AF581637

Sustainable Wastewater Treatment Systems

Sustainable Wastewater Treatment Systems

Special Issue Editor

José A. Herrera-Melián

MDPI • Basel • Beijing • Wuhan • Barcelona • Belgrade • Manchester • Tokyo • Cluj • Tianjin

Special Issue Editor
José A. Herrera-Melián
Universidad de Las Palmas de Gran Canaria
Spain

Editorial Office
MDPI
St. Alban-Anlage 66
4052 Basel, Switzerland

This is a reprint of articles from the Special Issue published online in the open access journal *Sustainability* (ISSN 2071-1050) (available at: https://www.mdpi.com/journal/sustainability/special_issues/Sustainable_Wastewater_Treatment_Systems).

For citation purposes, cite each article independently as indicated on the article page online and as indicated below:

LastName, A.A.; LastName, B.B.; LastName, C.C. Article Title. *Journal Name* **Year**, *Article Number*, Page Range.

ISBN 978-3-03928-971-4 (Pbk)
ISBN 978-3-03928-972-1 (PDF)

Cover image courtesy of José A. Herrera-Melián.

Contents

About the Special Issue Editor

José A. Herrera-Melián (Dr.) obtained a degree in Marine Science from the University of Las Palmas de Gran Canaria. His PhD studies were devoted to the electrochemical analysis of heavy metals (Ni and Co) in seawater and to the study of their biogeochemical cycle in the marine environment. During the last decades his research has been focused on wastewater treatment by means of advanced oxidation technologies, ponds and constructed wetlands. He has authored more than 60 papers and two chapters. His h-index is 27 (Google Scholar). Now he is a teacher of the Department of Chemistry of the University of Las Palmas de Gran Canaria and a member of the Institute of Environmental Studies and Natural Resources (i-UNAT).

Editorial

Sustainable Wastewater Treatment Systems (2018–2019)

José Alberto Herrera Melián

Department of Chemistry, i-Unat (Institute of Environmental Studies and Natural Resources) University of Las Palmas de Gran Canaria, 35017 Las Palmas de Gran Canaria, Spain; josealberto.herrera@ulpgc.es; Tel.: +34-928-45-44-38

Received: 19 February 2020; Accepted: 21 February 2020; Published: 3 March 2020

Abstract: An important part of the environmental degradation suffered by the planet is caused by the discharge of untreated or poorly treated wastewater. Industrial, urban, and agricultural wastewater contain many different types of pollutants such as biodegradable and nonbiodegradable organic matter, suspended solids, turbidity, nutrients, heavy metals, pesticides, pathogens, etc. All of these pose a threat to the environment and human health, so the selected treatment techniques must be adapted to their nature in order to optimize their removal. In addition to efficiency, wastewater treatment methods must be sustainable, not only from an environmental point of view, but also economically and ethically. As a result, no technological dependence should be generated in less developed countries or communities. Therefore, this Special Issue deals with improvements in various aspects of wastewater treatment including different aspects of water treatment such as the development of mathematical models, the application of life cycle techniques, or the experimental optimization of wastewater treatment methods. Thirteen articles were accepted covering some of the most relevant fields of wastewater treatment: activated sludge, nanoparticle treatment, constructed wetlands, energy–water nexus, nutrient recovery, eco-friendly sorbents, and reverse osmosis.

Keywords: water treatment; activated sludge; modeling; constructed wetland; advanced oxidation techniques; reverse osmosis; sorbents

---

Historically, water scarcity has been a major problem in many regions of the world. However, accessibility to water of sufficient quality is becoming an increasingly serious problem, mainly due to the pollution of aquifers and coastal areas, climate change, and overpopulation (Nguvava et al. [1], Ziadi et al. [2], Zhou et al. [3]).

Natural waters, and to a greater extent, sewage, must be treated before use, reuse, or being discharged. There are many pollutants that compromise water quality, and affect both the human being and the natural environment. Some of the most important threats come from industrial wastewater due to the toxicity of heavy metals, persistent organic compounds, etc. (Tayeb et al. [4]) However, urban wastewater, apparently easily treated by well-established methods for decades, is generating an increase in eutrophication events with direct consequences such as massive deaths of aquatic organisms, biodiversity reduction, red tides, etc. For example, the number of anoxic or sub-toxic zones in coastal waters has grown exponentially since 1960, with more than 400 hypoxic zones being reported worldwide (Diaz and Rosenberg [5]). The presence of pathogens (viruses, bacteria, and parasites) and their elimination remains an essential part in the treatment of wastewater (Chaukura et al. [6]). Hormonal disruptors are another of the many groups of pollutants that pose a health risk of unforeseeable consequences to humans and animals. These include a huge variety of substances including drugs, pesticides, additives of plastic products, bleaching agents, cleaning agents, etc. (Alvarez-Ruiz and Picó [7]).

The current environmental situation of the planet also requires that the methods of water treatment, like any other human activity, are sustainable, but what does sustainable mean? Many times, we think about environmental sustainability exclusively and forget that sustainability, as understood by the United Nations in its 17 Sustainable Development Goals, is much more than that, since it includes economics and ethical-social aspects such as gender equality or the eradication of poverty for all (United Nations [8]).

Water treatment methods include a wide variety of techniques of a different nature, ranging from physical methods such as dissolved air flotation or membrane techniques to chemical methods such as the advanced oxidation techniques and the great variety of biological methods (Holkar et al. [9]). Within this last group, the low-cost, decentralized, or ecological techniques such as lagoons, constructed wetlands, trickling filters, or biodiscs have undergone a remarkable development (Guedes-Alonso et al. [10]).

The objective of this Special Issue "Sustainable Wastewater Treatment Systems" has been to review the state-of-the-art of the latest advances in water management with an particular focus on sustainable methods of disinfection, grey water, constructed wetlands, ponds, membranes, reclaimed wastewater reuse, etc. Many submissions have been received with significant contributions for the main topics of interest in our Special Issue. However, only 13 high-quality papers were accepted after strict and rigorous review. In particular, these accepted papers mainly focused on various perspectives such as innovative applications and research covering the removal of nanoparticles, constructed wetlands, microbial aspects of activated sludge, adaptation to climate change in water-energy coupling, nutrient recovery from wastewater, ecofriendly sorbents, advanced oxidation processes, membrane technology, and modeling in reverse osmosis optimization.

All of the accepted articles (Contribution 1–Contribution 13) provide recent advances in the most active wastewater treatment research fields. Rizwan Khan, Muhammad Ali Inam, Saba Zam Zam, Muhammad Akram, Sookyo Shin, and Ick Tae Yeom explored the removal of CuO nanoparticles from water by coagulation at different pH values and dissolved organic matter concentrations. The media pH significantly affected the coagulation efficiency of the nanoparticles. They observed that the simultaneous effect of coagulants and charge neutralization at pH 6–8 enhanced the removal of CuO nanoparticles. Their study suggests that coagulation is effective in removing the nanoparticles from complex matrices in a wide pH range. Their findings provide insight into the coagulation and dissolution behavior of CuO nanoparticles during the water treatment process (Contribution 1). In the field of activated sludge, Jin Xu, Peifang Wang, Yi Li, Lihua Niu, and Zhen Xing studied the effect of different organic carbon/N ratii and dissolved oxygen (DO) levels and observed that the best treatment performance was achieved with a COD (Chemical Oxygen Demand)/N ratio of 7:1 or the DO levels of 2–2.5 mg/L. They observed evident microbial variance and changes in the richness and evenness of the microbial communities in the activated sludge. Their work provides valuable practical guidance for the operators of any wastewater treatment plant (Contribution 2). Research in constructed wetlands is continuously expanding and this Special Issue could attract different remarkable manuscripts. The sustainable application of constructed wetlands calls for the use of alternative materials to be used as substrates. Agro-forest wastes such as palm mulch can be a suitable alternative to gravel and sand. Thus, Marina Carrasco-Acosta, Pilar Garcia-Jimenez, José Alberto Herrera-Melián, Néstor Peñate-Castellano, and Argimiro Rivero-Rosales studied the effect of plants on relevant aspects of constructed wetland performance such as pollutant removal, substrate clogging, and bacterial community structure in organic-based vertical flow constructed wetlands. They observed that the presence of plants delayed the clogging of the reactors and reduced the biodiversity of Enterococci and *E. coli* as measured with terminal restriction fragment length polymorphism (T-RFLP) analysis (Contribution 3). Nowadays, water-energy management optimization is a key issue, but the threat of the climate change can make it imperative in the near future. Tuan-Viet Hoang, Pouya Ifaei, Kijeon Nam, Jouan Rashidi, Soonho Hwangbo, Jong-Min Oh, and Chang Kyoo Yoo proposed the optimization of a hybrid renewable energy system (HRES) coupled with a membrane bioreactor for the sustainable adaptation to climate change in Vietnam. The model-based HRES consisted of solar photovoltaic panels, wind turbines, and battery banks. The authors defined three scenarios, 101 sub-scenarios, and three management cases to optimize the system design. The results showed that the smallest environ-economic cost was obtained when 47% of the demand load of the membrane bioreactor was met using the HRES and the rest was supplied by the grid (Contribution 4). Another paper on constructed wetlands focused on the removal of heavy metals with the plant totora in the South American Altiplano region. Juan Blanco tested if the plant could be used in constructed wetlands treating mining wastewaters with high salinity and As and Pb concentrations. He compared the chemical composition of the leaves, rhizomes, and roots and observed that totora was a multi-hyperaccumulator for As, Fe, and Ni. These results, in addition with the plant's intrinsic high biomass production, slow decomposition, and usability as a raw material for local craftwork and industry, support the recommendation to use totora in wetlands to treat water polluted with heavy metals and/or with high salinity (Contribution 5). Rizwan Khan, Muhammad Ali Inam, Muhammad Mazhar Iqbal, Muhammad Shoaib, Du Ri Park, Kang Hoon Lee, Sookyo Shin, Sarfaraz Khan, and Ick Tae Yeom studied the influence of surfactant type in the removal of ZnO nanoparticles from natural waters by the coagulation–flocculation process. Anionic sodium dodecyl sulfate (SDS) and nonionic nonylphenol ethoxylate (NPEO) were employed as model surfactants. The adsorption of the nanoparticles, which was strongly pH-dependent, was studied with Freundlich and Langmuir models. The formation of mono-bilayer patches onto the nanoparticles was suggested. The cooperation of charge neutralization and adsorptive micellar flocculation might explain the coagulation mechanism. This study provides new insight into the behavior of ZnO nanoparticles and surfactants in water treatment processes (Contribution 6). Nitrogen and phosphorus play a key role in food production but their environmental impact can be devastating when they are discharged in the natural watercourses. Jan Peter Van der Hoek, Rogier Duijff, and Otto Reinstra studied nitrogen recovery from wastewater. The current N-based fertilizers have many drawbacks since energy requirements are high and in the wastewater treatment, N is lost to the atmosphere as N_2. The authors selected technologies for N recovery from wastewater considering four criteria: sustainability, the potential to recover N, the maturity of the technology, and the N concentration that can be handled by the technology. The most promising mature technologies that can be incorporated into existing wastewater treatment plants include struvite precipitation, the treatment of digester reject water by air stripping, vacuum membrane filtration, hydrophobic membrane filtration, and treatment of air from thermal sludge drying. Higher nitrogen recovery (60%) could be achieved by separate urine collection, but a completely new infrastructure for wastewater collection and treatment would be necessary. Different technologies in parallel are required to reach sustainable solutions (Contribution 7). Shuang Xu, Weiguang Yu, Sen Liu, Congying Xu, Jihui Li, and Yucang Zhang explored the adsorption of hexavalent chromium on a low cost banana pseudostem biochar. The biochar surface prepared at low temperature was rich in O-containing groups. The best results were obtained with the biochar prepared at 300 °C with a 125.44 mg/g maximum adsorption capacity. Pseudo-second-order kinetics and Langmuir model provided the best fit of the experimental data, indicating a monolayer chemi-adsorption. The adsorption of Cr(VI) was attributed to the reduction of Cr(VI) to Cr(III), ion exchange, and complexation (Contribution 8). Angela Gorgoglione and Vincenzo Torretta contributed with a revision of more than 120 constructed wetland (CWs) case studies with the goal of providing a tool for researchers and decision-makers considering using this green technology. The authors claim that although CWs are considered to be environmental-friendly and low cost, their sustainable management still remains a challenge. The study provides sustainable solutions for the performance and applications of CWs by means of the discussion of key aspects such as macrophyte species, media type, water level, hydraulic retention time, and hydraulic loading rate (Contribution 9). Additionally, very interesting research on the use of the green waste coffee silverskin in water treatment was performed by Angela Malara, Emilia Paone, Patrizia Frontera, Lucio Bonaccorsi, Giuseppe Panzera, and Francesco Mauriello. These authors assessed it for its suitability in the removal of Cu, Zn, and Ni divalent ions from water. The application of the Langmuir and Freundlich models demonstrated a monolayer-type adsorption. The results support the use of coffee silverskin as a new low cost adsorbent for metals in wastewater (Contribution 10). Reducing the effects of eutrophication on receiving waterbodies has many environmental, but also economic benefits. This way, Ben Morelli, Sarah Cashman, Xin (Cissy) Ma, Jay Garland, Jason Turgeon, Lauren Fillmore, Diana Bless, and Michael Nye applied life cycle and life cost assessments to determine the environmental benefits of upgrading a small community conventional activated sludge treatment process. The authors introduced biological nutrient removal, and enhanced primary settling and anaerobic digestion (AD) with co-digestion of high strength organic waste. The upgraded system significantly reduced eutrophication impact, global climate change potential, and cumulative energy demand relative to the legacy system (Contribution 11). Water treatment methods of different nature can also be combined to provide particularly suitable technologies. Hyun-Hee Jang, Gyu-Tae Seo and Dae-Woon Jeong proposed a combination of nanofiltration and ozone-hydrogen peroxide oxidation for the treatment of soy sauce waste. Currently, the application of ozone oxidation provides 34% color removal and 27% chemical oxygen demand reduction. The authors combined ozone with hydrogen peroxide and achieved color removal (52 %) and COD reduction (34 %) with the optimized method. When nanofiltration was used as a pre-treatment, the method was remarkably improved since color removal was 98% and COD removal was 98%. Thus, the NF-H_2O_2/O_3 process is one of the best methods to treat soy sauce waste (Contribution 12). Finally, with the goal of reducing the power consumption of the desalination industry in Kuwait, Bader S. Al-Anzi and Ashly Thomas developed a one-dimensional analytical model of pressure retarded osmosis in a parallel flow configuration. The model has been developed to "size" an osmotically-driven membrane process mass exchanger given the operating conditions and desired performance. The model has been used to determine mass transfer units as a function of mass flow rate ratio, recovery ratio, concentration factors, effectiveness, etc. The actual water permeation to the brine stream was related by the introduction of a new dimensionless dilution rate ratio and dilution rate, among others. A maximum power of 0.28 and 2.6 kJ can be produced by the system using seawater or treated wastewater effluent as the feed solution, respectively, which could help to reduce the power consumption of the desalination industry in Kuwait (Contribution 13).

List of Contributions:

1. Khan, R.; Inam, M.; Zam Zam, S.; Akram, M.; Shin, S.; Yeom, I. Coagulation and Dissolution of CuO Nanoparticles in the Presence of Dissolved Organic Matter under Different pH Values. *Sustainability* **2019**, *11*, 2825.
2. Xu, J.; Wang, P.; Li, Y.; Niu, L.; Xing, Z. Shifts in the Microbial Community of Activated Sludge with Different COD/N Ratios or Dissolved Oxygen Levels in Tibet, China. *Sustainability* **2019**, *11*, 2284.
3. Carrasco-Acosta, M.; García-Jiménez, P.; Herrera-Melián, J.; Peñate-Castellano, N.; Rivero-Rosales, A. The Effects of Plants on Pollutant Removal, Clogging, and Bacterial Community Structure in Palm Mulch-Based Vertical Flow Constructed Wetlands. *Sustainability* **2019**, *11*, 632.
4. Hoang, T.; Ifaei, P.; Nam, K.; Rashidi, J.; Hwangbo, S.; Oh, J.; Yoo, C. Optimal Management of a Hybrid Renewable Energy System Coupled with a Membrane Bioreactor Using Enviro-Economic and Power Pinch Analyses for Sustainable Climate Change Adaption. *Sustainability* **2019**, *11*, 66.
5. Blanco, J. Suitability of Totora (Schoenoplectus californicus (C.A. Mey.) Soják) for Its Use in Constructed Wetlands in Areas Polluted with Heavy Metals. *Sustainability* **2019**, *11*, 19.
6. Khan, R.; Inam, M.; Iqbal, M.; Shoaib, M.; Park, D.; Lee, K.; Shin, S.; Khan, S.; Yeom, I. Removal of ZnO Nanoparticles from Natural Waters by Coagulation-Flocculation Process: Influence of Surfactant Type on Aggregation, Dissolution and Colloidal Stability. *Sustainability* **2019**, *11*, 17.
7. Van der Hoek, J.; Duijff, R.; Reinstra, O. Nitrogen Recovery from Wastewater: Possibilities, Competition with Other Resources, and Adaptation Pathways. *Sustainability* **2018**, *10*, 4605.
8. Xu, S.; Yu, W.; Liu, S.; Xu, C.; Li, J.; Zhang, Y. Adsorption of Hexavalent Chromium Using Banana Pseudostem Biochar and Its Mechanism. *Sustainability* **2018**, *10*, 4250.

9. Gorgoglione, A.; Torretta, V. Sustainable Management and Successful Application of Constructed Wetlands: A Critical Review. *Sustainability* **2018**, *10*, 3910.
10. Malara, A.; Paone, E.; Frontera, P.; Bonaccorsi, L.; Panzera, G.; Mauriello, F. Sustainable Exploitation of Coffee Silverskin in Water Remediation. *Sustainability* **2018**, *10*, 3547.
11. Morelli, B.; Cashman, S.; Ma, X.; Garland, J.; Turgeon, J.; Fillmore, L.; Bless, D.; Nye, M. Effect of Nutrient Removal and Resource Recovery on Life Cycle Cost and Environmental Impacts of a Small Scale Water Resource Recovery Facility. *Sustainability* **2018**, *10*, 3546.
12. Jang, H.; Seo, G.; Jeong, D. Advanced Oxidation Processes and Nanofiltration to Reduce the Color and Chemical Oxygen Demand of Waste Soy Sauce. *Sustainability* **2018**, *10*, 2929.
13. Al-Anzi, B.; Thomas, A. One-Dimensional Analytical Modeling of Pressure- Retarded Osmosis in a Parallel Flow Configuration for the Desalination Industry in the State of Kuwait. *Sustainability* **2018**, *10*, 1288.

Funding: This research was supported by the collaboration of the Spanish Ministry of Science, Innovation and Universities, the University of Las Palmas de Gran Canaria (Grant CGL2016-78442-C2-2-R, GOBESP2017-04 ULPGC) and the Programa Innova Canarias 2020.

Conflicts of Interest: The author declares no conflict of interest.

References

1. Nguvava, M.; Abiodun, B.J.; Otieno, F. Projecting drought characteristics over East African basins at specific global warming levels. *Atmos. Res.* **2019**, *228*, 41–54. [CrossRef]
2. Ziadi, A.; Hariga, N.T.; Tarhouni, J. Mineralization and pollution sources in the coastal aquifer of Lebna, Cap Bon, Tunisia. *J. Afr. Earth Sci.* **2019**, *151*, 391–402. [CrossRef]
3. Zhou, Y.; Wang, L.; Zhou, Y.; Mao, X. Eutrophication control strategies for highly anthropogenic influenced coastal waters. *Sci. Total Environ.* **2020**, *705*, 135760. [CrossRef] [PubMed]
4. Tayeb, A.; Chellali, M.R.; Hamou, A.; Debbah, S. Impact of urban and industrial effluents on the coastal marine environment in Oran, Algeria. *Mar. Pollut. Bull.* **2015**, *98*, 281–288. [CrossRef] [PubMed]
5. Diaz, R.J.; Rosenberg, R. Spreading dead zones and consequences for marine marine ecosystems. *Science* **2008**, *321*, 926–929. [CrossRef] [PubMed]
6. Chaukura, C.; Marais, S.S.; Moyo, W.; Mbali, N.; Thakalekoala, L.C.; Ingwani, T.; Mamba, B.B.; Jarvis, P.; Nkambule, T.T.I. Contemporary issues on the occurrence and removal of disinfection byproducts in drinking water—A review. *J. Environ. Chem. Eng.* **2020**, *8*, 103659. [CrossRef]
7. Álvarez-Ruiz, R.; Picó, Y. Analysis of emerging and related pollutants in aquatic biota. *Trends Environ. Anal. Chem.* **2020**, *25*, e00082. [CrossRef]
8. United Nations. Transforming Our World: The 2030 Agenda for Sustainable Development. 2015. Available online: http://www.un.org/ga/search/view_doc.asp?symbol=A/RES/70/1&Lang=E (accessed on 3 March 2020).
9. Holkar, C.R.; Jadhav, A.J.; Pinjari, D.V.; Mahamuni, N.M.; Pandit, A.B. A critical review on textile wastewater treatments: Possible approaches. *J. Environ. Manag.* **2016**, *182*, 351–366. [CrossRef] [PubMed]
10. Guedes-Alonso, R.; Montesdeoca-Esponda, S.; Herrera-Melián, J.A.; Rodríguez-Rodríguez, R.; Ojeda-González, Z.; Landívar-Andrade, V.; Sosa-Ferrera, Z.; Santana-Rodríguez, J.J. Pharmaceutical and personal care product residues in a macrophyte pond-constructed wetland treating wastewater from a university campus: Presence, removal and ecological risk assessment. *Sci. Total Environ.* **2020**, *703*, 135596. [CrossRef] [PubMed]

Article

Coagulation and Dissolution of CuO Nanoparticles in the Presence of Dissolved Organic Matter Under Different pH Values

Rizwan Khan [1], Muhammad Ali Inam [1], Saba Zam Zam [1], Muhammad Akram [2], Sookyo Shin [1] and Ick Tae Yeom [1,*]

1 Graduate School of Water Resources, Sungkyunkwan University (SKKU), Suwon 16419, Korea; rizwankhan@skku.edu (R.K.); aliinam@skku.edu (M.A.I.); sabazamzam@skku.edu (S.Z.Z.); tkssk08@gmail.com (S.S.)

2 Shandong Key Laboratory of Water Pollution Control and Resource Reuse, School of Environmental Science and Engineering, Shandong University, Qingdao 266200, China; m.akramsathio@mail.sdu.edu.cn

* Correspondence: yeom@skku.edu; Tel.: +82-312-996-699

Received: 26 April 2019; Accepted: 15 May 2019; Published: 17 May 2019

Abstract: The increased use of copper oxide nanoparticles (CuO NPs) in commercial products and industrial applications raises concerns about their adverse effects on aquatic life and human health. Therefore, the current study explored the removal of CuO NPs from water via coagulation by measuring solubility under various pH values and humic acid (HA) concentrations. The results showed that the media pH significantly affected the coagulation efficiency of CuO NPs (30 mg/L) under various (0–0.30 mM) ferric chloride (FC) dosages. The concentration of dissolved Cu^{2+} ions at pH 3–6 was (16.5–4.8 mg/L), which was higher than at other studied pH (7–11). Moreover, the simultaneous effect of coagulants and charge neutralization at pH 6–8 enhanced the removal of CuO NPs. At a lower FC (0–0.05 mM) dosage, the higher HA concentration inhibited the aggregation of CuO NPs. However, at the optimum dose of (0.2 mM) FC, the efficiency of turbidity removal and solubility of CuO NPs between pH 8 and 11 was above 98% and 5%, respectively, probably due to coagulant enmeshment. Our study suggested that coagulation was effective in removing the CuO NPs from the complex matrices with pH values ranging from 8–11. The findings of the present study provide insight into the coagulation and dissolution behavior of CuO NPs during the water treatment process.

Keywords: aggregation; coagulation; CuO; dissolution; humic acid; nanoparticles

1. Introduction

Many commercial products containing metal-based nanoparticles (NPs) are currently available in the market [1]. It is reported that every year around 28–32% of the discharged NPs are released into the surface waters of the United States. Thus, a large number of released NPs pose a potential risk to human health and aquatic life [2]. Copper oxides (CuO) are among the most widely used NPs in textiles, wood preservatives, electronics, inks, films, coatings, and ceramics, because of their specific structural properties [3,4]. The global annual production of CuO NPs was around 570 tons in 2014 and an estimated 1600 tons by 2025. Upon production and application, CuO NPs enter different environments such as natural surface waters and sediments, thereby increasing the risk of exposure to organisms as well as affecting their life cycle [2]. CuO NPs may dissociate into Cu^{2+} and higher concentrations significantly affect the growth of aquatic organisms, especially *lymphocytes, Fagopyrum esculentum,* and *Pseudokirchneriella* [5]. A recent study reported that the discharged NPs substantially inhibit the capability of wastewater biofilms [6]. The toxic effects of NPs on humans via damage to DNA structure and cell membranes were also reported [7]. Thus, it is important to understand the dissolution phenomena of CuO NPs in the aquatic environment to minimize the related environmental and ecological risks of soluble engineered nanoparticles (ENPs) in general.

In the aquatic system, the dissolution of CuO NPs depends upon several factors such as particle size, shape, and surface charge, and physicochemical properties of the media, i.e., ionic strength (I.S), pH, and dissolved organic matter (DOM) [8]. A previous study reported that media pH strongly influenced the surface potential and dissolution of CuO NPs via protonation/deprotonation of surface hydroxyl groups [9]. For instance, a recent study showed that a decrease in the pH of aqueous environment enhanced the dissolution of ENPs [10]. Furthermore, natural waters also contain ubiquitous humic substances with concentrations of dissolved organic carbon ranging from 1–100 mg/L [11]. The DOM at a lower concentration (0.1 mg/L) might be adsorbed onto the NPs surface, resulting in a negative charge depending on various pH values to increase the stability of NPs suspension. Bian et al. reported that the polydentate structure of humic acid (HA) increases the dissolution of ZnO NPs at pH values ranging from 9–11 [12]. Furthermore, the rate of dissolution of NPs may be enhanced with increased HA concentration in the solution. Conversely, a recent study demonstrated that fulvic acid hindered the rate of carbon nanotubes dissolution and increased colloidal stability because of surface interaction [13]. A previous study showed that a coating of HA on the surface of CuO NPs resulted in particle disaggregation, and with further increase in HA concentration improved the dispersion of NPs [9]. Therefore, water contaminated with the CuO NPs enhances the risk of exposure to human and aquatic living organisms. Thus, it is important to consider the dissolution of CuO NPs in the treatment process, which is influenced by the solution chemistry.

Several advanced technologies, such as membrane filtration are used to remove NPs from water [14,15]. However, membrane fouling significantly affects performance and increases the cost of membrane treatment [15]. The particles can be removed via activated sludge, but most of the ENPs might be toxic to microorganisms as well as affect the overall sludge treatment by altering the sludge properties [16]. Coagulation is an efficient and simple process commonly used in water treatment to remove suspended solids, organic and inorganic substances from the water. Earlier studies [17–19] reported that NPs including cadmium telluride (CdTe), C(60), and multiwall carbon nanotubes (MWCNT) were efficiently removed from water by alum as well as polyaluminum chloride (PACl) coagulation. In addition, variable coagulation efficiencies of different ENPs, such as Ag (2–21%), ZnO (46–98%) and TiO_2 (2–9%) have also been reported [20]. The DOM has been found to hinder the agglomeration of TiO_2 NPs and affect the overall removal efficiency of the coagulation process [21]. For example, ZnO NPs coated with DOM such as HA and salicylic acid (SA) showed substantial adsorption capacities with increasing colloidal stability, thereby reducing the NPs removal because of electrostatic repulsion between organic molecules [22]. However, studies investigating the interactive behavior of CuO NPs in the presence of HA under different pH values were rarely investigated by environmental scholars. Furthermore, the studies also appear inadequate in elucidating the influence of HA on the dissolution and coagulation of CuO NPs. Thus, it is important to comprehensively understand the effect of HA on the removal of CuO NPs by coagulation in heterogenous water environments.

Accordingly, the present study investigated the effect of pH and HA on the removal of CuO NPs via coagulation from the water. In this study, we evaluated the removal efficiency by measuring the suspension turbidity and residual concentration of Cu^{2+} ions under different HA concentrations with varying pH values.

2. Materials and Methods

2.1. Chemicals Reagents and Stock Solution Preparation

The CuO NPs (CAS No: 1317380, purity ≥99.0%) with vendor reported diameter <50 nm were obtained from Sigma-Aldrich (St. Louis, MO, USA) and used without additional purification (Supplementary Materials [SM] Table S1). The CuO NPs stock suspension was prepared by weighing

3 mg of CuO powder and dispersion into 100 mL of 1 mM $NaHCO_3$ solution. The effect of the probe sonication on the turbidity of CuO NPs suspension was determined (Figure S1A). All suspensions were sonicated according to the optimized settings using an ultrasonicator (Bio-Safer 1200/90, Â 12 mm, Nanjing, China) for 30 min prior to coagulation experiments. The suspension pH was adjusted to 9 by 0.1 M HCl or 0.1 M NaOH solutions to reduce the effect of dissolution of CuO NPs during the sonication. Subsequently, 1 mL of the NPs suspension was transferred into a DTS0012 cuvette and placed in the Zetasizer sample chamber for hydrodynamic (HDD) size measurements, while the results were reported in intensity-volume % distribution. Humic acid (HA) with (99+% purity) was purchased from Sigma-Aldrich (St. Louis, MO, USA) and used as a model for dissolved organic matter (DOM). The stock solution of HA was prepared by dissolving 100 mg of HA powder in 100 mL of deionized (DI) water and adjusting the solution pH to 10 using 0.1 M NaOH in order to ensure complete dissolution of HA. The solution was stirred at 600 rpm for 24 h to increase the stability and then filtered with a 0.45 μm glass fiber filter followed by pH adjustment to 7. The total organic carbon (10 mg/L HA) was 4.5 mg/L. The coagulant iron (III) chloride hexahydrate ($FeCl_3{\cdot}6H_2O$) with (98+% purity) was purchased from local suppliers. Coagulant stock solution, i.e., 0.1 M FC was prepared by dissolving $FeCl_3{\cdot}6H_2O$ into DI water.

2.2. Coagulation and Dissolution

Coagulation experiments were performed in a jar tester (Model: SJ-10, Young Hana Tech Co., Ltd. Gyeongsangbuk-Do, Korea) as follows: Rapid mixing at 200 rpm within 1 min followed by slow mixing at 40 rpm for 15 min and settling for 30 min. The solubility of CuO NPs significantly increases at pH below 7 and above 11 as previously reported [23]. Systematic experiments were performed at pH values 3–11 to evaluate the effect of pH on the dissolution phenomena of CuO NPs. The performance was evaluated via measurement of supernatant turbidity after coagulation using a turbidimeter (Hach Benchtop 2100 N, Loveland, CO, USA). An aliquot of ~30 mL was then collected to measure the concentration of the dissolved Cu^{2+} and ferric (Fe^{3+}) ions via inductively coupled plasma optical emission spectrometry (ICP-OES: Model Varian, Agilent technologies, Sana Clara, CA, USA). In order to simulate the natural water conditions, the concentration of HA was maintained in the range of 0–10 mg/L (ref). The Visual MINTEQ 3.1 (KTH, Stockholm, Sweden) was used to determine the speciation of Cu and Fe(III) in aqueous media after coagulation with different HA concentrations. To further explore the impact of carbon dioxide (CO_2) on Cu and Fe(III) species, parallel tests were conducted with and without CO_2 in the system. Furthermore, all of the experiments were performed in triplicate, and the relative standard deviations (RSD) were reported.

2.3. Additional Characterization of CuO NPs

The specific surface area of CuO NPs was analyzed using the N_2-Brunauer Emmett Teller (BET) method (ASAP 2020, Micromeritics, Norcross, GA, USA). The zeta potential of CuO NPs was determined from their electrophoretic mobility using a Zetasizer analyzer (Malvern Nano-ZS, Worcestershire, UK) at 25 ± 1 °C. The particle size distribution of the CuO NPs in solution was measured using dynamic light scattering (DLS) with a He-Ne laser (λ = 632.8 nm).

3. Results and Discussion

3.1. Characteristics of CuO NPs

The CuO NPs were characterized in water immediately after sonication. As shown in Figure S1A, the optimal CuO NPs dispersion was obtained at a sonication time of 30 min; however, no significant difference in suspension turbidity was detected upon further increase in sonication time. Further, as shown in Figure S1B, the diameter of most particles was in the range of 150–250 nm suggesting a much larger size of CuO NPs in water compared with the reported primary particle size (<50 nm). The change in particle size may be due to an increase in van der Waals (vdW) forces among the NPs, resulting in the formation of larger aggregates in solution [24]. The BET specific surface area of CuO NPs was found to be 29.2 m^2/g. The impact of pH on the dissolution and zeta potential of CuO (30 mg/L) suspension were investigated (Figure 1). It was obvious that the pH range of 2.0–6.9 facilitated the dissolution of CuO NPs, which was consistent with previous studies [9,23]. The Cu^{2+} released from CuO at pH range 2.0–5.0 was up to 19.8–8.4% of the total Cu. However, the solubility of CuO NPs declined above pH 7, and less than 1.3% of the total Cu was measured at pH 8. Under acidic environment, the interaction of H^+ ions with CuO NPs significantly released a large amount of Cu^{2+} ions, whereas at alkaline pH, the formation of $Cu(OH)^+$ complexes occurred due to the presence of high concentrations of OH^- ions [12]. Therefore, considering the dissolution of CuO NPs in highly acidic conditions, the coagulation experiments were conducted at pH 9 to minimize the NPs dissolution. Further, based on the speciation of Fe(III), large amounts of $Fe(OH)_2{}^+$ were generated around pH 8, which may facilitate the coagulation process (Figure S1C). The average zeta potential of CuO at pH 8 was +6.14 mV, which declined gradually with a pH increasing to more alkaline values. As shown in Figure 1, CuO NPs exhibit a stable and positive charge (+27.9 ± 1.3 mV) from pH 3.0–5.0. Subsequently, the zeta potential sharply decreased to the Isoelectric point (IEP) at a pHiep of approximately ~8.2 ± 0.1, which is consistent with earlier studies [9,23,25] investigating CuO NPs. Further increase in pH resulted in charge reversal with zeta potential values of −24.76 ± 1.3 mV at pH 11 ± 0.1. These results were consistent with previous study findings [25], suggesting that the surface potential was positive when pH < pHiep and negative at pH > pHiep. In addition, NPs with zeta potential below ±15 mV were supposed to be unstable, while NPs with zeta potential higher than ±30 mV were considered stable in the solution.

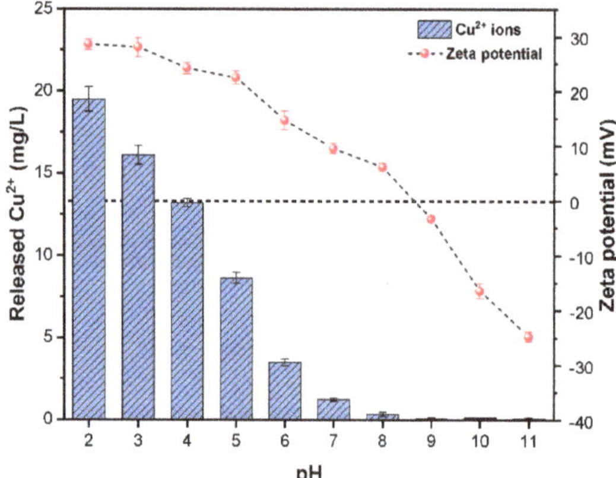

Figure 1. Cu^{2+} released (mg/L); and zeta potential of CuO NPs as a function of pH.

3.2. *Effect of Coagulant Dose on CuO NPs Removal*

Figure 2 illustrates the turbidity removal efficiency of CuO NPs (30 mg/L) at various FC dosages at pH 9. It can be seen that at a lower FC dosage (0–0.05 mM), the removal efficiency was below 60%, which may be attributed to fewer active Fe sites available in the solution. In addition, the inadequate compression of the electrical double layer (EDL) of NPs induced weak interparticle bridging among the colloidal flocs [18]. The turbidity removal efficiency was remarkably enhanced with increasing FC dose and the highest removal up to ~98% was obtained with 0.2 mM FC dosage. However, a further increase of coagulant dose (up to 0.30 mM) slightly decreased the turbidity removal efficiency of CuO NPs probably due to the charge inversion and restabilization of NPs flocs following oversaturation of polyelectrolytes [26]. In addition, as Figure 2 further suggests the addition of FC hinders the dissolution of CuO NPs at pH 9 and the FC dosage had a slight impact on the solubility of CuO NPs. The measured solubility of CuO NPs at optimum FC dosage (0.2 mM) was around 2%, which can be ignored compared with the higher removal of NPs. In general, the coagulation was found to be an efficient process in removing the CuO NPs from aqueous solution at pH 9.

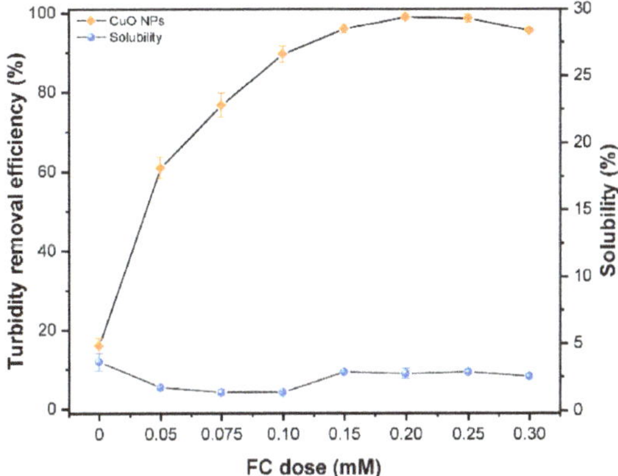

Figure 2. Effect of various FC (0–0.30 mM) dose on turbidity removal efficiency and solubility of CuO NPs (30 mg/L) at pH 9.

3.3. *Effect of HA Concentration on CuO NPs Removal*

The influence of HA concentration (10 mg/L) on the turbidity removal efficiency of CuO NPs (30 mg/L) was initially investigated at 0.05 mM FC dosage as a function of pH (Figure 3). The results indicated that the addition of HA decreased the coagulation efficiency of CuO NPs. The zeta potential of CuO at various HA concentration is shown in (Figure S2A,B), which suggests that the steric hindrance among NPs is further enhanced at higher HA concentration. The HA impeded the aggregation of CuO NPs and stabilized the negatively charged CuO NPs in solution. Mohd Omar et al. reported that the sorption of HA resulted in the disaggregation of NPs because of the Van der Walls (vdW) interaction between particles and HA molecules [27]. Furthermore, DOM has been known to stabilize the NPs by altering the zeta potential of the suspension from positive to negative. In order to further elucidate the effect of various HA concentrations on coagulation efficiency, an optimum dose of 0.2 mM was used in the following batch experiments.

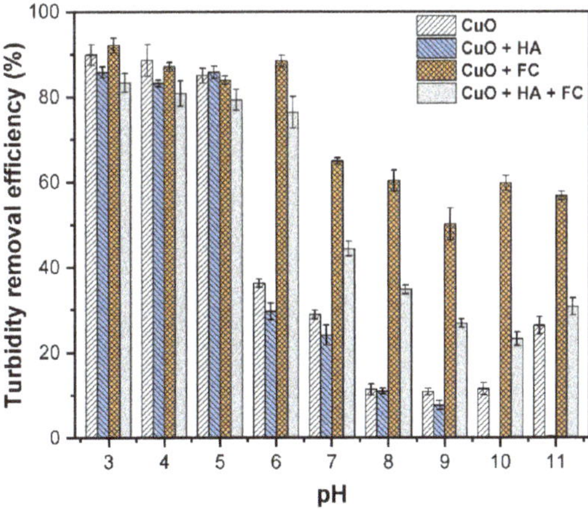

Figure 3. Turbidity removal efficiency of CuO NPs (30 mg/L) under different pH (3–11) values of FC (0.05 mM) and HA (10 mg/L).

It is noteworthy that in the absence of humic substance, the CuO NPs were dissolved under the highly acidic aqueous environment. Thus, CuO NPs were easily removed by coagulation at the pH range between 8 and 11 (Figure 4). In the absence of HA, the highest turbidity removal efficiency (99.1%) was obtained at pH 8 and 9, while the average removal efficiency was higher than 95%. At pH 8, the interaction between dominant Fe(III) species ($Fe(OH)_2{}^+$) and CuO NPs was increased via adsorption, resulting in higher turbidity removal [28]. Moreover, the coagulation efficiency increased with increased solution pH (9–10), which might be related to the enmeshed FC coagulants (Figure S1C). However, under highly alkaline pH (above 10), the effect of enmeshment was weaker due to the increased concentrations of $Fe(OH)_4{}^-$ species in the solution, which increased the electrostatic repulsion between negatively charged CuO NPs and $Fe(OH)_4{}^-$ (Figures S3A and S4A). Therefore, a slight decrease in the removal of CuO NPs above pH 10 was observed during the coagulation.

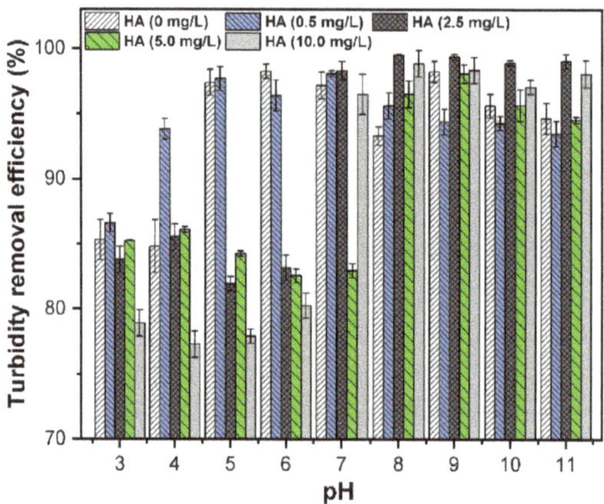

Figure 4. Efficiency of turbidity removal by CuO NPs (30 mg/L) under various pH (3–11) values and concentrations of HA (0, 0.5, 2.5, 5, 10 mg/L) at optimum FC (0.2 mM) dosage.

In addition, the CuO NPs were effectively removed even in the presence of HA when pH ranged between 8 and 11 (Figure 4). However, the higher concentration of HA significantly reduced the removal of CuO NPs at pH 3 and 6 compared with a lower concentration. The higher concentration of HA reverses the zeta potential from +6.14 to −45.2 mV (Figure S2A). However, the surface charge plays an insignificant role in the efficient removal of turbidity. The measured zeta potential of CuO NPs after coagulation experiment is shown in Figure S2B, which further explains the removal phenomena of CuO NPs from the heterogeneous environment. The possible removal mechanism might be related to the sufficient dosage of coagulant (0.2 mM FC), since the aqueous environment with pH greater than 8 was conducive to the formation of Fe(III) flocs (Figure S1C). Nonetheless, negatively charged CuO NPs, HA and $Fe(OH)_4{}^-$ (Figures S3A and S4A) were present in the solution. Therefore, the enmeshment of FC was stronger compared with the electrostatic repulsive forces among the NPs, which played a critical role in the removal of CuO NPs from aqueous solution. Moreover, the speciation of Cu and Fe(III) based on the presence of CO_2 after the completion of experiments to elucidate the role of CO_2 in the removal process (Figures S3 and S4). The results suggested that the presence of CO_2 at pH 6 and 11 enhanced the concentrations of $[Fe(OH)^2]^+$ and $[Fe(OH)_4]^-$, respectively. Moreover, ferric species preferably combine with the hydroxyl group. Based on these observations, it can be concluded that the combined effects of the enmeshment of coagulants and charge neutralization mechanism contributed to robust removal of CuO NPs from the water.

3.4. Effect of pH and HA on the Dissolution of CuO NPs

In the natural environment, the fluctuations in pH and the interaction between CuO NPs and humic substances affect the dissolution phenomena and increase their bioavailability in aquatic organisms. It has been proposed that after coagulation, the solution can be divided into three different segments including Cu^{2+}, suspended CuO NPs and CuO NPs flocs. Thus, the solubility of CuO NPs was determined via measurement of residual Cu^{2+} concentration in the solution (Figure 5). The results indicated that in the absence of HA, the solubility of CuO NPs was reduced with increasing solution pH and significantly decreased when the pH value was changed from 6–7. Figures S3A and S4A represent the dominant Cu species in the preferred pH range (3–11).

When pH ranged between 3–6, the concentration of Cu^{2+} ions in the supernatants was around 13 mg/L, which suggested that about 54.10% of NPs were dissolved in the aqueous solution because of the effect of protonation on the surface of CuO [12]. Hence, under highly acidic conditions, the turbidity removal of CuO NPs resulted from dissolution rather than coagulation. However, the measured concentration of Cu^{2+} at pH 9 was approximately 0.52 mg/L, while the total amount of copper species in the supernatant was 23.87 mg/L. The average solubility and turbidity removal at pH 9 and 11 were nearly 3.54% and 96.87%, respectively, which further indicated that coagulation is an effective method to remove CuO NPs from aqueous solution. These results are consistent with a previous study [29], which reported that at pH below 7, the dominant forms of copper were Cu^{2+} and $Cu(OH)^{+}$. However, at pH values above 9, the dissolution of CuO NPs occurred due to soluble hydroxy and hydroxide complexes. Therefore, the solubility of CuO NPs at pH 9 was less compared with highly acidic conditions. The possible chemical reactions of CuO NPs that may occur in highly acidic and alkaline conditions are as follows [30].

$$CuO\ (s) + 2H^{+}(aq) \leftrightarrows Cu^{+}(aq) + H_2O\ (l) \tag{1}$$

$$CuO(s) + H^{+}(aq) \leftrightarrows Cu(OH)^{+}(aq) \tag{2}$$

$$CuO(s) + OH^{-}(aq) + H_2O(l) \leftrightarrows CU(OH)_3^{-}(aq) \tag{3}$$

$$CuO(s) + 2OH^{-}(aq) + H_2O(I) \rightleftarrows Cu(OH)_4^{2-}(aq) \tag{4}$$

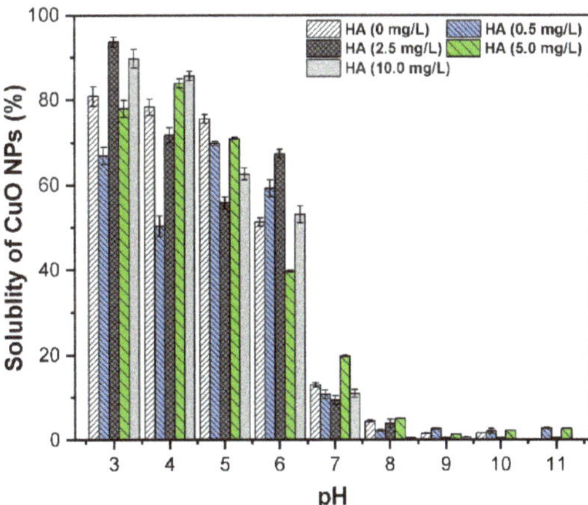

Figure 5. Solubility of CuO NPs (30 mg/L) under various pH (3–11) values and concentrations of HA (0, 0.5, 2.5, 5, and 10 mg/L) at FC (0.2 mM) dosage.

In addition, at a pH of 6–9, the presence of humic substances reduced the solubility with increasing pH, and the major speciation of copper ions is shown in Figures S3B and S4B. The lower concentration of HA inhibited the dissolution of CuO NPs suspension due to adsorption of organic molecules on the surface of NPs [13,22]. Moreover, the change in pH was also monitored before and after the coagulation experiment, which indicated a slight variation in pH, and an insignificant role in the dissolution of CuO NPs (Figure S5A,B). These findings have practical implications in that the presence of HA decreases the release of Cu^{2+} ions and reduces the associated risk and toxicity to aquatic life. Furthermore, HA has an insignificant effect on the coagulation efficiency of CuO NPs under alkaline conditions. In general, these results indicate that the coagulation process effectively removes the CuO NPs from the solution at a pH range of 8–11.

4. Conclusions

This study demonstrated that coagulation is an effective method to remove CuO NPs from the aqueous environment. However, the removal efficiency depends on the coagulant dosage used and the media pH. The results indicate that at a pH range of 8–11, the removal efficiency of CuO NPs (30 mg/L) at the optimum dose FC (0.2 mM) dose was higher than 98%, whereas the removal decreased under highly acidic pH conditions. Furthermore, when the pH ranged from 3–6, the dissolution of CuO NPs mainly occurred and was followed by the transformation of the released Cu into soluble hydroxide, hydroxy complex, and Cu^{2+} in the supernatant. However, at pH 6–8, the removal of CuO NPs resulted from the combined effects of charge neutralization as well as adsorption. In addition, the effect of the enmeshment of coagulants was stronger when the pH was between 8 and 11, which played a vital role in the removal of CuO NPs from the water. These results indicated that media pH and HA concentration influence the fate, mobility, and coagulation performance of ENPs in water/wastewater treatment.

Supplementary Materials: The following additional materials are available online at http://www.mdpi.com/2071-1050/11/10/2825/s1, **Figure S1**. (A) Effects of sonication time (5–40 min) on the turbidity of CuO NPs stock suspension (30 mg/L) at pH 9; (B) size distribution by volume (%) of CuO NPs at the optimized time (30 min) of sonication; (C) speciation of Fe (III) as a function of solution pH. **Figure S2**. Measured zeta potential of CuO NPs at different pH values (3–11) and HA concentrations (0.5, 2.5, 5, and 10 mg/L); (A) without FC; (B) with (0.2 mM) FC dosage. **Figure S3**. The species of Fe(III) and Cu ions in the supernatant after coagulation without CO_2 showing (A) HA = 0 mg/L; (B) HA = 10 mg/L. **Figure S4**. The species of Fe(III) and Cu ions in the supernatant after coagulation with CO_2 (A) HA = 0 mg/L; (B) HA = 10 mg/L. **Figure S5**. Change in pH before and after coagulation experiments of CuO NPs (30 mg/L) with 0.2 mM FC dosage (A) HA = 0 mg/L; (B) HA = 10 mg/L. **Table S1**: Physicochemical properties of CuO NPs used in the current study.

Author Contributions: R.K. and I.T.Y. designed the study; R.K. and M.A.I. performed the experiment and analyzed the data; and M.A. S.Z.Z, M.A. and S.S. provided critical feedback and directed the research program; R.K. wrote the final version of the manuscript.

Funding: The BK21 plus program supported this work through the National Research Foundation of Korea (NRF), funded by the Ministry of Education of Korea (Grant No. 22A20152613545).

Acknowledgments: The authors acknowledge the help of Adeela Hanif and Muhammad Haroon for their support in measurements and provision of critical feedback. The staff members of Zero Emission Center at Sungkyunkwan University are also acknowledged for their support in measurements and data reduction.

Conflicts of Interest: The authors declare the absence of any conflict of interest.

References

1. Vance, M.E.; Kuiken, T.; Vejerano, E.P.; McGinnis, S.P.; Hochella Jr, M.F.; Rejeski, D.; Hull, M.S. Nanotechnology in the real world: Redeveloping the nanomaterial consumer products inventory. *Beilstein J. Nanotechnol.* **2015**, *6*, 1769–1780. [CrossRef] [PubMed]
2. Keller, A.A.; Vosti, W.; Wang, H.; Lazareva, A. Release of engineered nanomaterials from personal care products throughout their life cycle. *J. Nanopart. Res.* **2014**, *16*, 2489. [CrossRef]
3. Gawande, M.B.; Goswami, A.; Felpin, F.-X.; Asefa, T.; Huang, X.; Silva, R.; Zou, X.; Zboril, R.; Varma, R.S. Cu and Cu-Based Nanoparticles: Synthesis and Applications in Catalysis. *Chem. Rev.* **2016**, *116*, 3722–3811. [CrossRef] [PubMed]

4. Ju-Nam, Y.; Lead, J.R. Manufactured nanoparticles: an overview of their chemistry, interactions and potential environmental implications. *Sci. Total Environ.* **2008**, *400*, 396–414. [CrossRef] [PubMed]
5. Aruoja, V.; Dubourguier, H.-C.C.; Kasemets, K.; Kahru, A. Toxicity of nanoparticles of CuO, ZnO and TiO_2 to microalgae Pseudokirchneriella subcapitata. *Sci. Total Environ.* **2009**, *407*, 1461–1468. [CrossRef] [PubMed]
6. Zheng, X.; Wu, R.; Chen, Y. Effects of ZnO nanoparticles on wastewater biological nitrogen and phosphorus removal. *Environ. Sci. Technol.* **2011**, *45*, 2826–2832. [CrossRef]
7. Dreher, K.L. Health and environmental impact of nanotechnology: toxicological assessment of manufactured nanoparticles. *Toxicol. Sci.* **2004**, *77*, 3–5. [CrossRef] [PubMed]
8. Son, J.; Vavra, J.; Forbes, V.E. Effects of water quality parameters on agglomeration and dissolution of copper oxide nanoparticles (CuO-NPs) using a central composite circumscribed design. *Sci. Total Environ.* **2015**, *521*, 183–190. [CrossRef]
9. Peng, C.; Shen, C.; Zheng, S.; Yang, W.; Hu, H.; Liu, J.; Shi, J. Transformation of CuO Nanoparticles in the Aquatic Environment: Influence of pH, Electrolytes and Natural Organic Matter. *Nanomaterials* **2017**, *7*, 326. [CrossRef]
10. Zhang, H.; Chen, B.; Banfield, J.F. Particle size and pH effects on nanoparticle dissolution. *J. Phys. Chem. C* **2010**, *114*, 14876–14884. [CrossRef]
11. Wall, N.A.; Choppin, G.R. Humic acids coagulation: influence of divalent cations. *Appl. Geochem.* **2003**, *18*, 1573–1582. [CrossRef]
12. Bian, S.W.; Mudunkotuwa, I.A.; Rupasinghe, T.; Grassian, V.H. Aggregation and dissolution of 4 nm ZnO nanoparticles in aqueous environments: Influence of pH, ionic strength, size, and adsorption of humic acid. *Langmuir* **2011**, *27*, 6059–6068. [CrossRef]
13. Yang, K.; Xing, B. Adsorption of fulvic acid by carbon nanotubes from water. *Environ. Pollut.* **2009**, *157*, 1095–1100. [CrossRef]
14. Zhong, Z.; Li, W.; Xing, W.; Xu, N. Crossflow filtration of nanosized catalysts suspension using ceramic membranes. *Sep. Purif. Technol.* **2011**, *76*, 223–230. [CrossRef]
15. Springer, F.; Laborie, S.; Guigui, C. Removal of SiO_2 nanoparticles from industry wastewaters and subsurface waters by ultrafiltration: Investigation of process efficiency, deposit properties and fouling mechanism. *Sep. Purif. Technol.* **2013**, *108*, 6–14. [CrossRef]
16. Tiede, K.; Boxall, A.B.A.; Wang, X.; Gore, D.; Tiede, D.; Baxter, M.; David, H.; Tear, S.P.; Lewis, J. Application of hydrodynamic chromatography-ICP-MS to investigate the fate of silver nanoparticles in activated sludge. *J. Anal. At. Spectrom.* **2010**, *25*, 1149–1154. [CrossRef]
17. Zhang, Y.; Chen, Y.; Westerhoff, P.; Crittenden, J.C. Stability and removal of water soluble CdTe quantum dots in water. *Environ. Sci. Technol.* **2007**, *42*, 321–325. [CrossRef]
18. Hyung, H.; Kim, J.-H.H. Dispersion of C60in natural water and removal by conventional drinking water treatment processes. *Water Res.* **2009**, *43*, 2463–2470. [CrossRef]
19. Holbrook, R.D.; Kline, C.N.; Filliben, J.J. Impact of source water quality on multiwall carbon nanotube coagulation. *Environ. Sci. Technol.* **2010**, *44*, 1386–1391. [CrossRef]
20. Abbott Chalew, T.E.; Ajmani, G.S.; Huang, H.; Schwab, K.J. Evaluating nanoparticle breakthrough during drinking water treatment. *Environ. Health Perspect.* **2013**, *121*, 1161–1166. [CrossRef]
21. Wang, H.T.; Ye, Y.Y.; Qi, J.; Li, F.T.; Tang, Y.L. Removal of titanium dioxide nanoparticles by coagulation: effects of coagulants, typical ions, alkalinity and natural organic matters. *Water Sci. Technol.* **2013**, *68*, 1137–1143. [CrossRef] [PubMed]
22. Khan, R.; Inam, M.; Park, D.; Zam Zam, S.; Shin, S.; Khan, S.; Akram, M.; Yeom, I. Influence of Organic Ligands on the Colloidal Stability and Removal of ZnO Nanoparticles from Synthetic Waters by Coagulation. *Processes* **2018**, *6*, 170. [CrossRef]
23. Khan, R.; Inam, M.A.; Park, D.R.; Khan, S.; Akram, M.; Yeom, I.T. The Removal of CuO Nanoparticles from Water by Conventional Treatment C/F/S: The Effect of pH and Natural Organic Matter. *Molecules* **2019**, *24*, 914. [CrossRef] [PubMed]
24. Keller, A.A.; Wang, H.; Zhou, D.; Lenihan, H.S.; Cherr, G.; Cardinale, B.J.; Miller, R.; Zhaoxia, J.I. Stability and aggregation of metal oxide nanoparticles in natural aqueous matrices. *Environ. Sci. Technol.* **2010**, *44*, 1962–1967. [CrossRef]

25. Miao, L.; Wang, C.; Hou, J.; Wang, P.; Ao, Y.; Li, Y.; Lv, B.; Yang, Y.; You, G.; Xu, Y. Effect of alginate on the aggregation kinetics of copper oxide nanoparticles (CuO NPs): bridging interaction and hetero-aggregation induced by Ca^{2+}. *Environ. Sci. Pollut. Res.* **2016**, *23*, 11611–11619. [CrossRef]
26. Zhang, L.; Mao, J.; Zhao, Q.; He, S.; Ma, J. Effect of AlCl3 concentration on nanoparticle removal by coagulation. *J. Environ. Sci.* **2015**, *38*, 103–109. [CrossRef]
27. Omar, F.M.; Aziz, H.A.; Stoll, S. Aggregation and disaggregation of ZnO nanoparticles: influence of pH and adsorption of Suwannee River humic acid. *Sci. Total Environ.* **2014**, *468*, 195–201. [CrossRef] [PubMed]
28. Tang, H.X. Inorganic polymer flocculation theory and flocculants. *China Build. Ind. Press. Peking* **2006**.
29. Zirino, A.; Yamamoto, S. A pH-dependent model for the chemical speciation of copper, zinc, cadmium, and lead in seawater. *Limnol. Oceanogr.* **1972**, *17*, 661–671. [CrossRef]
30. Albrecht, T.W.J.; Addai-Mensah, J.; Fornasiero, D. Effect of pH, concentration and temperature on copper and zinc hydroxide formation/precipitation in solution. In Proceedings of the Chemeca 2011: Engineering a Better World, Sydney Hilton Hotel, NSW, Australia, 18–21 September 2011; pp. 2100–2110.

Article

Removal of ZnO Nanoparticles from Natural Waters by Coagulation-Flocculation Process: Influence of Surfactant Type on Aggregation, Dissolution and Colloidal Stability

Rizwan Khan [1], Muhammad Ali Inam [1], Muhammad Mazhar Iqbal [1], Muhammad Shoaib [1], Du Ri Park [1], Kang Hoon Lee [2], Sookyo Shin [1], Sarfaraz Khan [3] and Ick Tae Yeom [1,*]

1. Graduate School of Water Resources, Sungkyunkwan University (SKKU) 2066, Suwon 16419, Korea; rizwankhan@skku.edu (R.K.); aliinam@skku.edu (M.A.I.); mazhar0559@skku.edu (M.M.I.); changezi@skku.edu (M.S.); enfl8709@skku.edu (D.R.P.); tkssk08@gmail.com (S.S.)
2. Center for Built Environment, Sungkyunkwan University, (SKKU) 2066, Suwon 16419, Korea; diasyong86@gmail.com
3. Key Laboratory of the Three Gorges Reservoir Region Eco-Environment, State Ministry of Education, Chongqing University, Chongqing 400045, China; Sfk.jadoon@yahoo.com

* Correspondence: yeom@skku.edu

Received: 30 November 2018; Accepted: 19 December 2018; Published: 20 December 2018

Abstract: The zinc oxide nanoparticles (ZnO NPs) and surfactants that are widely used in commercial and industrial products lead to the likelihood of their co-occurrence in natural water, making it essential to investigate the effect of surfactants on the fate and mobility of ZnO NPs. The present study seeks to elucidate the effect of an anionic sodium dodecyl sulfate (SDS) and a nonionic nonylphenol ethoxylate (NPEO), on ZnO NPs adsorption, aggregation, dissolution, and removal by the coagulation process. The results indicate that the presence of SDS in ZnO NPs suspension significantly reduced the ζ-potential and hydrodynamic diameter (HDD), while the effect of NPEO was found not to be significant. The sorption of SDS and NPEO by ZnO NPs were fitted with Langmuir model, but the Freundlich isotherm was more suitable for SDS at pH 9.0. Moreover, the adsorption was strongly pH-dependent due to the formation of mono-bilayer patches onto the NPs. The SDS remarkably affect the dissolution and aggregation phenomena of ZnO NPs in natural waters as compared to NPEO. Finally, the coagulation results showed that the removal efficiency of ZnO, Zn^{2+} and the surfactant in synthetic and wastewaters at optimum ferric chloride (FC) dosage reached around 85–98% and 20–50%, respectively. Coagulation mechanism investigation demonstrated that the cooperation of charge neutralization and adsorptive micellar flocculation (AMF) might play an important role. In summary, this study may provide new insight into the environmental behavior of coexisting ZnO NPs and surfactants in water treatment processes, and it may facilitate their sustainable use in commercial products and processes.

Keywords: adsorption; aggregation; coagulation; dissolution; surfactants; wastewater treatment; ZnO NPs

1. Introduction

Zinc oxide nanoparticles (ZnO NPs) are used in various industrial applications, such as material sciences, environmental remediation, cosmetics, medicine, foods, and biosensors [1]. The annual production of ZnO NPs is estimated to be (31,500–34,000 t/y), which is anticipated to increase over time. It has been demonstrated that around 10–25% of produced ZnO NPs may release into the environment and enter freshwater sources [2,3]. A recent study [3] has reported the ZnO NPs concentration up to

0.05–10 μg/L in United States (U.S) surface waters; thus, NPs accumulate in aquatic organisms and may consequently cause undesirable environmental and health risks. Several studies [4,5] have reported the toxicity of ZnO NPs to aquatic organisms, such as bacteria, algae, cladocerans, soil organisms, rotifers, fish, and daphnia. Moreover, the genotoxicity of NPs in human mesenchymal stem cells has been demonstrated in previous studies as well [5,6]. The Zn-based NPs can penetrate the cells and it may cause DNA fragmentations as well as chromosomal aberrations. Thus, environmental transportation and transformation of ZnO NPs in the water matrices are essential to be studied in order to control the associated potential environmental risks.

When ZnO NPs are released into aquatic environments, their transport within the system depends on multiple environmental aspects as well as physiochemical properties (i.e., size, shape, concentration, solubility). Some studies [7,8] recently reported that factors, such as pH, dissolved organic matter (DOM), electrolyte type, and ionic strength (IS) significantly affect the NPs agglomeration process. The ZnO NPs is stable at pH value between 6.5 to 8.5, while the higher IS enhance the NPs destabilization with decreasing tendency in the order of $Ca^{2+} > Mg^{2+} > Na^{+} \approx K^{+}$ [7]. The presence of DOM remarkably enhanced the stability of ZnO NPs due to the effect of steric hindrance and electrostatic repulsion [8]. It has been reported that ZnO NPs coated with DOM (humic and salicylic acid) hinders the removal of NPs during the coagulation process [9]. In addition, surfactants can typically be detected in municipal and industrial wastewater in concentrations ranging from 1–1000 mg/L [10,11]. The surfactants are harmful to the aquatic organism and humans, while they also affect the quality of water due to froth formation [12]. Besides their strong adsorption capacity, surfactants can form highly stable colloids in solution, even at low concentration, by lowering the interfacial surface tension. Therefore, the co-presence of ZnO NPs and surfactants in the system may alter their transportation behavior as well as increase their bioavailability in aqueous environments.

The nature of the interaction between the ZnO NPs and surfactants depends on environmental water matrices. The surfactants, such as Triton X-100 and dodecyl benzene sulfonate (SDBS), were found to inhibit the uncontrollable aggregation of titanium dioxide (TiO_2) by enhancing the electro steric repulsive forces between the NPs [13]. Xuankun Li showed that ionic sodium dodecyl sulfate (SDS) significantly reduce the aggregation of TiO_2 NPs and stabilize the colloidal suspension at higher concentration [14,15]. Moreover, it was also demonstrated that at pH 9.0, increasing surfactant concentration led to the enhancement of transport distance of TiO_2 in porous media [15,16]. The significant effect of low concentrations (0.004% (w:v)) of SDBS and cetyl trimethyl ammonium bromide (CTAB) was found on the retention and transport properties of graphene NPs in saturated porous media [17,18]. Some researchers [18] have used SDBS to disperse and stabilize several NPs to produce surfactant-solubilized NPs that were highly mobile in saturated sandy as well as porous media. Several functional groups of surfactants may readily be adsorbed on the ZnO NPs surface and affect the aggregation behaviors by changing the surface properties, thereby increasing the colloidal stability and release of Zn^{2+} ions in receiving water bodies [19]. Thus, it is important to explore the removal of ZnO NPs and potential surfactants ions from water to reduce the NPs associated environmental risks.

To date, many studies [10,11,19,20] have focused on the effect of surfactant on the aggregation behaviors of NPs in simulated natural waters. To address the current limitations in knowledge regarding the effect of surfactants on ZnO NPs behavior in natural surface waters and its fundamental mechanism, we thoroughly studied the effect of two surfactants, i.e., anionic sodium dodecyl sulfate (SDS) and nonionic nonylphenol ethoxylates (NPEO), on the stability, dissolution, and sedimentation of ZnO in different aquatic chemistry conditions. In particular, the effects of SDS and NPEO on sorption ability and ZnO NPs stability at various pH ranges were assessed. In addition, we also studied the simultaneous removal of ZnO NPs, Zn^{2+}, and surfactants from synthetic as well as natural waters by coagulation.

2. Materials and Methods

2.1. Chemical Reagents

The commercial ZnO nanopowder (CAS # 1314-13-2) that was used in this study has a purity of 97% with vendor reported transmission electron microscope (TEM) primary size (50 nm) Table 1 for detail properties; SDS (99%) and NPEO (97%) surfactants were procured from Sigma Aldrich (St. Louis, MO, USA). The iso-electric point (pHiep) of ZnO was determined, see details in Supplementary Material (SM) (Figure S2). Sodium hydroxide (NaOH), iron (III) chloride hexahydrate ($FeCl_3$-$6H_2O$), and hydrochloric acid (HCl) were procured from local suppliers. The Synergy water system (Milli-Q, Millipore, Burlington, MA, USA) was used to produce ultra-pure water and it was used in the preparations of the stock solutions as well as synthetic waters.

Table 1. Properties of zinc oxide nanoparticles (ZnO NPs) used in the current study.

Nanomaterials Parameter	Unit	Technique	Value
Manufacturer-reported size	nm	TEM	<50
Bulk Density	$g(cm^3)^{-1}$	-	5.60
Solubility			High
BET specific surface area	m^2g^{-1}	BET	12.2 ± 0.4
Iso-electric point (pH iep, see Figure S2B)	-	Zetasizer	9.2
Zeta potential in pure water (at pH7)	mV		$+14 \pm 2.1$
HDD measured in pure water (n = 50)	nm	DLS	280 ± 35
Purity/moisture content (see Figure S1D)	wt.%	TGA/ICP-OES	96.52/1.85
Crystalline structure (Figure 1C)	-	XRD	Hexagonal
Shape			Polyhedral roughly round
Hamaker Constant	J	-	1.9×10^{-20}
Net energy barrier in pure water (IS 5×10^{-6} M)	kT [(a)]	-	42.8

[a] 1 KT = 4.1142 × 10–21 J at 25 °C.

2.2. Preparation of NPs and Surfactants Stock Solutions

First, 100 mg of ZnO powder was weighed using a microbalance (Mettler Toledo Ag Model XP26DR) and dispersed into 1000 mL water. The dispersion was subjected to probe sonicate for a cycle of 30 min using an ultrasonicator (Bio safer 1200-90, 400 W, 20 kHz, 40% amplitude-continuous mode, Nanjing, China). The properties of the surfactants that were used in this study are shown in Table 2, while 0.1% (w:v) stock solutions of both surfactants were prepared in ultra-pure water. The coagulant stock solution, i.e., 0.1 M ferric chloride (FC) concentration solution was prepared by dissolving a specific amount of $FeCl_3$-$6H_2O$ into the pure water. The stock solutions were kept in the dark at 4 °C before diluting it to the required experimental concentration.

Table 2. Properties of surfactants used in the present study.

Surfactant Type	Formula	Structure	Molecular Weight (g/mol)
SDS (Anionic)	$CH_3(CH_2)_{11}OSO_3Na$	O-S(=O)(=O)-O⁻ Na⁺	288.38
NPEO (Nonionic)	$C_9H_{19}C_6H_4(OCH_2CH_2)_9OH$	C_9H_{19} … OH, n	616.82

2.3. Preparation and Characteristic of Environmental Water Samples

Six water samples were used to assess the impact of different surfactants on the environmental behavior of ZnO NPs. Four synthetic waters (W1–W4) with different concentrations of both surfactants were prepared in the lab, as shown in Table 3. The other two samples were tap water (TW) (Sungkyunkwan University, Suwon Campus, Gyeonggi-do, Korea). The wastewater (WW) samples was obtained from metal processing industry (LS-Nikko Copper Inc., Onsan National Industrial Complex, Ulsan, Korea). The wastewater sample was diluted to reduce the concentration of total organic carbon (TOC) and heavy metal ions. The details of several parameters are shown in Table S1.

Table 3. Characteristics of synthetic and environmental waters.

Surfactant Type	Concentration (w:v) %		Water Code	Released Zn^{2+} (mg/L)	ZnO NPs (mg/L)	ζ-potential (mV)	HDD (nm)	pH
Control	0	-	Control	0.893 ± 0.10	8.56 ± 0.41	18.0 ± 1.3	458 ± 56	7.06 ± 0.02
Anionic	0.030	SDS	W1	1.524 ± 0.04	6.78 ± 0.50	−16.4 ± 0.5	280 ± 35	6.94 ± 0.04
	0.050		W2	1.680 ± 0.16	6.12 ± 0.31	−28.3 ± 0.8	205 ± 70	6.76 ± 0.17
Nonionic	0.030	NPEO	W3	0.684 ± 0.28	8.05 ± 0.14	16.4 ± 0.6	520 ± 60	7.02 ± 0.01
	0.050		W4	0.921 ± 0.33	7.57 ± 0.24	12.1 ± 0.4	485 ± 48	6.85 ± 0.13
Anionic	0.050	SDS	(WW+SDS)	3.587 ± 0.02	5.027 ± 0.08	−25.6 ± 0.1	190 ± 35	7.87 ± 0.21
			(TW+SDS)	2.423 ± 0.01	7.41 ± 0.10	−14.0 ± 0.3	235 ± 62	6.95 ± 0.34
Nonionic	0.050	NPEO	(WW+NPEO)	2.719 ± 0.02	6.09 ± 0.23	−17.8 ± 0.5	265 ± 79	7.56 ± 0.15
			(TW+NPEO)	0.769 ± 0.01	8.26 ± 0.08	8.6 ± 0.2	584 ± 98	7.02 ± 0.01

The pH of water samples was measured by (HACH-HQ40d portable multi-parameter meter, MA, USA). The TOC in water samples were analyzed using TOC analyzer (TOC-5000A, Shimadzu Corporation, Japan). Moreover, the concentrations of F^-, $SO_4{}^{2-}$, Cl^-, HCO_3 −, K^+, Ca^{2+}, Na^+, and Mg^{2+} ions in the natural waters were measured with ion chromatography (861-Advanced Compact IC, Switzerland) using standard methods [21]. All of the collected water samples were filtered through 0.45 μ glass fiber membrane and stored in 5 L amber reagent bottle before being used for the experiment.

2.4. Laboratory Batch Experiments

2.4.1. Adsorption Study and Isotherm Modelling

In order to fully understand the adsorption behaviors of the different surfactants on ZnO NPs surface, batch adsorption kinetics of SDS and NPEO (10 mg/L) on ZnO NPs (30 mg/L) at various pH ranges were determined. The pHs of the solutions were adjusted to 5.0, 7.0, and 9.0 with NaOH (0.1 M) and HCl. Subsequently, the suspensions vials were shaken at 150 rpm in a shaker (SK, 300 companions, Korea) and were sampled at consistent intervals of time from 0.5 to 72 h. An aliquot of each mixture was taken and centrifuged for 30 min at 10,000 rpm using (Hettich Centrifuge Universal 320R) in order to measure the residual surfactant concentration using TOC analysis. The kinetics data of each surfactant at different pH onto ZnO surface was fitted by pseudo-first order (PFO), as well as pseudo-second-order (PSO), models Equations (1) and (2), respectively.

$$\log(q_e - q_t) = \log q_e - \frac{k_1 t}{2.303} \tag{1}$$

$$\frac{t}{q_t} = \frac{1}{k_2 q_e^2} + \frac{t}{q_e} \tag{2}$$

where qe and qt are the adsorbed amounts of SDS and NPEO at equilibrium or any time (mg/g), respectively; k_1 (1/h) and k_2 (g/mg/h) represent the PFO and PSO rate constants, respectively.

Furthermore, stock solutions of both surfactants were diluted into vials in order to form 50 mL adsorption experiment suspension. The concentrations in the vials ranged between 0 to 50 mg /L, and the pHs of all experimental solutions were adjusted to 5.0, 7.0, and 9.0. Afterwards, suspensions were shaken at 150 rpm for 24 h, so as to equilibrate the suspension, then centrifuged at 10,000 rpm for 30 min. Moreover, in order to explore the adsorption mechanism and bond analysis onto NPs surface at different pH values, the Fourier transform infrared spectroscopy (JASCO, FT-IR-4700, USA) of ZnO NP surfactants complexes was conducted. The experimental data was further fitted with nonlinear forms of Langmuir and Freundlich isotherm models using Equations (3) and (4), respectively.

$$q_e = \frac{q_m K_L C_e}{1 + K_L C_e} \tag{3}$$

$$q_e = K_F\, C_e^{\frac{1}{n}} \tag{4}$$

where q_e (mg/g) is the adsorbed amount of SDS and NPEO on the ZnO surface; C_e (mg/L) is the equilibrium concentration of SDS and NPEO in suspension; q_m and K_L (L/mg) shows a maximum saturated adsorption amount and Langmuir constant relates to binding strength; and, n and K_F (L/mg) in the Freundlich equation relates to the intensity of heterogeneity and adsorption capacity, respectively.

2.4.2. Sedimentation and Dissolution Kinetics in Different Waters

The aggregation and sedimentation experiment of ZnO NPs was carried out in synthetic and environmental waters. The ZnO NPs stock suspension was spiked in water samples in the glass vials, so as to accomplish the desired concentrations, and was then immediately sonicated. The vials containing the mixtures were placed on a shaker and shaken at 150 rpm for 24 h to simulate the circulation conditions within natural waters. After 24-h of equilibration, the ζ-potential and intensity weighted average hydrodynamic diameter (HDD) of each suspension were measured with a dynamic light scattering (DLS) (Zeta-sizer NanoZS, Malvern, Worcestershire, UK) analyzer. Subsequently, the vials were left undisturbed, and the sedimentation rate of the NPs was measured with varying times through a turbidimeter (Hach Benchtop 2100 N, Loveland, CO, USA). Since turbidity increases with NPs concentration, the sedimentation rate $\delta(C/C_0)/\delta t$ can be related to the normalized NPs turbidity C/C_0, where C and C_0 are the turbidity at times t and 0, respectively. The sedimentation rate can be estimated from the initial 10% decrease in normalized turbidity, which tends to occur within the first 2 h for fast sedimentation and within 24 h for slow sedimentation. For the dissolution experiment, the stock solution of ZnO NPs was diluted into the 10 mg/L and dispersed through the ultrasonication. The ZnO NPs suspension was added in the equal volume of pure water as a control, tap water, and wastewaters suspension (Table 3). The initial concentration of Zn^{2+} ions in each water sample was recorded after the start of the experiments and left unattended for 10 days, as shown in Table 2. Subsequently, at different time intervals, an aliquot of 2 mL from each mixture was obtained and centrifuged at 10,000 rpm for 30 min prior to the Zn^{2+} ions concentration analysis.

2.4.3. Coagulation-Flocculation Experiments

Prior to the coagulation-flocculation experiment, the ZnO NP suspension that was prepared in 1 mM $NaHCO_3$ solution was sonicated, as described previously in Section 2.2, and transferred into 250 mL beakers. The coagulant was dosed and pH was adjusted to 7.0 using 0.1 M HCl or NaOH solution. The coagulation experiments were performed using a jar tester (SJ10, Young Hana, Korea). The coagulation was carried out in three steps, as described in our earlier study [9], and precisely described as: (1) rapid mixing at 200 rpm for 2 min to initialize the coagulation, (2) slow mixing at 40 rpm for 20 min to increase flocculation, (3) and sedimentation for 30 min. After sedimentation, 100 mL aliquot was collected from 2 cm below the surface for the analysis of various water quality parameters. The effect of FC dosage between 0 and 10 mg/L was studied on synthetic and natural waters samples, where 0 shows the control experiment (no coagulant) for each water sample. Each experiment was performed in triplicate and relative standard deviations (RSD) were reported, while the removal efficiency (%) was calculated using Equation (5).

$$\text{Removal efficiency } (\%) = \frac{C_o - C_f}{C_o} \times 100 \quad (5)$$

where C_o (mg/L) is the initial and C_f (mg/L) is the final concentration of ZnO NPs, Zn^{2+}, SDS, and NPEO, respectively.

2.5. Characterization and Measurement of ZnO NPs Suspension

Following the coagulation experiment, 100 mL sample was immediately collected and centrifuged at 10,000 rpm for 30 min in order to analyze the Zn^{2+} and ZnO NPs in the solution. The residual concentration of metals in the supernatant was measured by Inductively Coupled Plasma Optical Emission Spectroscopy (ICP-OES, Agilent Technologies). The analysis of SDS and NPEO surfactant were quantified using liquid chromatography and then a mass spectrometer (MS 4500 Qtrap, AB SCIEX, USA). The optical properties of ZnO NPs were determined through absorption spectra using a UV Vis spectrophotometer (Optizen-2120, Mecasys, Korea) at 25 $\pm$ 1 °C in a 10 mm quartz cuvette from the 250–800 nm wavelength range. The structural characterization of the ZnO powder, such as XRD, XPS, and Brunauer Emmett Teller (BET) specific surface area were investigated by Rigaku D max C III, X-ray diffractometry (Rigaku Corporation, Japan), and X-ray photoelectron spectroscopy (XPS) using (XSAM HS, KRATOS) and (ASAP 2020, Micromeritics, USA). Furthermore, Raman spectroscopy was acquired with a high-resolution confocal Lab Ram HR Evolution Horiba Jobin Yvon microscope.

3. Results and Discussions

3.1. NPs Characterization and Properties

The morphology and surface properties of pristine ZnO NPs were investigated through various characterization techniques. The UV−Vis spectra of ZnO NPs (Figure 1A) indicated that the strong absorption peak at 370 nm, which should be derived from the $\pi-\pi^*$ transition of the Zn moiety [9]. The XPS spectra of ZnO NPs showed the Zn $2p_{1/2}$ and $2p_{3/2}$ binding energy peaks at 1042 and 1021 eV, respectively (Figure 1B). The O 1s spectrum that is shown in the inset of (Figure 1B) also indicates that the binding energy of O 1s is resolved into two major peaks, i.e., 530 eV ascribed to surrounding Zn atoms and the peak at 531.2 eV is attributed to the formation of (Zn-OH) with chemisorbed oxygen species [22]. The XRD pattern in (Figure 1C) shows the peaks at diffraction angles, i.e., 31.76°, 34.37°, 36.24°, 47.54°, 56.58°, 62.85°, and 67.97°, corresponding to the planes (1 0 0), (0 0 2), (1 0 1), (1 0 2), (1 1 0), (1 0 3), and (1 1 2), respectively. Therefore, it is indexed to a wurtzite hexagonal ZnO and a highly crystalline structure. The FT-IR spectra further revealed various chemically reactive functional groups and a capping agent on the outermost surface of ZnO NPs, as shown in (Figure 1D). The Raman spectra and specific surface area of ZnO NPs, as well as FT-IR analysis of both surfactants, were also conducted, as shown in Figure S1A–C.

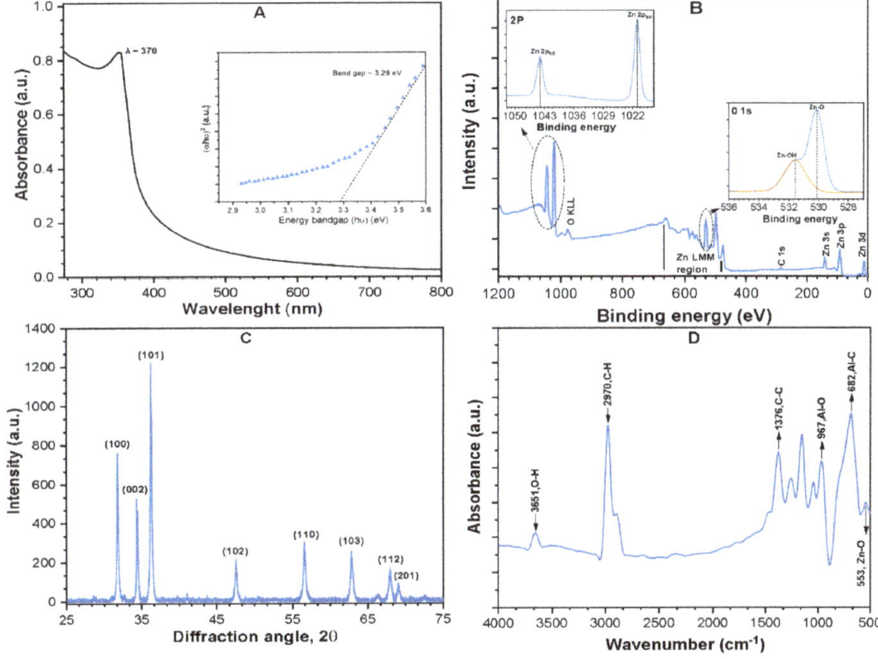

Figure 1. (**A**) UV−Vis spectra of ZnO NPs (10 mg/L) in ultra-pure water by full-wave scanning at pH 7; (**B**) High-resolution X-ray photoelectron spectroscopy (XPS) survey scan of ZnO; (**C**) XRD pattern of commercial ZnO NPs powder; and, (**D**) FT-IR spectra of ZnO. The inset in (Figure 1A,B) shows the band gap, 2p peak, and O 1s peak, respectively.

3.2. Effects of pH and Surfactants on ζ-potential and HDD of ZnO NPs

The influences of surfactants type and concentration on ζ-potential and HDD of ZnO NPs suspension were determined at various pHs 5.0, 7.0, and 9.0 (Figure 2A,B). As shown in Figure 2A, SDS presented greater ability for strengthening the surface charge and promote the stability of ZnO suspension than NPEO in all pH ranges. For instance, as the SDS concentration increased from 0 to 0.05%, the absolute value of ζ-potential at pH 5.0, 7.0, and 9.0 significantly decreased from 12.32 to −10.51, 14.05 to −21.14, and 5.32 to −35.96 ± 2.39 mV, respectively. The NPEO exhibited no interference effect on ζ-potential, even at higher concentrations in solution (Figure 2A), which was in accordance with the earlier study by Shimada et al. [19]. The low concentration of 0.015% of SDS at pH 5.0 and 7.0 sharply increases the HDD of ZnO NPs, with a maximum value of 2354.1 ± 150.3 and 1822.6 ± 153.8 nm, respectively (Figure 2B), which corresponds to the neutralization of the ZnO NPs surface charge [23]. It has been proposed that at the appropriate concentration, the SDS efficiently compresses the double layer (DL) of NPs, which later decreases the electrostatic repulsive force [17]. However, the HDD of ZnO NPs rapidly declined and became much smaller upon an increase of SDS concentration above 0.02% (Figure 2B). A probable explanation might be that, when the ζ-potential value is higher than ±15 to ±20 mV, NPs tend to repel each other and remain stable in the suspension [24]. The impact of SDS on ZnO NPs suspension was intense at pH 7.0 and 9.0, which was consistent with the findings of previous studies [25,26] that positive charges on the surface of NPs provide favorable attachment sites for surfactant molecules. In addition, with the further increase of SDS concentration up to 0.05%, the ZnO NPs HDD (280.2 ± 35.1 nm) significantly reduces and becomes stable. This reversed the ζ-potential from positive to negative and caused a significant increase in the absolute surface charge of ZnO NPs (Figure 2A). This might be due to the electro steric stabilization of ZnO NPs and shielding of SDS molecules by forming outer-sphere surface complexation [27].

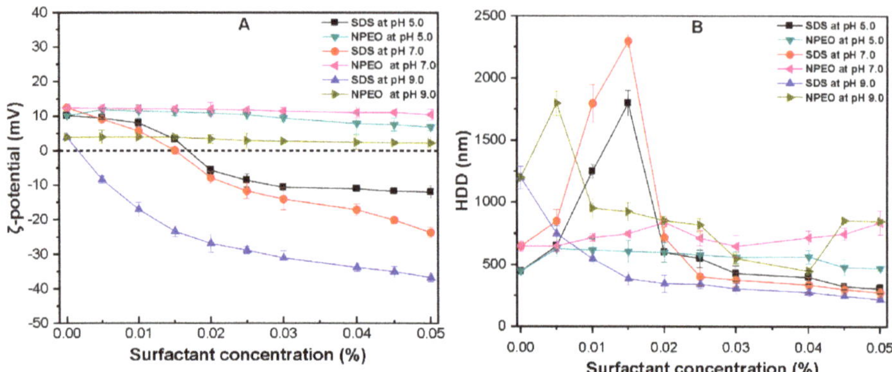

Figure 2. At various pH and concentration of surfactant showing (**A**) ζ-potential; and, (**B**) hydrodynamic diameter (HDD) of ZnO NPs.

By contrast, the presence of NPEO surfactant reduced the HDD of the ZnO NPs aggregate without exerting any noticeable impact on the ζ-potential throughout the pH range. Since non-ionic surfactants are non-charged and their hydrophilic part usually ester, phenol is non-dissociable, while the hydrophilic part contains the alkyl or alkylbenzene group [17,19]. Thus, due to the adsorption of non-ionic surfactant onto ZnO NPs through the process that reduced the agglomeration of the particle due to steric inhibition effect of surfactants macro-molecules. NPEO surfactant contains multiple phenol ring structures. As the NPEO concentration reached 0.05%, the ZnO aggregate size reduced to 536.5 $\pm$ 47.2, 790 $\pm$ 35.64, and 800 $\pm$ 50.84 nm at pH 5.0, 7.0, and 9.0, respectively. These results indicated that high concentrations of both surfactants significantly enhanced the stability of ZnO NPs. Moreover, the variations in charge reversal and HDD might be related to their specific adsorption mechanisms onto ZnO NPs. Thus, adsorption kinetics and isotherms modeling studies were conducted.

3.3. Surfactants Adsorption onto NPs Surface

Figure 3A,B show the adsorption kinetics of SDS and NPEO on the surface of ZnO NPs at various pH. The adsorption rates of SDS and NPEO were rapid in the first 5 h in all studied pH, and they reached 90.2 and 80.4% of their maximum adsorptions, respectively. However, it took an additional 17 h to reach 98% of their maximum adsorption capacity. This excellent sorption ability of surfactant might be attributable to the relatively high surface area of ZnO NPs. It has been reported previously [28,29] that more interstices between NPs enhance the surfactants molecules binding to active sites of ZnO [28,29]. It can also be observed that the adsorption efficiencies of SDS and NPEO on ZnO NPs were pH dependent with excellent sorption capacities at pH 9.0 and 7.0. With increasing pH, the surface charge of ZnO NPs gradually reversed from positive to negative, while the Zn-O^- on the surface at pH 9.0 was gradually protonated to form Zn-OH^{2+} [30,31]. Moreover, the sulfate group that was present in SDS surfactant could bind to ZnO NPs to form outer-sphere complexes by replacing the previously adsorbed OH group [32]. Such phenomena could be weakened with decreasing pH due to the protonation of hydroxyl groups on NPs surface.

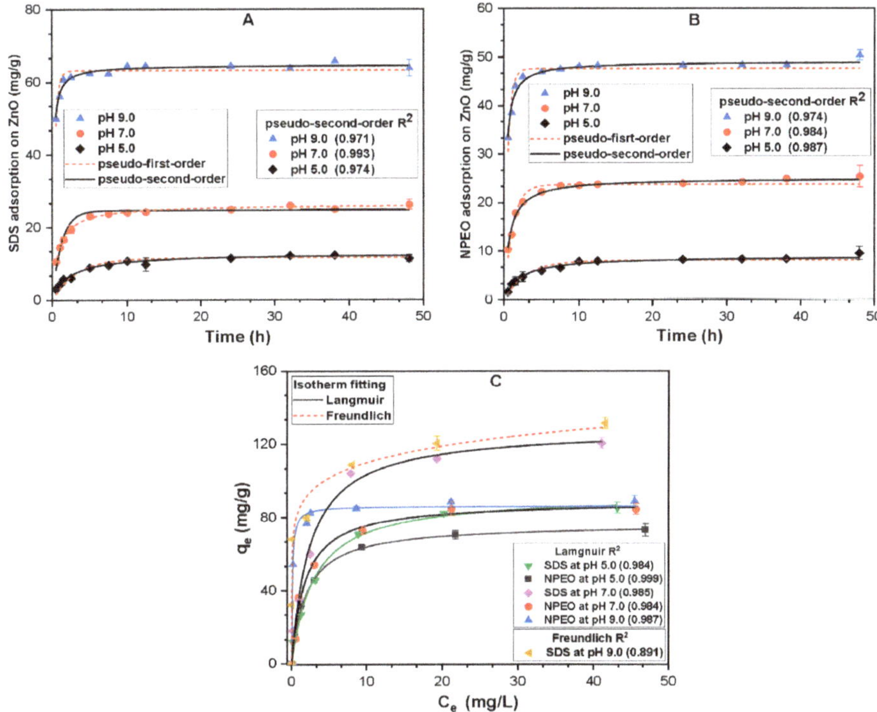

Figure 3. Adsorption kinetics study of (10 mg/L of each surfactant); (**A**) sodium dodecyl sulfate (SDS); (**B**) nonionic nonylphenol ethoxylate (NPEO); and, (**C**) isotherms modeling (0–50 mg/L of SDS and NPEO surfactants) at different pH onto ZnO NPs (30 mg/L).

Our results are consistent with previous studies [14,17,19,24], which suggested that adsorption of SDS onto TiO_2 NPs was surface bond via H-Bond interaction between the sulfate head group and oxide surface. The adsorption of SDS was higher and faster than that of NPEO based on rate constants (*k*) in all pH ranges (Table 4). The SDS is considered relatively more hydrophilic as compared with NPEO due to the higher contents of sulfate species [14,33]. Under the influence of electrostatic attractive and hydrophilic forces, SDS molecules could favorably be adsorbed onto the ZnO NPs surface. These results serve as a reconciliation with the previous observation [24] that anionic surfactants have a higher affinity to TiO_2 NPs than nonionic surfactants due to their bilayer adsorption property on the colloid surface. It is noteworthy that the adsorptions of both surfactants were even observed at pH 5.0, but to a lesser extent, which might be ascribed to surfactant self-assembly interaction at the solid/aqueous interface. It was reported [11,34] that the H^+ possessed a greater affinity for functional groups of ZnO NPs or negatively charged functional groups of surfactant molecules, both of which form a bridge with surfactant molecules, thereby bonding to ZnO NPs by competitive adsorption. Similar phenomena were also observed [34,35] for goethite NPs, where at lower pH organic pollutant adsorbed onto the surface by forming H-Bond between the Fe-O^- and O-H group of pollutant. The PSO model was fitted better due to its correlation coefficients being higher than PFO (Table 4). The results in Table 4 demonstrate that the attachment of surfactants onto ZnO NPs favors the chemisorption process that might be rate limiting in the sorption step of both surfactants.

Table 4. The pseudo-first and second-order kinetics constants for adsorption of SDS and NPEO onto ZnO NPs.

		Pseudo-First-Order			Pseudo-Second-Order		
Surfactant	pH	q_e (mg/g)	k_1 (1/h)	R^2	q_e	k_2 (g/mg/h)	R^2
SDS	5	11.58	0.745	0.914	12.76	0.0354	0.974
	7	24.77	1.881	0.928	26.33	0.0471	0.993
	9	63.27	6.582	0.858	64.72	0.1105	0.971
NPEO	5	8.134	0.806	0.950	8.956	0.0535	0.987
	7	23.76	2.099	0.966	25.09	0.0573	0.984
	9	47.66	4.740	0.896	49.09	0.0887	0.974

In order to obtain a comprehensive understanding of adsorption behavior, the isotherm study was conducted under varying surfactant concentration (0–50 mg/L) with ZnO NPs concentration (30 mg/L), as shown in Figure 2C. As expected, the NPEO surfactant shows less adsorption capacity than SDS. The isotherm of both surfactants reached a plateau at a concentration up to 25 mg/L, indicating the nearly complete surface coverage (Figure 2C). The result of adsorption isotherms for binding of NPEO onto ZnO NPs at various pH was better fitted by the Langmuir model (Table 4). This observation indicated that electrostatic cross-linking forces might be responsible for the monolayer adsorption of NPEO onto ZnO due to the relatively high binding affinity of Zn^{2+} with a phenol group [24]. Similar results have been observed in previous studies [36,37], where the preferential adsorption of nonionic surfactant TX-100 was observed on the surfaces of TiO_2. Similarly, the bridging effect on TiO_2 may occur as nonionic surfactant micelles that would adsorb on surface, owing to the availability of the thermodynamically-favorable surface sites [37]. The fitted parameters of both surfactants sorption on ZnO NPs surface are shown in Table 5.

Table 5. Langmuir and Freundlich fitting parameters for SDS and NPEO adsorption onto ZnO NPs.

		Langmuir Fitting			Freundlich Fitting		
Surfactant	pH	K_L	q_{max}	R^2	K_F	n	R^2
SDS	**5.0**	0.343 ± 0.06	92.67 ± 4.15	0.984	32.753 ± 4.43	3.569 ± 0.57	0.947
	7.0	0.413 ± 0.07	128.52 ± 5.26	0.985	46.528 ± 6.98	3.55 ± 0.60	0.930
	9.0	0.827 ± 1.05	130.11 ± 26.58	0.527	85.073 ± 9.46	8.89 ± 3.10	0.891
NPEO	**5.0**	0.516 ± 0.01	76.40 ± 0.62	0.999	31.206 ± 4.39	4.05 ± 0.77	0.921
	7.0	0.569 ± 0.09	88.91 ± 3.28	0.984	36.49 ± 5.75	4.00 ± 0.85	0.904
	9.0	9.731 ± 2.02	87.47 ± 1.96	0.987	70.52 ± 2.46	12.34 ± 2.07	0.981

By contrast, the adsorptions of SDS were fitted better to Langmuir model at pH 5.0 and 7.0, but the Freundlich model was more suitable at pH 9.0. This discrepancy suggests that the multilayer adsorption of SDS might be developed at pH 9.0, while the monolayer adsorption was formed under lower pH conditions. The possible mechanism for this would be that SDS presented high hydrophilicity at basic conditions, which mainly promoted multilayer adsorption of SDS on the NPs besides electrostatic interaction forces [27]. The adsorption trend of both surfactants suggests the uptake of surfactant-enhanced at higher pH values. This might be related to H^+ ion competition at lower pH value and the weakly acidic nature of the ZnO NPs active sites that favors the molecules bonding [38]. It can be concluded that ionic surfactant (SDS) adsorbed to ZnO NPs more effectively than nonionic surfactant (NPEO) due to the presence of higher levels of adsorptive interaction and electrostatic force.

Characteristics of ZnO-Surfactant Complexes

The FT-IR spectra for ZnO-surfactant complexes as a function of various pH were analyzed in order to illustrate the adsorption mechanisms, as shown in Figure 4. The appearance of some bands and shift in peaks of the ZnO-surfactant complexes confirms that the surfactant made enough impact on the metal surface. The IR intensities are enhanced with increasing pH in both surfactants (Figure 3A,B). These results are consistent with pH-dependent surface adsorption, with a predominance of outer-sphere complexation, as indicated by adsorption study (Figure 2). The peaks in the 3000–2800 cm^{-1} region in Figure 4A is ascribed to the asymmetric and symmetric stretching of CH_3 and CH_2 of the hydrocarbon tail [39,40]. This strong adsorption could be attributed to electrostatic interactions between the negatively charged sulfonate head and the positively charged hydroxyls groups in NPs [41]. The bands at 1461 and 1377 cm^{-1} correspond to the bending of CH_2 and CH_3 deformation, respectively [42]. The asymmetric doublet peak for OSO_3 stretching in pristine SDS (Figure S1C) shifted down in frequency at 1198 cm^{-1} and it appeared in the spectrum of SDS adsorbed ZnO complexes [43]. This significant shift in wavenumber indicated that the primary adsorption mechanism might involve hydrophobic and electrostatic interactions.

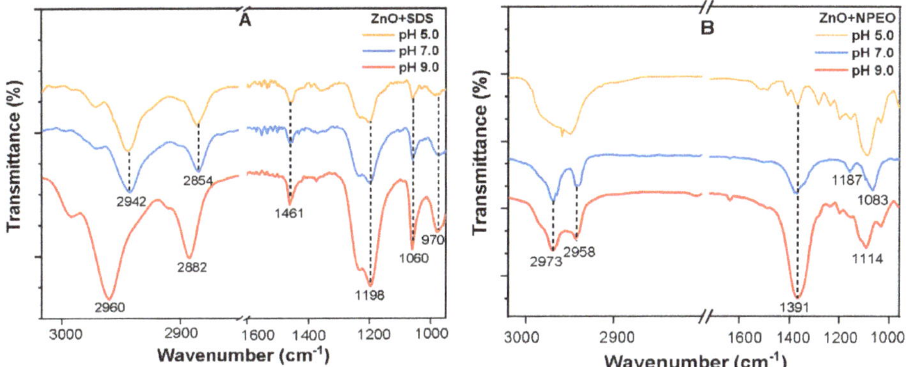

Figure 4. FT-IR spectra of ZnO NPs surfactant complexes at various pH and (10 mg/L) of each surfactant showing; (**A**) ZnO + SDS; (**B**) ZnO + NPEO.

Earlier reports [33] indicated that SDS adsorbed to the Al_2O_3 via OSO_3^- group and reduced remaining groups in the solution. A similar shift was previously reported [42], that, at appropriate SDS concentration, H-bond formation between one or two charged sulfonate oxygens and surface of Zn-OH groups may occur. Therefore, during the adsorption of SDS onto ZnO NPs, the sulfate head group of SDS may directly interact with ZnO NPs surface. The possible adsorption mechanism at pH 5.0 may be that under these particular conditions (pH < pHiep), SDS micelles are weakly, but positively adsorbed to the ZnO NPs surface via counterion (Na^+) bridging and attractive electrostatic forces [44,45].

Figure 4B illustrates the NPEO spectra that were collected at different pH values. The two major visible spectral regions correspond to the hydrophobic tail (2800–3000 cm^{-1}) and hydrophilic head (1250–1050 cm^{-1}), respectively. A noticeable change in the ratio of the intensity of the hydrophilic tail relative to the phenol head was observed in all spectra. As indicated in Figure S2, the ZnO NPs were highly positively charged at pHs 5.0 and 7.0; thus, the adsorption may involve the formation of a surficial monolayer due to electrostatic attraction between the phenol head group with NPs surface [44]. Conversely, when pH is near the pHiep of ZnO, the hydrophilic surface of NPs may still serve as a suitable substrate for adsorption of the surfactant tail, forming more complex cylindrical zinc micelles with the phenol head group toward the solution [45]. In addition, adsorption mechanisms between the NPEO molecules onto the ZnO NPs surface are thought to involve head group electrostatics and tail group lateral hydrophobic interactions. In conclusion, the specific adsorption characteristics of both surfactants affected the ζ- potential and the HDD of ZnO NPs. These finding might be instructive for assessing the colloidal stability of ZnO NPs in real wastewater.

3.4. Sedimentation and Aggregation of ZnO NPs in Different Waters

The sedimentation of ZnO aggregates in different kinds of waters over 24 h is presented in Figure 5. Notable aggregation and highest settling rate (0.08, 0.035, and 0.32/h) were observed for 24 h for ZnO NPs in control water, W3 and W4 (Figure 5A,C), thus leaving only 20–50% of the suspended aggregate in the above solution. In the absence or little steric hindrance, NPs flocculates form more prominent aggregates with higher settling rate. However, the sedimentation is less distinct, and consequently the sediment rate was lowest (0.01 and 0.005/h) in W1, W2 waters containing the SDS surfactant (Figure 5A,C). It may be noted that over 80 and 95% of the ZnO NPs were still suspended after the experimental period. This could be attributed to the stronger steric hindrance generated by the surface attached anionic surfactant molecules that restricted the NPs aggregation [46]. Moreover, it might also be correlated well with the particle size of ZnO NPs in SDS containing waters, which was found to be at its lowest (280 ± 35 and 205 ± 70 nm) in W1 and W2, respectively (Table 3). Many previous studies [8,16,34] have indicated that smaller size NPs have a larger coefficient of diffusion in comparison to larger size NPs, which agglomerates rapidly in solution due to the effect of Brownian motion.

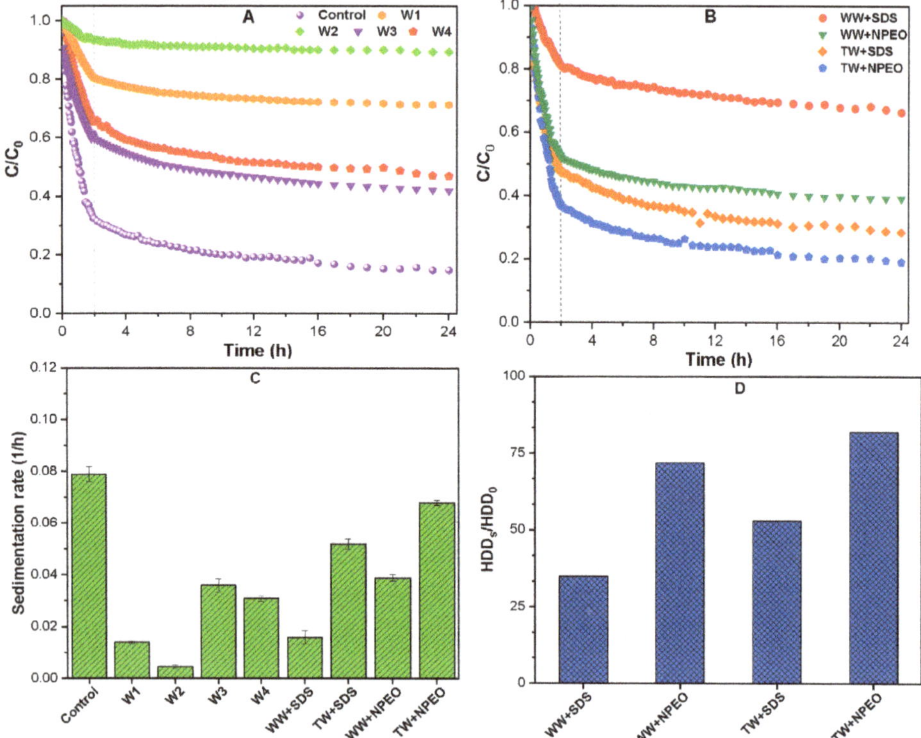

Figure 5. Aggregation kinetics (**A**) synthetic water; (**B**) environmental waters; (**C**) settling rate of ZnO NPs; and, (**D**) particle size ratio with and without surfactants.

As shown in Figure 5B, the ZnO NPs aggregates exhibited a pronounced difference with most stable in wastewater containing SDS and least suspended in remaining waters. The suspended ZnO

NPs aggregates in TW+NPEO, and WW+NPEO waters were found to be 20.8 and 32.2%, respectively. This discrepant behaviors of ZnO NPs in both environmental waters might be related to the IS. It has been reported [7] that an increase in IS will lead to effective compression of the electrical double layer (EDL), thereby reducing the interparticle repulsion forces and thus enhancing the rate of aggregation. In the presence of high electrolyte contents, the DOM may also increase the aggregation of ZnO NPs, through intermolecular bridging between DOM and ZnO NPs. Omar and Aziz [47,48] reported that parameters, such as pH value, IS, and DOMs can affect the aggregation phenomena of ZnO NPs in the aqueous system. In addition, the presence of SDS in TW+SDS and WW+SDS waters increase the degree of suspension and around (40.10 and 70.96%) NPs were found to be suspended (Figure 5B). These results showed that ZnO dispersed in SDS has lesser agglomeration, indicating better dispersion due to the higher level of adsorption of SDS molecules onto the ZnO surface (Figure 2). The slightly higher aggregation in TW+SDS as compared to WW+SDS water may be related to the increasing interaction of SDS molecules with positively charged sites of electrolytes that can eliminate the energy barrier of NPs. Some studies have [49] reported that a high concentration of monovalent ions, such as NaCl (10 mM), increased the sedimentation of hematite NPs and significantly reduced the critical micellar concentrations (CMC) of anionic surfactants in solution. The reduced agglomeration is WW+SDS water might be related to the high TOC concentration (15.68 mg/L) in wastewater samples, thus suggesting the enhanced sorption competition between DOMs and ionic surfactants. Moreover, the presence of metal ions, such as antimony (Sb) and arsenic (As), may have some possible interference and contribution in the stability of NPs in such waters. However, further investigations are necessary to understand the complex aggregation phenomena in real wastewater containing heavy metals ions.

The size ratio of ZnO agglomerates after sedimentation are shown in Figure 5D, where HDD_s and HDD_0 are the sizes of ZnO agglomerates in the presence and absence of surfactants in the environmental waters, respectively. The presence of both surfactants reduced the aggregate size of the ZnO NPs in all studied waters. In general, the retarding effect of surfactants on the growth of aggregates was found to be most remarkable in waters containing SDS, in comparison to sample containing NPEO surfactant. The presence of SDS and NPEO in tap water reduced the agglomerates size by 51.67% and 25.91%, respectively. Previous studies [50–55] have found that when the IS exceeds the critical coagulation concentration (CCC) value, the energy barriers are overcome, thereby increasing the agglomeration. The DOMs, such as humic acid, amino acid, and polyethylene glycol may generate the steric hindrance and electrostatic repulsion by adsorption onto NPs [31,51]. Moreover, the heavy metal ions that are present in wastewater might induce the cross-linking of NPs through the metal coordinating bonding, resulting in stable ZnO NPs suspension [8,34]. It has been proposed that the metal complexes be stabilized by a combination of ion-dipole, hydrophobic, and electrostatic interactions.

Dissolution of ZnO NPs in Environmental Waters

The time-dependent dissolution of ZnO NPs was measured for a 10-day period, as shown in Figure 6. The concentration of Zn^{2+} in pure water was 0.89 mg/L at the beginning, which increased to 4.82 mg/L at the last day (Figure 6). These results are consistent with the previous study report [38] that observed the enhanced released of Zn^{2+} up to 7.5–10 mg/L at pH 7. The solubility of ZnO NPs was found to be at its lowest in pH range 7–11, while it increased under highly acidic and alkaline conditions [9]. By contrast, the measured concentration of Zn^{+2} in WW+SDS (3.587 mg/L) and WW+NPEO (2.719 mg/L) waters reduced to 1.40 and 2.01 mg/L, respectively. The wastewater contains high concentrations of IS, TOC, and several anions, as well as metal ions. Therefore, a reduction in dissolution may account for the preferential formation of Zn complexes. Our observations are comparable with previous studies, which showed that the formation of hydrozincite ($Zn_5(CO_3)_2(OH)_6$) in synthetic waters lead to a reduction of solubility of ZnO NPs [56,57]. In the heterogeneous aqueous environment, ZnO may form various Zn complexes, such as Zn phosphate, Zn iron hydroxides, and

Zn sulfide [50–53]. These suspended complexes are less soluble as compared to pristine ZnO. Hence, the solubility might reduce with the formation of Zn complexes.

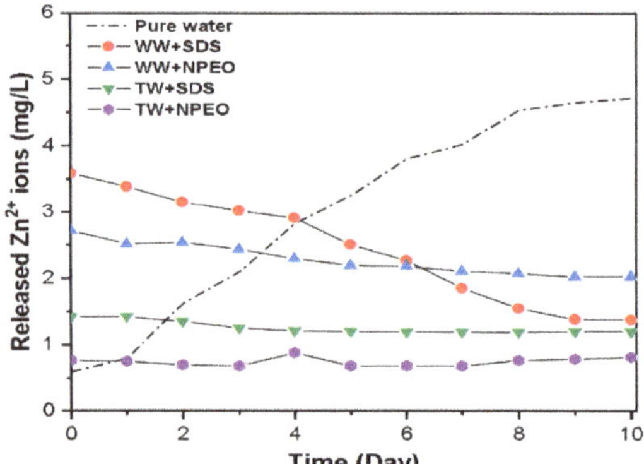

Figure 6. Time-dependent dissolution of ZnO NPs in surfactants containing environmental waters.

However, the Zn^{2+} concentration in TW+SDS (2.423 mg/L) and TW+NPEO (0.769 mg/L) waters remained stable within the whole experiment period. Our results are comparable with the dissolution patterns shown in previous studies [19,23,31], which observed that dissolution reached the maximum value within the initial 2–4 days. In this study, the concentration of Zn^{2+} ions in other waters was lower than that in the pure water, which might be due to the inhibition effect from high electrolytes content and the low solubility effect of aqueous pH. The phenomena behind the reduced solubility of ZnO NPs in WW+SDS water in comparison to other three waters was complicated. However, as stated previously, a new series of experiments similar to current conditions are required to understand the reduced solubility in this complex's matrix, which was beyond the scope of this work. In general, when combining the high colloidal stability and enhanced solubility of ZnO NPs in waters containing SDS, it is worth noting that the stable colloid may decrease/influence the coagulation efficiency of NPs in the wastewater treatment process.

3.5. *Removal of ZnO NPs and Zn^{2+} from Synthetic Waters*

The removal of ZnO NPs and Zn^{2+} from synthetics and environmental waters after coagulation with various FC dosages are shown in Figure 7. In the absence of coagulant, only 30–40% ZnO NPs removal was observed in synthetic waters (W1–W4), while in environmental waters, the ZnO NPs removal was found to be higher (50–65%) in (TW+SDS, TW+NPEO) waters than in (WW+SDS, WW+NPEO) waters (20–25%). The probable explanation for the higher removal rate in TW samples might be related to the presence of metal cations, i.e., Mg^{2+} and Ca^{2+}, which could effectively compress the EDL and weaken the steric inhibition effect, thus enhancing NPs flocculation (ref). The results also indicated that the coagulation efficiencies of Zn^{2+} and surfactants were low, at (10–20%) and (<40%), respectively, in all the studied waters.

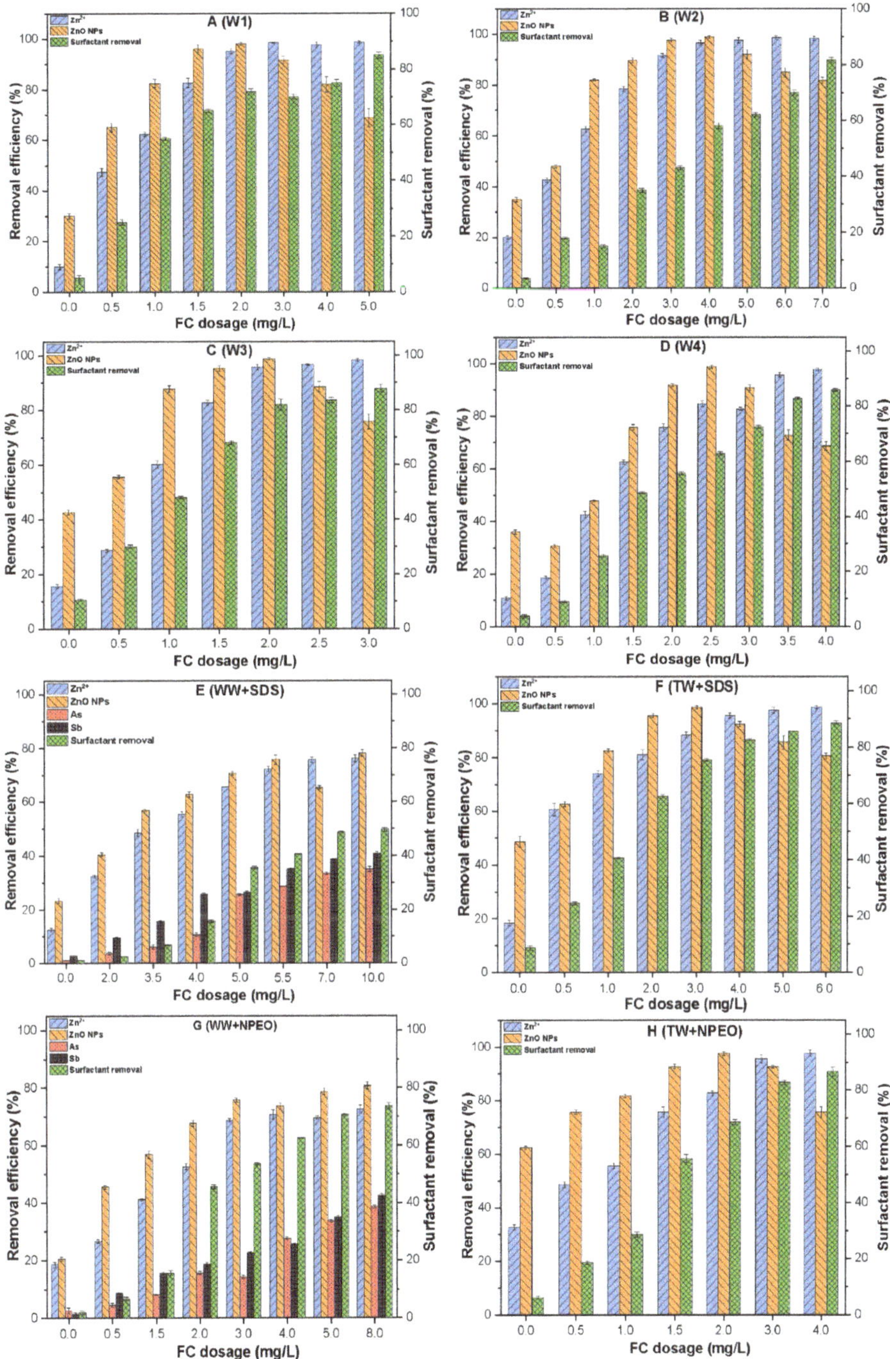

Figure 7. Removal of ZnO NPs, Zn^{2+}, surfactant, As, and Sb from synthetic waters (**A**(W1)–**D**(W4)), and environmental waters (**E**–**H**), respectively, as a function of FC coagulant dosage.

The removal efficiencies of ZnO NPs and Zn^{2+} enhanced with increasing FC dosage in all synthetic waters (Figure 7A–D). The presence of FC results in ampler NP flocs formation until reaching a plateau, i.e., where ~98% of removal is achieved. The optimum FC dosage significantly varies among the surfactant type, such as the NPEO surfactant containing (W3, W4) waters that required lower optimum doses between 2.0 to 4.0 mg/L (Figure 7C,D). In addition, waters containing the SDS (W1, W2) required larger FC dosage between 3.0 to 5.0 mg/L to obtain maximum removal efficiency up to 95% of both contaminants. Since the ZnO NPs that were dispersed in the NPEO surfactants waters showed a larger size than waters containing SDS surfactants (Table 2, and Figure 7). The results suggested that NPs size plays an influential role in the estimation of optimum dosage (OD), as the smaller NPs have a larger diffusion coefficient, and thus required higher coagulant dosage then larger particles [9]. Many previous studies [54,55] have indicated that anionic surfactant forms stable aggregates, thus requiring more a coagulant dose to achieve a higher removal rate. The refractory nature of SDS might be responsible for the difference in OD, because the presence of sulfonate groups in structure might decrease the electrostatic neutralization with the cationic Fe species of the coagulant [55]. In all studied waters, after OD, an excess of coagulant is counterproductive since it enhances the turbidity of suspension as ζ-potential of ZnO NPs became more positive. Earlier reports [9,18] indicated that the charge inversion and restabilization of coagulated colloids occur due to the excess adsorption of polyelectrolytes chains onto stable NPs in solution.

In all tested waters, the comparatively lower removal rate of Zn^{2+} ions was observed at a lower FC dosage, because of the lack of available FC sites in the aqueous phase. Interestingly, the higher removal efficiencies of Zn^{2+} up to 90–95% were observed at OD of FC in all synthetic waters (Figure 7A–D). It was assumed that heavy metals in cationic form might be associated with the precipitate matrix particle via the mechanism known as occlusion [56]. This result may explain the effect of electrolyte on Zn^{2+} removal from the heterogeneous aqueous environment and is consistent with previous literature [35,51]. The result confirms the finding of Sousa VS, et al. that, at lower coagulant dose, the interparticle bridging did not form due to the inadequate compression of EDL of stabilized NPs [57]. Further, increasing FC dosage induces the destabilization of colloid by neutralizing the electro steric forces through hydrolyzed Fe^{3+} species in suspension. The ZnO NPs that were spiked in SDS containing synthetic waters (W1, W2) have a negative surface charge (-16.4 ± 0.5 and -28.3 ± 0.8) as compared to NPEO containing (W3, W4) waters with a slightly larger particle size (Table 2). In aquatic environments, the Fe hydrolysis product strongly adsorbs on the ZnO surface, thus reducing the ζ-potential of suspension. Previous studies have reported [58] that NPs colloids might be practically neutralized when the ζ-potential ranges between $\pm$ 20 mV. Our result implies that at OD, colloids are destabilized and bridging flocculation occurs due to an increase in attractive electrostatic charges among the NPs (Figure 7A–D). The surfactant may bind to ZnO NPs colloids through steric stabilization or electrostatic interaction and it enhanced the NPs mobility, as observed in aggregation and sedimentation experiment (Figure 5). The enmeshment effect of coagulant at the OD was stronger than the electro steric forces among the NPs and might play a dominant role in the removal of NPs from these waters.

3.5.1. Removal of ZnO NPs and Zn^{2+} from Environmental Waters

Numerous jar tests experiments in environmental waters, such as tap water and wastewater, were conducted in order to establish a practical understanding of removal behavior of ZnO NPs and Zn^{2+} in the presence of both surfactants. In addition, the residual concentration of metal ions (Sb, As) in wastewater was measured to evaluate the competitive inhibition effect of these ions on the removal process. The efficiency of ZnO NPs and Zn^{2+} removal was significantly influenced by the characteristics of treated waters, as shown in Figure 7E–H. The addition of lower dosage of FC in (TW+SDS and TW+NPEO) waters resulted in the enhanced removal rate of ZnO NPs and Zn^{2+}, while no significant effect was observed in (WW+SDS and WW+NPEO) waters (Figure 7E–H). This is apparently due to the presence of high molecular weight (HMW) compounds in wastewater, thus hindering the electrostatic interactions between the ZnO NPs and FC coagulant [54]. The tap water containing both surfactants shows the similar removal phenomena of ZnO NPs and Zn^{2+} with complete removal at the optimum dosage of 3.0–4.0 mg/L (Figure 7F,H). An earlier study [57] reported that divalent metal cations might cause the ZnO NPs agglomeration in complex waters, owing to their specific adsorption behavior. By contrast, the complete removal of both pollutants from wastewater was not achieved, even at higher coagulant dosage (8.0–10.0 mg/L) (Figure 7E,G), and around (25% Zn^{2+} and ZnO NPs) remains in the solution. The probable explanation might be that wastewater contains highly reactive species, which was elucidated through its high surface potential in aqueous media (Table 2). The simultaneous influence of other parameters such as the high metal concentration of As, Sb has an adverse effect on ZnO NPs and Zn^{2+} removal ratio. Thus, the removal rate decreased with the increase in competition adsorption between the species [56]. These results showed that higher concentration of other ions resulted in less neutralization, absorption, and interparticle bridging among NPs colloids and FC coagulant [43]. Overall, the removal efficiencies of metal ions increased with higher FC dosage as Zn^{2+} (70–75%), Sb (30–40%), and As (30–35%) were observed in WW+SDS and WW+NPEO waters due to the availability of more FC attachment sites. A previous study [59] reported that coagulation efficiencies significantly decreased in a complex matrix such as domestic and industrial wastewaters because of their variable origin. Regardless, the results indicate that the competitive inhibition effect of As, Sb species is due to strong adsorption potential on FC the surface [60,61]. Moreover, charge neutralization and adsorption to FC are the primary pollutant removal mechanisms that may act simultaneously during the water treatment by the coagulation-flocculation process.

3.5.2. Removal of Surfactant from Synthetic and Environmental Waters

The presence of different surfactant in real wastewater may affect the overall colloidal stability of ZnO NPs. The reduction of surfactant content reached up to 80–85% in all synthetic and environmental waters, except for (WW+SDS), where less than 50% removal was observed, even at higher FC dosages, while further increases in dosage did not improve the removal efficiency of the surfactants. The hydrophilic micropollutants, such as SDS with (pKa = 5.7), exist as negative ions at neutral pH [62]. Therefore, it can be effectively removed with a Fe^{3+} ion from synthetic waters by the coagulation process. The lower removal of surfactant from (WW+SDS) water might be related to the solubilization capacity of surfactants and NOM, which increase the dissolution effect in the solution. Similar results were obtained by Stackelberg et al., which shows the low removal <40% of micropollutant ats the coagulation step in water treatment plant [55]. The presence of surfactants in these waters further stabilized the NPs-NOMs composite due to their ionization state and surface-active properties, and thus produced thermodynamically stabilized suspension. In addition, the extended H-Bonding might enhance the steric stabilization effect due to the change in the surfactant layer configuration [62]. It was assumed that the polarized groups of surfactant interacted predominantly with the hydroxyl group of FC via hydrophobic interactions, producing an increasingly hydrophilic surface [9]. The measured ζ-potential during the coagulation further explained the removal phenomena, such as the higher removal, were obtained at the ζ-potential around ± 10 mV, while the removal rate decreases at ζ-potential ($>\pm$ 30 mV). It has been reported [57] that at ζ-potential (+3 and -10 mV), the removal of most stable colloids can be easily achieved. The results also suggest that the organic fractions mostly composed of the aromatic compound (UV_{254}) and dissolved organic carbon (DOC) were efficiently removed from wastewaters. These conclusions are consistent with previous studies [9,14,51,58] that HMW aromatic compounds are removed more easily than other fractions. At OD, the high charge metal cations Fe^{3+} may bind to the surface of the micellar and creates a region of high cation concentration across the Stern and diffuse layer around the micelle. Shielding of surface charge leads to flocculation of the micelles, which form a large disordered structure, amorphous with fragments of the liquid crystal inlaid [54]. Moreover, organic pollutants may adsorb as complexes within the Stern-diffuse layer due to the high concentration of polyvalent cations. This process reduces the ζ-potential of the residual suspension, thereby removing the pollutant through the adsorptive micellar flocculation (AMF) [59]. The conditions for the coagulation of the surfactant are similar to organic matter, thus understanding the AMF process is needed to help explain this sequestering behavior in wastewater and its applicability limits. It can be concluded that, in addition to the charge neutralization process, the AMF mechanism may also contribute to the removal of surfactants and organic matters from wastewater. Based on our finding, surfactant type, concentration, and other metal ions appear to be a crucial factor in influencing the treatability in environmental waters and it should be included in any mechanistic approach to water treatment.

3.6. Study Significance and Limitation

This study is the first-time approach to provide some insight into the aggregation, dissolution, and removal behavior of ZnO NPs in synthetic and real wastewater containing various surfactants, as well as heavy metal ions. Previous studies by [10,14,17,40,41] have described the effect of surfactant on the individual parameters of ENPs. No study has yet been conducted that describes the influence of different surfactants on the removal of ZnO NPs from real industrial wastewater. The current study was conducted on four synthetic and two real waters with different characteristics and surfactant concentration. The results highlight that the sorption of anionic surfactant onto ZnO NPs restricted the aggregation through steric hindrance and enhanced the solubility. Thus, the transformation of ZnO NPs into heavy metal ions may alter their fate, bioavailability, and the potential effect on biota. However, the agglomeration of NPs decreases the rate of dissolution in environmental waters by reducing surface area for the sorption of surfactant molecules. The major limitation of this study is the unavailability of examined data of other wastewater quality parameters, such as Cu metal ions and organic fractions of compounds, which might play a critical role in influencing the removal efficiency of Zn^{2+} ions. Moreover, it should also be noted that a further investigation is needed to study the competitive effect of heavy metals on NPs removal. In this study, we only considered the higher concentration of ions, such as As and Sb, with maximum possible discussion, providing insight into the overall effect of surfactant on coagulation behavior. Here, we demonstrated that the coagulation process could completely remove ZnO NPs, Zn^{2+}, and other micropollutants from synthetic waters; however, it shows a limited removal performance in real wastewater under the studied dosage. This study underscores the importance of understanding the effect of different surfactants on the transport and fate of ZnO NPs in the wastewater environment.

4. Conclusions

In this study, we investigated the influence of ionic (SDS) and nonionic (NPEO) surfactants on ZnO NPs stability and removal behavior from synthetic and natural waters by the coagulation-flocculation process. The presence of SDS significantly decreased the ζ-potential and aggregate size, thereby hindering the NPs aggregation rate. This study found that both surfactants could be adsorbed onto ZnO NPs surface with the strong sorption capacity of SDS as compared to NPEO. The FT-IR analysis of ZnO-surfactants composite indicated the strong pH-dependence with increasing IR intensities due to the formation of mono-bilayer patches onto the ZnO surface. The presence of both surfactants in environmental waters results in a distinct adverse effect on dissolution and aggregate size of ZnO NPs, while the effect of SDS was more pronounced than NPEO. The results showed that, at optimum FC dosage, the removal efficiencies of ZnO NPs, Zn^{2+}, and surfactant were more than 98, 95, and 85%, respectively, in all synthetic waters. However, the presence of higher concentration of metals ions in wastewater significantly inhibited the coagulation efficiency of Zn^{2+}, ZnO, and surfactant up to 20, 25, and 50%, respectively. The mechanisms, such as charge neutralization and adsorptive micellar flocculation (AMF), might be involved in the removal of NPs composite pollutant by coagulation. The future research and endeavors shall focus on polymerized coagulants, such as polyaluminum chloride, polyferric chloride, and polyferric sulphate, which might enhance the coagulation efficiency of ZnO NPs under competitive conditions.

Supplementary Materials: The following are available online at http://www.mdpi.com/2071-1050/11/1/17/s1, **Figure S1.** (**A**) BET surface Area; (**B**) Raman spectra of ZnO powder; and (**C**) FT-IR spectra of SDS and NPEO surfactants; (**D**) TGA% purity of ZnO NPs. **Figure S2.** ζ-potential of ZnO (10 mg/L) in absence and presence of SDS/NPEO at various pH ranges. **Table S1.** The composition of environmental waters.

Author Contributions: R.K. and I.T.Y. designed the study; R.K., M.A.I., performed the experiment and analyzed the data; M.M.I., M.S.; D.R.P., K.H.L., S.S. and S.K. provided critical feedback and helped shape the research; R.K. wrote final version of the manuscript.

Acknowledgments: The BK21 plus program supported this research through the National Research Foundation of Korea (NRF) funded by the Ministry of Education of Korea (Grant No. 22A20152613545).

Conflicts of Interest: The authors declare no conflict of interest.

References

1. Wang, Z.L. Zinc oxide nanostructures: Growth, properties and applications. *J. Phys. Condens. Matter* **2004**, *16*. [CrossRef]
2. Gottschalk, F.; Nowack, B. The release of engineered nanomaterials to the environment. *J. Environ. Monit.* **2011**, *13*, 1145. [CrossRef]
3. Keller, A.A.; McFerran, S.; Lazareva, A.; Suh, S. Global life cycle releases of engineered nanomaterials. *J. Nanopart. Res.* **2013**, *15*. [CrossRef]
4. Adam, N.; Leroux, F.; Knapen, D.; Bals, S.; Blust, R. The uptake and elimination of ZnO and CuO nanoparticles in Daphnia magna under chronic exposure scenarios. *Water Res.* **2015**, *68*, 249–261. [CrossRef] [PubMed]
5. Dreher, K.L. Health and environmental impact of nanotechnology: Toxicological assessment of manufactured nanoparticles. *Toxicol. Sci.* **2004**, *77*, 3–5. [CrossRef] [PubMed]
6. Zhou, D.; Keller, A.A. Role of morphology in the aggregation kinetics of ZnO nanoparticles. *Water Res.* **2010**, *44*, 2948–2956. [CrossRef]
7. Peng, Y.-H.; Tso, C.; Tsai, Y.; Zhuang, C.; Shih, Y. The effect of electrolytes on the aggregation kinetics of three different ZnO nanoparticles in water. *Sci. Total Environ.* **2015**, *530*, 183–190. [CrossRef]
8. Khan, R.; Inam, M.; Zam, S.; Park, D.; Yeom, I. Assessment of Key Environmental Factors Influencing the Sedimentation and Aggregation Behavior of Zinc Oxide Nanoparticles in Aquatic Environment. *Water* **2018**, *10*, 660. [CrossRef]
9. Khan, R.; Inam, M.; Park, D.; Zam Zam, S.; Shin, S.; Khan, S.; Akram, M.; Yeom, I. Influence of Organic Ligands on the Colloidal Stability and Removal of ZnO Nanoparticles from Synthetic Waters by Coagulation. *Processes* **2018**, *6*, 170. [CrossRef]
10. Narkis, N.; Ben-David, B. Adsorption of non-ionic surfactants on activated carbon and mineral clay. *Water Res.* **1985**, *19*, 815–824. [CrossRef]
11. Bouchard, D.; Zhang, W.; Powell, T.; Rattanaudompol, U. Aggregation kinetics and transport of single-walled carbon nanotubes at low surfactant concentrations. *Environ. Sci. Technol.* **2012**, *46*, 4458–4465. [CrossRef] [PubMed]
12. Lewis, M.A. Chronic and sublethal toxicities of surfactants to aquatic animals: A review and risk assessment. *Water Res.* **1991**, *25*, 101–113. [CrossRef]
13. Godinez, I.G.; Darnault, C.J.G. Aggregation and transport of nano-TiO_2 in saturated porous media: Effects of pH, surfactants and flow velocity. *Water Res.* **2011**, *45*, 839–851. [CrossRef] [PubMed]
14. Li, X.; Yoneda, M.; Shimada, Y.; Matsui, Y. Effect of surfactants on the aggregation and stability of TiO_2 nanomaterial in environmental aqueous matrices. *Sci. Total Environ.* **2017**, *574*, 176–182. [CrossRef] [PubMed]
15. Zhang, Y.; Chen, Y.; Westerhoff, P.; Crittenden, J. Impact of natural organic matter and divalent cations on the stability of aqueous nanoparticles. *Water Res.* **2009**, *43*, 4249–4257. [CrossRef] [PubMed]
16. Keller, A.A.; Wang, H.; Zhou, D.; Lenihan, H.S.; Cherr, G.; Cardinale, B.J.; Miller, R.; Zhaoxia, J.I. Stability and aggregation of metal oxide nanoparticles in natural aqueous matrices. *Environ. Sci. Technol.* **2010**, *44*, 1962–1967. [CrossRef]
17. Liu, L.; Gao, B.; Wu, L.; Sun, Y.; Zhou, Z. Effects of surfactant type and concentration on graphene retention and transport in saturated porous media. *Chem. Eng. J.* **2015**, *262*, 1187–1191. [CrossRef]

18. Tkachenko, N.H.; Yaremko, Z.M.; Bellmann, C.; Soltys, M.M. The influence of ionic and nonionic surfactants on aggregative stability and electrical surface properties of aqueous suspensions of titanium dioxide. *J. Colloid Interface Sci.* **2006**, *299*, 686–695. [CrossRef]
19. Li, X.; Yoneda, M.; Shimada, Y.; Matsui, Y. Effect of surfactants on the aggregation and sedimentation of zinc oxide nanomaterial in natural water matrices. *Sci. Total Environ.* **2017**, *581*, 649–656. [CrossRef]
20. Sun, P.; Zhang, K.; Fang, J.; Lin, D.; Wang, M.; Han, J. Transport of TiO_2 nanoparticles in soil in the presence of surfactants. *Sci. Total Environ.* **2015**, *527*, 420–428. [CrossRef]
21. Apha, A. *WPCF, Standard Methods for the Examination of Water and Wastewater*; American Public Health Association: Washington, DC, USA, 1995.
22. Wagner, C.D.; Gale, L.H.; Raymond, R.H. Two-dimensional chemical state plots: A standardized data set for use in identifying chemical states by x-ray photoelectron spectroscopy. *Anal. Chem.* **1979**, *51*, 466–482. [CrossRef]
23. Rupasinghe, R.A. Dissolution and Aggregation of Zinc Oxide Nanoparticles at Circumneutral pH: A Study of Size Effects in the Presence and Absence of Citric Acid. Master's Thesis, University of Iowa, Iowa City, IA, USA, 2011.
24. Chibowski, E.; Holysz, L.; Terpilowski, K.; Wiacek, A.E. Influence of ionic surfactants and lecithin on stability of titanium dioxide in aqueous electrolyte solution. *Croat. Chem. Acta* **2007**, *80*, 395–403.
25. Saleh, N.; Kim, H.-J.; Phenrat, T.; Matyjaszewski, K.; Tilton, R.D.; Lowry, G.V. Ionic strength and composition affect the mobility of surface-modified Fe0 nanoparticles in water-saturated sand columns. *Environ. Sci. Technol.* **2008**, *42*, 3349–3355. [CrossRef] [PubMed]
26. Nevskaia, D.M.; Guerrero-Ruız, A.; de D. López-González, J. Adsorption of polyoxyethylenic nonionic and anionic surfactants from aqueous solution: Effects induced by the addition of NaCl and $CaCl_2$. *J. Colloid Interface Sci.* **1998**, *205*, 97–105. [CrossRef] [PubMed]
27. Shao, D.; Jiang, Z.; Wang, X. SDBS Modified XC-72 Carbon for the Removal of Pb (II) from Aqueous Solutions. *Plasma Process. Polym.* **2010**, *7*, 552–560. [CrossRef]
28. Ho, Y.-S. Review of second-order models for adsorption systems. *J. Hazard. Mater.* **2006**, *136*, 681–689. [CrossRef]
29. Qiu, H.; Lv, L.; Pan, B.; Zhang, Q.; Zhang, W.; Zhang, Q. Critical review in adsorption kinetic models. *J. Zhejiang Univ. A* **2009**, *10*, 716–724. [CrossRef]
30. Illés, E.; Tombácz, E. The effect of humic acid adsorption on pH-dependent surface charging and aggregation of magnetite nanoparticles. *J. Colloid Interface Sci.* **2006**, *295*, 115–123. [CrossRef]
31. Hu, J.-D.; Zevi, Y.; Kou, X.-M.; Xiao, J.; Wang, X.-J.; Jin, Y. Effect of dissolved organic matter on the stability of magnetite nanoparticles under different pH and ionic strength conditions. *Sci. Total Environ.* **2010**, *408*, 3477–3489. [CrossRef]
32. Filius, J.D.; Meeussen, J.C.L.; Lumsdon, D.G.; Hiemstra, T.; van Riemsdijk, W.H. Modeling the binding of fulvic acid by goethite: The speciation of adsorbed FA molecules. *Geochim. Cosmochim. Acta* **2003**, *67*, 1463–1474. [CrossRef]
33. Piccolo, A.; Conte, P.; Spaccini, R.; Chiarella, M. Effects of some dicarboxylic acids on the association of dissolved humic substances. *Biol. Fertil. Soils* **2003**, *37*, 255–259.
34. Philippe, A.; Schaumann, G.E. Interactions of dissolved organic matter with natural and engineered inorganic colloids: A review. *Environ. Sci. Technol.* **2014**, *48*, 8946–8962. [CrossRef] [PubMed]
35. Filius, J.D.; Lumsdon, D.G.; Meeussen, J.C.L.; Hiemstra, T.; Van Riemsdijk, W.H. Adsorption of fulvic acid on goethite. *Geochim. Cosmochim. Acta* **2000**, *64*, 51–60. [CrossRef]
36. Romero-Cano, M.; Martín-Rodríguez, A.; de las Nieves, F. Electrokinetic behaviour of polymer colloids with adsorbed Triton X-100. *Colloid Polym. Sci.* **2002**, *280*, 526–532.
37. Sharma, K.P.; Aswal, V.K.; Kumaraswamy, G. Adsorption of nonionic surfactant on silica nanoparticles: Structure and resultant interparticle interactions. *J. Phys. Chem. B* **2010**, *114*, 10986–10994. [CrossRef] [PubMed]
38. Bian, S.W.; Mudunkotuwa, I.A.; Rupasinghe, T.; Grassian, V.H. Aggregation and dissolution of 4 nm ZnO nanoparticles in aqueous environments: Influence of pH, ionic strength, size, and adsorption of humic acid. *Langmuir* **2011**, *27*, 6059–6068. [CrossRef] [PubMed]
39. Bai, B.; Hankins, N.P.; Hey, M.J.; Kingman, S.W. In situ mechanistic study of SDS adsorption on hematite for optimized froth flotation. *Ind. Eng. Chem. Res.* **2004**, *43*, 5326–5338. [CrossRef]

40. Dobson, K.D.; Roddick-Lanzilotta, A.D.; McQuillan, A.J. In situ infrared spectroscopic investigation of adsorption of sodium dodecylsulfate and of cetyltrimethylammonium bromide surfactants to TiO_2, ZrO_2, Al_2O_3, and Ta_2O_5 particle films from aqueous solutions. *Vib. Spectrosc.* **2000**, *24*, 287–295. [CrossRef]
41. Gao, X.; Chorover, J. Adsorption of sodium dodecyl sulfate (SDS) at ZnSe and α-Fe_2O_3 surfaces: Combining infrared spectroscopy and batch uptake studies. *J. Colloid Interface Sci.* **2010**, *348*, 167–176. [CrossRef]
42. Hug, S.J. In situ fourier transform infrared measurements of sulfate adsorption on hematite in aqueous solutions. *J. Colloid Interface Sci.* **1997**, *188*, 415–422. [CrossRef]
43. Somasundaran, P. *Reagents in Mineral Technology*; Routledge: Abingdon-on-Thames, UK, 2018; ISBN 1351419625.
44. Bhagat, R.P. Kinetics of sodium dodecyl benzene sulfonate adsorption on hematite and its interaction with polyacrylamide. *Colloid Polym. Sci.* **2001**, *279*, 33–38. [CrossRef]
45. Tan, G.; Ford, C.; John, V.T.; He, J.; McPherson, G.L.; Bose, A. Surfactant solubilization and the direct encapsulation of interfacially active Phenols in mesoporous silicas. *Langmuir* **2008**, *24*, 1031–1036. [CrossRef] [PubMed]
46. Brunelli, A.; Pojana, G.; Callegaro, S.; Marcomini, A. Agglomeration and sedimentation of titanium dioxide nanoparticles (n-TiO_2) in synthetic and real waters. *J. Nanopart. Res.* **2013**, *15*, 1684. [CrossRef]
47. Omar, F.M.; Aziz, H.A.; Stoll, S. Aggregation and disaggregation of ZnO nanoparticles: Influence of pH and adsorption of Suwannee River humic acid. *Sci. Total Environ.* **2014**, *468*, 195–201. [CrossRef] [PubMed]
48. Chowdhury, I.; Walker, S.L.; Mylon, S.E. Aggregate morphology of nano-TiO_2: Role of primary particle size, solution chemistry, and organic matter. *Environ. Sci. Process. Impacts* **2013**, *15*, 275–282. [CrossRef]
49. Han, Y.; Kim, D.; Hwang, G.; Lee, B.; Eom, I.; Kim, P.J.; Tong, M.; Kim, H. Aggregation and dissolution of ZnO nanoparticles synthesized by different methods: Influence of ionic strength and humic acid. *Colloids Surf. A Physicochem. Eng. Asp.* **2014**, *451*, 7–15. [CrossRef]
50. Gelabert, A.; Sivry, Y.; Ferrari, R.; Akrout, A.; Cordier, L.; Nowak, S.; Menguy, N.; Benedetti, M.F. Uncoated and coated ZnO nanoparticle life cycle in synthetic seawater. *Environ. Toxicol. Chem.* **2014**, *33*, 341–349. [CrossRef] [PubMed]
51. Lombi, E.; Donner, E.; Tavakkoli, E.; Turney, T.W.; Naidu, R.; Miller, B.W.; Scheckel, K.G. Fate of zinc oxide nanoparticles during anaerobic digestion of wastewater and post-treatment processing of sewage sludge. *Environ. Sci. Technol.* **2012**, *46*, 9089–9096. [CrossRef]
52. Ma, R.; Levard, C.; Judy, J.D.; Unrine, J.M.; Durenkamp, M.; Martin, B.; Jefferson, B.; Lowry, G.V. Fate of zinc oxide and silver nanoparticles in a pilot wastewater treatment plant and in processed biosolids. *Environ. Sci. Technol.* **2013**, *48*, 104–112. [CrossRef]
53. Rasool, K.; Lee, D.S. Effects of sulfidation on ZnO nanoparticle dissolution and Aggregation in sulfate-containing suspensions. *J. Nanosci. Nanotechnol.* **2015**, *15*, 7334–7340. [CrossRef]
54. Beltrán-Heredia, J.; Sánchez-Martín, J.; Solera-Hernández, C. Anionic surfactants removal by natural coagulant/flocculant products. *Ind. Eng. Chem. Res.* **2009**, *48*, 5085–5092. [CrossRef]
55. Stackelberg, P.E.; Gibs, J.; Furlong, E.T.; Meyer, M.T.; Zaugg, S.D.; Lippincott, R.L. Efficiency of conventional drinking-water-treatment processes in removal of pharmaceuticals and other organic compounds. *Sci. Total Environ.* **2007**, *377*, 255–272. [CrossRef] [PubMed]
56. Duan, J.; Gregory, J. Coagulation by hydrolysing metal salts. *Adv. Colloid Interface Sci.* **2003**, *100*, 475–502. [CrossRef]
57. Sousa, V.S.; Corniciuc, C.; Teixeira, M.R. The effect of TiO_2 nanoparticles removal on drinking water quality produced by conventional treatment C/F/S. *Water Res.* **2017**, *109*, 1–12. [CrossRef] [PubMed]
58. Sun, J.; Gao, B.; Zhao, S.; Li, R.; Yue, Q.; Wang, Y.; Liu, S. Simultaneous removal of nano-ZnO and Zn^{2+} based on transportation character of nano-ZnO by coagulation: Enteromorpha polysaccharide compound polyaluminum chloride. *Environ. Sci. Pollut. Res.* **2017**, *24*, 5179–5188. [CrossRef] [PubMed]
59. Czech, B. Surfactants removal from water and wastewater using Co modified TiO_2/Al_2O_3 photocatalysts. *Ann. UMCS Chem.* **2011**, *66*, 81–93. [CrossRef]
60. Inam, M.; Khan, R.; Park, D.; Lee, Y.-W.; Yeom, I. Removal of Sb(III) and Sb(V) by Ferric Chloride Coagulation: Implications of Fe Solubility. *Water* **2018**. [CrossRef]

61. Inam, M.A.; Khan, R.; Park, D.R.; Ali, B.A.; Uddin, A.; Yeom, I.T. Influence of pH and Contaminant Redox Form on the Competitive Removal of Arsenic and Antimony from Aqueous Media by Coagulation. *Minerals* **2018**, *8*, 574. [CrossRef]
62. Paton-Morales, P.; Talens-Alesson, F.I. Effect of ionic strength and competitive adsorption of Na^+ on the flocculation of lauryl sulfate micelles with Al^{3+}. *Langmuir* **2001**, *17*, 6059–6064. [CrossRef]

Article

Advanced Oxidation Processes and Nanofiltration to Reduce the Color and Chemical Oxygen Demand of Waste Soy Sauce

Hyun-Hee Jang [1], Gyu-Tae Seo [1,2,*] and Dae-Woon Jeong [1,2,*]

1 Department of Environmental and Chemical Engineering, Eco-Friendly Offshore Plant FEED Engineering Course, Changwon National University, 20 Changwondaehak-ro, Uichang-gu, Changwon-si, Gyeongsangnam-do 51140, Korea; jang_@changwon.ac.kr
2 School of Civil, Environmental and Chemical Engineering, Changwon National University, 20 Changwondaehak-ro, Uichang-gu, Changwon-si, Gyeongsangnam-do 51140, Korea
* Correspondence: gts@changwon.ac.kr (G.-T.S.); dwjeong@changwon.ac.kr (D.-W.J.); Tel.: +82-55-213-3746 (G.-T.S.); +82-55-213-3743 (D.-W.J.)

Received: 20 July 2018; Accepted: 15 August 2018; Published: 17 August 2018

Abstract: Currently, the ozone (O_3) oxidation efficiency in the treatment of waste soy sauce provides 34.2% color removal and a 27.4% reduction in its chemical oxygen demand (COD). To improve the O_3 oxidation efficiency, hydrogen peroxide (H_2O_2) is used to cause a H_2O_2/O_3 process. In H_2O_2/O_3 process experiments, a previously optimized pH of 11 and applied O_3 dose of 50 mg L^{-1} were used and the H_2O_2/O_3 ratio was varied between 0.1 and 0.9 in intervals of 0.2. The results show that an H_2O_2/O_3 ratio of 0.3 results in the highest efficiencies in terms of color removal (51.6%) and COD reduction (33.8%). Nanofiltration (NF) was used to pretreat the waste soy sauce to improve color removal and COD reduction. The results showed that NF with an NE-70 membrane results in 80.8% color removal and 79.6% COD reduction. Finally, the combination of NF and H_2O_2/O_3 process resulted in the best treatment efficiency: 98.1% color removal and 98.2% COD reduction. Thus, NF & H_2O_2/O_3 process can be considered as one of the best treatment methods for waste soy sauce, which requires high intrinsic color removal and COD reduction efficiencies.

Keywords: waste soy sauce; AOPs; H_2O_2/O_3 process; oxidation; nanofiltration; membrane

1. Introduction

Waste soy sauce has a high chemical oxygen demand (COD). It has an intense dark brown color due to the presence of caramel pigments and the occurrence of the Maillard reactions producing melanin or melanoidin [1,2]. Several processes are typically used to treat soy sauce wastewater including nitrification and denitrification, an activated sludge process, and biological treatments. However, it is difficult to control the stability of the soy sauce treatment processes because of annual fluctuations in the amount of waste soy sauce generated (47,914 tons were produced in 2016 compared to 37,232 tons in 2015 and 40,461 tons in 2014) [3]. Biological treatment removes only a small fraction of the organic matter that contributes to the color of the effluents [4–6]. Therefore, the wastewater discharged to the environment is colored, which is aesthetically displeasing; moreover, the wastewater could affect the ecosystem due to its non-biodegradability and recalcitrance [7]. Therefore, it is necessary to decolorize waste soy sauce before it is discharged.

In our earlier study, overcoming the limitations of ozone (O_3) oxidation to reduce the color and COD of waste soy sauce was intensively investigated [8]. Up to 34.2% color removal and 27.4% COD reduction were achieved by O_3 oxidation at a pH of 11 with an applied O_3 dose of 50 mg L^{-1} [8]. However, the color removal and COD reduction were not thoroughly investigated as the salt water regenerated from the waste soy sauce was reused. Therefore, a combination of different techniques must be considered as an alternative treatment method for removing the color and reducing the COD of waste soy sauce as completely as possible. In addition, a high concentration of salt can be recovered from waste soy sauce through the use of advanced technologies.

Many researchers conduct O_3 oxidation by adding hydrogen peroxide (H_2O_2) to boost oxidation to raise efficiency of the color removal and COD reduction in wastewater. Fahmi, M. R et al. reported that advanced oxidation processes (AOPs) better facilitate the removal of reactive red 120 than ozonation [9]. This indicates that H_2O_2 accelerates O_3 activation by forming hydroxyl radicals, which quickly oxidize reactive red 120 [9]. Rollon, A. A. et al. reported that the AOPs combination of H_2O_2 and O_3 are most effective, followed by UV/H_2O_2-treatment, UV/O_3-treatment, O_3-treatment, and UV-C treatment (which is the least effective) [10]. However, it is difficult to improve the efficiency of AOPs without creating oxidative conditions. This is because the O_3 oxidation at the optimum applied O_3 dose involves a large number of hydroxyl radicals reacting instantaneously with O_3-demanding species (the target substances) [8]. In other words, there is a high instantaneous ozone demand (IOD) in high concentration COD products such as waste soy sauce. Therefore, a pretreatment is required to overcome this IOD that is generated during oxidation.

Other researchers have proposed the use of membrane-based filtration techniques to remove color and reduce the COD of high-organic-content wastewater. Zheng, Y. et al. reported that under optimum conditions (pressure of 0.8 bar and volume-concentrating factor of 4.0), a submerged nanofiltration (NF) system exhibited a steady permeate flux of 5.15 L m^{-2} h, a color reduction of 99.3%, and a COD reduction of 91.5% [11]. Abid, M. F. et al. (2012) achieved 93.77%, 95.67%, and 97% removal of red, black and blue dyes using a NF membrane under a pressure of 8 bar [12]. However, it is difficult to remove the color and organic matter completely using typical membrane processes other than reverse osmosis because chromophoric dissolved organic matter generally has various molecular weights [13–15]. Furthermore, despite research progress to date, membrane filtration has yet to be applied in removing the intense color to reduce the high COD of waste soy sauce. Membrane systems have been considered promising for the pretreatment of waste soy sauce before oxidation. However, there remain some technological challenges to be solved for this application, including developing a physicochemical technique of membrane filtration and an oxidation method that can overcome the limitations of biological treatments. In this study, we applied H_2O_2/O_3 process to enhance color removal and COD reduction by oxidation and the H_2O_2/O_3 ratio was optimized. In addition, NF was combined with H_2O_2/O_3 process to maximize the color removal and COD reduction. Herein, we identify the best method for treating waste soy sauce from five different treatment methods that will guide future wastewater treatment strategies to allow the saline wastewater to be reused and mitigate the impact on the environment when it is discharged.

2. Materials and Methods

2.1. Sampling Procedure

The waste soy sauce used in this study was provided by Monggo Foods, Inc. (Changwon-si, Gyeongsangnam-do, Korea).

2.2. Characterization

All samples were analyzed after removing the impurities with a 1.2 μm GF/C filter (Whatman). A seven compact™ pH/Ion S220 instrument (Mettler-Toledo GmbH, Greifensee, Switzerland) was used to measure the pH. The biochemical oxygen demand (BOD_5), chemical oxygen demand (COD_{cr}), total nitrogen (TN), and total phosphorus (TP) were measured using a US/DR3900 (320−1100 nm, Hach Company, Düsseldorf, Germany) kit assay according to the methods for testing water pollution set forth by the Ministry of Environment of Korea. The total organic carbon (TOC) was analyzed using a Shimadzu JP/TOC−5000A (Kyoto, Japan). The color was calculated as the transmittance of wavelength at 390, 400, 456 nm as measured using an Agilent Cary60 UV−VIS spectrophotometer (Santa Clara, CA, USA). The salinity was measured using an SB1500PRO instrument (0%−10%, HM Digital, Seoul, Korea) and a table salinometer.

2.3. *Size Exclusion Chromatography (SEC)*

SEC analysis was performed with high performance liquid chromatography (HPLC, LC600 Shimadzu) with UVA (SPD-6A Shimadzu) and DOC (Modified Sievers Turbo total organic carbon analyzer) detectors. The columns employed were a TSK-50S column, a Polyacrylamide Bio-Gel P-6 column, and a Waters Protein-Pak 125 silica column (Table 1) [16].

Table 1. Characteristics of tested columns.

Type	Column Packing	Particle Size (μm)	Separation Range (Da)	Column Size (cm)
Hydroxylated organic	Protein PAK 125	10	1000–30,000	0.78 × 30
Polyacrylamide	TSK-50S	30	<5 × 10^6	2 × 25
Silica	Bio-Gel P-6	90–180	1000–6000	0.5 × 90

Molecules that were larger than the pore size of the packing material were excluded and eluted first, at the void volume. Smaller molecules were able to penetrate through the porous infrastructure and were attenuated, corresponding to a higher retention time [16].

2.4. *H_2O_2/O_3 Process*

A schematic of the benchtop-scale reactor system used herein is shown in Figure 1. Experimental runs were performed in a 1 L Pyrex reactor with 500 mL of sample. The reactor was filled with 500 mL of waste soy sauce and agitated with a magnetic stirrer at 150 rpm. The optimum conditions for color removal and COD reduction by O_3-based oxidation were a pH of 11 and an applied O_3 concentration of 50 mg L^{-1} [8]. The O_3-containing gas was supplied continuously for 30 min through a gas diffuser at the bottom of the reactor. An O_3 trap containing a 2% potassium iodide solution was connected in series with the reactor to verify the O_3 gas concentration in the outlet gas stream. A 0.1 N $Na_2S_2O_3$ (sodium thiosulfate) solution was used as the reducing agent for the reverse titration in the trap. Subsequently, 0.1 N H_2SO_4 (sulfuric acid) was used to facilitate the reaction of the O_3 in the liquid phase with the I_2 [8]. All experiments were performed in duplicate at room temperature under a fume hood (for safety due to the presence of O_3 gas). The pH, applied O_3 concentration, and reaction time were kept constant while the H_2O_2/O_3 ratio was varied between 0.1 and 0.9 in intervals of 0.2. Table 2 summarizes the parameters for the H_2O_2/O_3 process experiments. In addition, the calculated vent gas of ozone by Equation (1) is shown in Table 3. The resulting color change and COD reduction were observed in 50 mL samples collected from the supernatant.

$$\frac{[a]\mathrm{eqNa_2S_2O_3}}{\mathrm{L}}\ \frac{1\mathrm{eqO_3}}{2\mathrm{eqNa_2S_2O_3}}\ \frac{48\mathrm{gO_3}}{1\mathrm{eqO_3}}\ \frac{[b]\mathrm{mL}}{[c]\mathrm{min}}\ \frac{1}{\mathrm{Reactor\ volume\ (L)}} \tag{1}$$

where [a]: $Na_2S_2O_3$ concentration, [b]: $Na_2S_2O_3$ consumption, [c]: Contact time.

Table 2. The experimental conditions for the H_2O_2/O_3 process.

Parameter	Value
pH	11
Applied O_3 conc. (mg L^{-1})	50
H_2O_2/O_3 ratio (wt/wt)	0.1, 0.3, 0.5, 0.7, and 0.9
Reaction time (min)	30
Mixing speed (rpm)	150
Sample volume (mL)	500

Table 3. Quantitative assessment of O_3 consumption.

Parameter	1st Experiment	2nd Experiment
$Na_2S_2O_3$ conc.	0.1	
$Na_2S_2O_3$ consumption	1.4	2.9
Vent O_3 conc. (mg L^{-1} min^{-1})	1.12	2.32

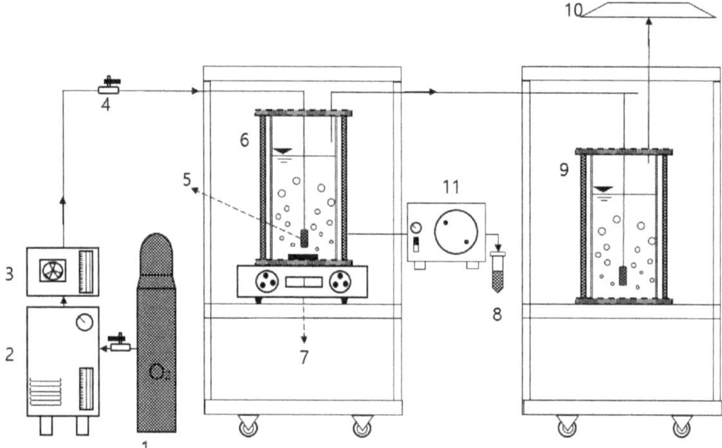

1. Oxygen gas
2. Ozone generator
3. Injection ozone monitor
4. Ozone injection valve
5. Diffuser
6. Oxidation reactor
7. Magnetic stirrer
8. Sampling port
9. Trap solution (KI)
10. Fume hood
11. Peristaltic pump

Figure 1. Schematic of the O_3-oxidation experimental system.

2.5. Nanofiltration (NF) System

The NF membranes used in this study were thin-film composite NF membranes, NE-70 and NE-90 (Toray Chemical, Korea). These membranes have different top layers, zeta potentials, wettabilities, and roughnesses as detailed in Table 4. The characteristics of the NF membrane were investigated in previous study [17].

Table 4. The characteristics of the NF membrane.

Membrane	Material	MWCO (Da)	Zeta Potential at pH 7 (mV)	Contact Angle (°)	Roughness (nm)
NE-70	Sulfonated polyethersulfone	350	−47.2	22.6 ± 1.9	8.69
NE-90	Meta-phenylene diamine	210	−38.7	41.5 ± 3.7	48

A laboratory-scale dead-end NF membrane was used in this experiment. The membrane surface area was 0.015 m^2. Waste soy sauce was fed into the filtration module by a gear pump. The filtration experiments were performed with a commercial NF module. Figure 2 shows a schematic of the NF system. In all experiments, a pressure of 1.5 MPa (15 bar) was applied at room temperature. The evolution of flux and rejection progressed over a period of 6 hours. For analysis and comparison, the measurements were taken after 1 h of filtration.

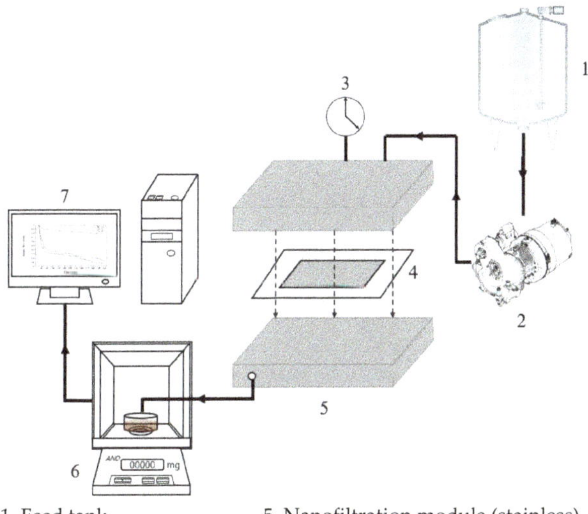

1. Feed tank
2. High-pressure pump
3. Pressure gauge
4. Nano-sized membrane
5. Nanofiltration module (stainless)
6. Electronic balance
7. Data collection computer

Figure 2. Schematic of the NF experimental system.

3. Results and Discussion

3.1. Characteristics of Waste Soy Sauce

The chemical properties of the sample were evaluated according to the guidelines for testing water pollution developed by the Ministry of Environment of Korea [18]. Table 5 shows the characteristics of the waste soy sauce. The pH value of 4.6 indicates that the waste soy sauce was acidic. The color was found to be 3810 TCU and the COD was measured to be 231.5 g/L. These results are similar to those reported in a previous study: 4038 TCU color and 229.1 g/L COD [8]. The T-N and T-P concentrations were measured to be 10.4 and 2.8 g/L, respectively. The salinity, 16.4%, was much higher than that found in other types of organic wastewater. In general, the high-salinity wastewater does not exhibit high removal efficiencies with biological treatment systems. This is because the performances of biological treatment processes are adversely affected by the negative effects of salt on microbial flora [19]. Moreover, the TOC, which is used as an indicator of water pollution, was found to be 57.6 g/L.

Table 5. Chemical characteristics of waste soy sauce.

Parameter	Value
pH	4.4 ± 0.2
[1] COD_{cr} (g/L)	231.5 ± 0.9
[2] BOD_5 (g/L)	129.4 ± 6.6
[3] TN (g/L)	10.4 ± 0.7
[4] TP (g/L)	2.8 ± 0.3
[5] TOC (g/L)	57.6 ± 2.7
Salinity (%)	16.4 ± 0.2
Color (TCU)	3810 ± 130

[1] COD: Chemical oxygen demand; [2] BOD: Biochemical oxygen demand; [3] TN: Total nitrogen; [4] TP: Total phosphorus; [5] TOC: Total organic carbon.

3.2. Optimization of H_2O_2/O_3 Ratio

Our previous study revealed that a pH of 11 and an applied O_3 concentration of 50 mg L^{-1} are optimal for color removal and COD reduction; these conditions were used in this study. H_2O_2 was combined with O_3 to accelerate the oxidation of the organic molecules present in the wastewater. The H_2O_2/O_3 ratio was varied between 0.1 and 0.9 to optimize the color removal by H_2O_2/O_3 process; the results of this experiment are shown in Figure 3. As the H_2O_2/O_3 ratio increased from 0.1 to 0.3, the color removal increased gradually from 41.2 to 51.6%. The H_2O_2/O_3 ratio of 0.3 resulted in the greatest color removal. The color removal decreased sharply from 51.6 to 17.1% as the H_2O_2/O_3 ratio increased above 0.3.

The lower residual color after oxidation with H_2O_2/O_3 as the H_2O_2/O_3 ratio increased was attributed to the increased formation of OH radicals [20–23]. Other studies have reported increased H_2O_2/O_3 ratios and biodegradability after wastewater treatment with O_3 and H_2O_2 [24,25]. The decreased color removal with H_2O_2/O_3 ratios above 0.3 was attributed to the strong scavenging effects of carbonate (CO_3^{2-}) and bicarbonate (HCO_3^-). The results showed that an H_2O_2/O_3 ratio of 0.3 was optimal for color removal (Yielding 51.6% color reduction) because of the high oxygenation capacity resulting from the suitable amount of hydroxyl radicals. Thus, it was confirmed that the H_2O_2/O_3 process are more efficient (51.6%) in removing color than in O_3-based oxidation (34.2%) [8].

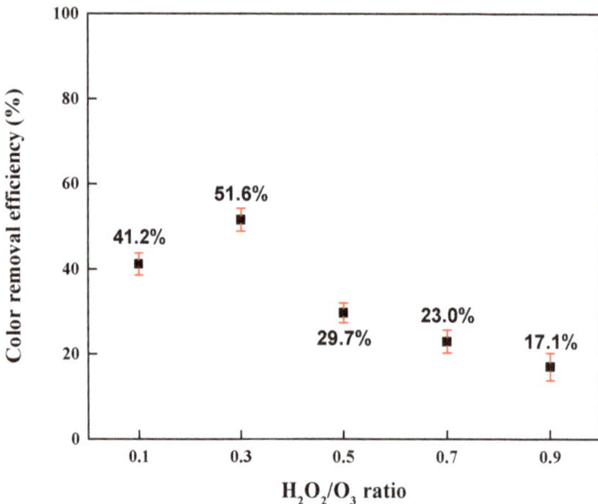

Figure 3. Removal efficiency of color with various H_2O_2/O_3 ratios during H_2O_2/O_3 process.

Figure 4 shows the influence of the H_2O_2/O_3 ratio between 0.1 and 0.9 on the COD reduction due to the O_3 treatment of waste soy sauce. The results show that the COD reduction as a function of the H_2O_2/O_3 ratio followed a similar trend to that of the color removal shown in Figure 3. The COD reduction was optimal (a 73.6 g/L reduction from 217.73 g/L) at an H_2O_2/O_3 ratio of 0.3. This result confirms that a sufficient amount of OH radicals were produced at an H_2O_2/O_3 ratio of 0.3. The H_2O_2 addition facilitated the production of OH radicals, resulting in a synergistic effect between the applied O_3 concentration and the COD reduction [26]. The COD reduction effect decreased distinctly from 33.8 to 16.6% as the H_2O_2/O_3 ratio increased above 0.3 as observed with the color removal. This was likely due to the OH radicals being consumed by the excessive amounts of H_2O_2 [27–29].

The results shown in Figures 3 and 4 confirm that the color removal and COD reduction strongly depend on the H_2O_2/O_3 ratio. However, the H_2O_2/O_3 ratio influences color removal more than COD reduction. Moreover, an H_2O_2/O_3 ratio of 0.3 was the most effective in terms of both color removal and COD reduction.

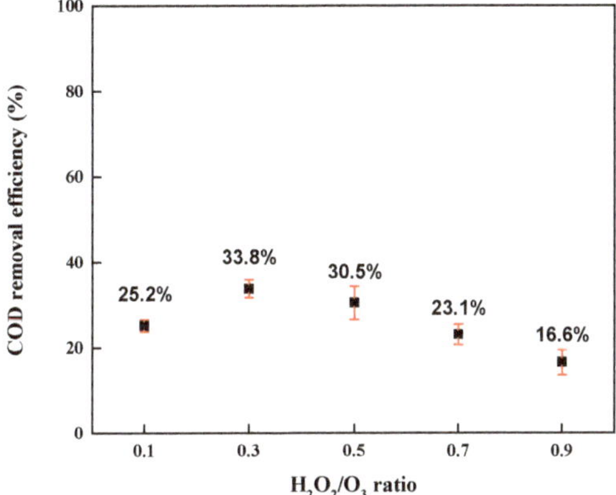

Figure 4. Removal efficiency of COD with various H_2O_2/O_3 ratios during H_2O_2/O_3 process.

3.3. *Nanofiltration (NF) System*

3.3.1. Size Exclusion Chromatography Analysis

The COD reduction was not significant despite the use of a H_2O_2/O_3 process because the COD of waste soy sauce is considerably higher than that of other types of waste and wastewater. In addition, there is a high IOD due to the high COD of the waste soy sauce. Thus, the initial COD of waste soy sauce makes it unsuitable for the oxidation treatment [30]. In addition, it is known that H_2O_2 interferes with the COD reduction by this process by consuming oxidation agents including potassium dichromate [31]. Thus, the molecular weight distribution of the total organic carbon was measured to determine the best membrane-based filtration pretreatment in the interest of improving the COD reduction effect of the H_2O_2/O_3 process.

Figure 5 shows the molecular weight distribution in the waste soy sauce as measured via size-exclusion chromatography (SEC) using a method for detecting organic matter wherein the adsorption of the sample was measured at an ultraviolet (254 nm) wavelength [32]. The main peak shows the molecular weight of the organic matter ranged 750–2200 dalton.

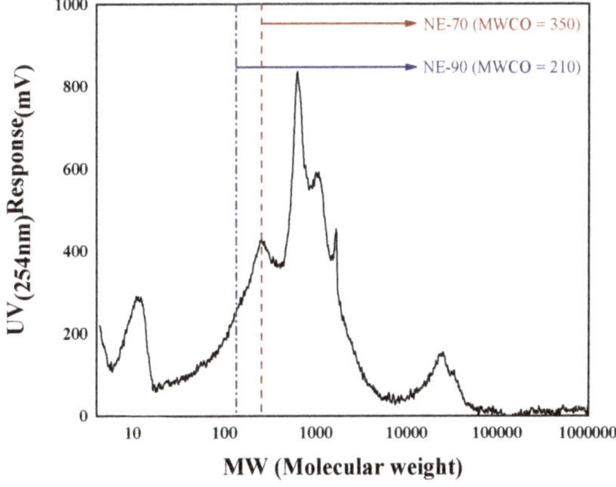

Figure 5. Molecular weight distribution of waste soy sauce.

From this data, the appropriate molecular weight cutoff (MWCO) for the NF membranes (NE-70 and NE-90) was selected to approve the COD reduction effect by H_2O_2/O_3 process. To facilitate stable treatment by oxidation, the COD reduction by the membrane pretreatment was calculated to be over 70%. Thus, experiments were conducted using two membranes, NE-70 (71.8% of the expected COD removal) and NE-90 (73.2% of expected COD removal), confirming the effectiveness of the pretreatment in terms of color removal and COD reduction.

3.3.2. Color Removal and COD Reduction by Nanofiltration (NF)

In our earlier study, O_3-based oxidation exhibited low color removal (34.2%) and COD reduction (27.4%) efficiencies at pH 11 with an O_3 injection dose of 50 mg L^{-1} [8]. In this study, we used H_2O_2/O_3 process and found that the removal efficiency was higher than that with than O_3-based oxidation under the same conditions. However, the color removal and COD reduction were not complete even with H_2O_2/O_3 process. Thus, we applied NF as a pretreatment to enhance the color removal and COD reduction by oxidation method.

Table 6 shows that color removal and COD reduction by NF was similar with the NE-70 and NE-90 membranes. The results show that the color removal and COD reduction were similar with both membranes even though the MWCO of the NE-90 membrane (210 daltons) is lower than that of the NE-70 membrane (350 daltons). The NE-90 membrane, which yielded a color reduction of 81.3% and a COD reduction of 80.7%, was slightly more effective for removing color and reducing the COD than the NE-70 membrane, which yielded a color reduction of 80.8% and a COD of 79.6%. However, the NE-70 is of sustainability usable than NE-90 for experiment. It can further be concluded that the NE-70 membrane is suitable as a pretreatment for waste soy sauce treatment as it improves the color removal and COD reduction.

Table 6. Color removal and COD reduction by nanofiltration (NF).

Classification	Type	Amount of Removal (Removal Efficiency)	
		Color (TCU)	COD (g/L)
Treated waste soy sauce (Removal efficiency)	NE-70	2908.8 (80.8%)	173.3 (79.6%)
	NE-90	2926.8 (81.3%)	175.7 (80.7%)

3.3.3. Flux Variation

To predict the lifetime of the nanofiltration (NF) membranes for waste soy sauce treatment, flux experiments were conducted. Figure 6 shows the flux decline of the two NF membranes during 4 hours of operation. The flux decreased sharply initially then steadily decreased after 45 min of operation. The flux of the NE-70 membrane was lower than that of the NE-90 membrane. This was attributed to the properties of the membranes: NE-90 has a higher contact angle ($41.5 \pm 3.7°$) than NE-70 ($22.6 \pm 1.9°$), showing that its surface is more hydrophobic [17,33,34]; this would cause hydrophobic organic compounds to adsorb onto the surface [17]. In addition, NE-90 has a higher surface roughness than NE-70, which results in a greater repulsive force between the membrane surface and the solute and, thus, a lower solute permeability [17].

These results show that NE-70 is more suitable for this application as its flux decline is lesser during the waste soy sauce treatment than that of NE-90 although the color removal and COD reduction were similar for the two membranes.

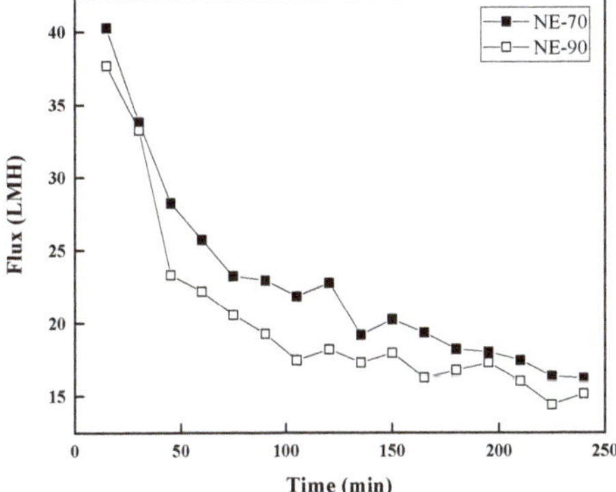

Figure 6. Flux decline of NE-70 and NE-90 in filtration of waste soy sauce.

3.3.4. Comparison of Treatment Methods

To enhance the effectiveness of the O_3 oxidation, we aimed to increase the OH radical production by adding H_2O_2 and applying NF as a pretreatment to overcome the limitation (i.e., the high COD concentration) of the H_2O_2/O_3 process. Figure 7 and Table 7 show comparisons of color removal and COD reduction obtained by various treatment methods for waste soy sauce treatment. The results show that the NF & H_2O_2/O_3 process resulted in the greatest color removal and COD reduction (98.1% and 98.2%, respectively) of the five methods considered. However, the residual color was 74.4 TCU and the residual COD was 4.2 g/L, which are considerably higher than those in ordinary wastewater. Therefore, additional treatment methods should be investigated in order to further improve the process.

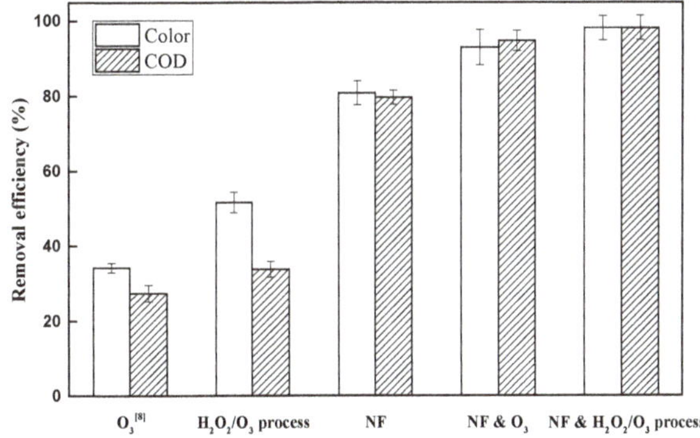

Figure 7. Comparison of removal efficiencies from five treatment methods.

Table 7. Color removal and COD reduction by various treatment methods.

Parameter	Amount of Removal (Removal Efficiency)	
	Color (TCU)	COD (g/L)
O_3 [8]	1333.8 (34.2%)	63.3 (27.4%)
H_2O_2/O_3 process	1857.6 (51.6%)	73.6 (33.8%)
NF (NE-70)	2908.8 (80.8%)	173.3 (79.6%)
NF & O_3	3344.4 (92.9%)	206.2 (94.7%)
NF & H_2O_2/O_3 process	3531.6 (98.1%)	213.8 (98.2%)

4. Conclusions

In this study, the H_2O_2/O_3 process were optimized for removing the color and reducing the COD of waste soy sauce. The H_2O_2/O_3 process were conducted under optimized conditions (H_2O_2/O_3 ratio or 0.3, pH of 11.0, and applied O_3 dose of 50 mg L^{-1}), resulting in 51.6% color removal and 33.8% COD reduction. This was primarily due to the high oxidation capability of O_3 in the presence of the hydroxyl radicals introduced by the addition of H_2O_2. Moreover, an appropriate membrane was selected for waste soy sauce pretreatment based on the molecular weight distribution of the wastewater. The addition of this pretreatment resulted in 98.1% color removal and 98.2% COD reduction. Comparing with alternative methods, NF & H_2O_2/O_3 process can be considered one of the best treatment methods for waste soy sauce, which requires particularly high degrees of color removal and COD reduction. These results can ultimately guide future research into the best wastewater treatment techniques that would allow the wastewater to be reused and mitigate the environmental impacts when it is discharged.

Author Contributions: H.-H.J. conceived and designed the experiments, analyzed the data, participated literature review, and preparation of the manuscript. G.-T.S. provided methodology of the research, participated in analyzed the data, and preparation of the manuscript. D.-W.J. offered suggestions on the concept and preparation of the manuscript.

Funding: This research was funded by the Gyeongnam Green Environment Center (16-4-20-22-8).

Acknowledgments: This work was supported by the research grant of the Gyeongnam Green Environment Center (16-4-20-22-8).

Conflicts of Interest: The authors declare no conflict of interest.

References

1. Lertsiri, S.; Maungma, R.; Assavanig, A.; Bhumiratana, A. Roles of the Maillard reaction in browning during moromi process of Thai soy sauce. *J. Food Process. Preserv.* **2001**, *25*, 149–162. [CrossRef]
2. Nursten, H.E. *The Maillard Reaction: Chemistry, Biochemistry, and Implications*; Royal Society of Chemistry: London, UK, 2005.
3. KOrean Statistical Information Service (KOSIS). Available online: http://kosis.kr/statHtml/statHtml.do?orgId=145&tblId=TX_14503_B016&conn_path=I3 (accessed on 13 April 2018).
4. Pooja, R.N. Removal of Organic Matter from Wastewater by Activated Sludge Process—Review. *Int. J. Sci. Eng. Technol. Res.* **2014**, *3*, 1260–1263.
5. Zheng, C.; Zhao, L.; Zhou, X.; Fu, Z.; Li, A. *Treatment Technologies for Organic Wastewater*; Intech: London, UK, 2013; pp. 249–286.
6. Pipe-Martin, C. Dissolved organic carbon removal by biological treatment. In Proceedings of the Ninth International Conference on Water Pollution: Modelling, Monitoring, and Management, Alicante, Spain, 9–11 June 2008; pp. 445–452.
7. Yoon, Y.; Hwang, Y.; Kwon, M.; Jung, Y.; Hwang, T.M.; Kang, J.W. Application of O_3 and O_3/H_2O_2 as post-treatment processes for color removal in swine wastewater from a membrane filtration system. *J. Ind. Eng. Chem.* **2014**, *20*, 2801–2805. [CrossRef]
8. Jang, H.H.; Seo, G.T.; Jeong, D.W. Investigation of Oxidation Methods for Waste Soy Sauce Treatment. *Int. J. Environ. Res.* **2017**, *14*, 1190. [CrossRef] [PubMed]
9. Fahmi, M.R.; Abidin, C.Z.A.; Rahmat, N.R. Characteristic of colour and COD removal of azo dye by advanced oxidation process and biological treatment. In Proceedings of the International Conference on Biotechnology and Environment Management, Singapore, 2–4 September 2011; pp. 8–13.
10. Rollon, A.A.; Dionisio, M.G.G. Treatment of Distillery Wastewater using Combinations of Advanced Oxidation Processes: UV/O_3, UV/H_2O_2 & H_2O_2/O_3. In Proceedings of the 14th International Conference on Environmental Science and Technology, Rhodes, Greece, 3–5 September 2015.
11. Zheng, Y.; Yu, S.; Shuai, S.; Zhou, Q.; Cheng, Q.; Liu, M.; Gao, C. Color removal and COD reduction of biologically treated textile effluent through submerged filtration using hollow fiber nanofiltration membrane. *Desalination* **2013**, *314*, 89–95. [CrossRef]

12. Abid, M.F.; Zablouk, M.A.; Abid-Alameer, A.M. Experimental study of dye removal from industrial wastewater by membrane technologies of reverse osmosis and nanofiltration. *Iranian J. Environ. Health Sci. Eng.* **2012**, *9*, 17. [CrossRef] [PubMed]
13. Shahata, A.; Omata, T.; Urase, T. Removal of Color from Molasses Wastewater Using Membrane Bioreactor with Acidic Condition. *J. Water Environ. Technol.* **2013**, *11*, 539–546. [CrossRef]
14. Sharma, P.; Joshi, H. MF as Pretreatment of RO for Tertiary Treatment of Biologically Treated Distillery Spentwash. *Int. J. Environ. Sci. Technol.* **2016**, *7*, 172. [CrossRef]
15. Guo, W.D.; Huang, J.P.; Hong, H.S.; Xu, J.; Deng, X. Resolving excitation emission matrix spectroscopy of estuarine CDOM with parallel factor analysis and its application in organic pollution monitoring. *J. Environ. Sci.* **2010**, *31*, 1419–1427.
16. Her, N.; Amy, G.; Foss, D.; Cho, J.; Yoon, Y.; Kosenka, P. Optimization of method for detecting and characterizing NOM by HPLC-size exclusion chromatography with UV and on-line DOC detection. *Environ. Sci. Technol.* **2002**, *36*, 1069–1076. [CrossRef] [PubMed]
17. Kaewsuk, J.; Seo, G.T. A Study on Application of NF Membrane Process for Enhanced Drinking Water Quality. Ph.D. Thesis, Changwon National University, Gyeongsangnam-do, Korea, 2011.
18. Ministry of Environment. Available online: http://www.me.go.kr/home/web/index/do?menuId=70 (accessed on 10 March 2018).
19. Ching, Y.C.; Redzwan, G. Biological Treatment of Fish Processing Saline Wastewater for Reuse as Liquid Fertilizer. *Sustainability* **2017**, *9*, 1062. [CrossRef]
20. Al−Kdasi, A.; Idris, A.; Saed, K.; Guan, C.T. Treatment of textile wastewater by advanced oxidation processes—A review. *Glob. Nest J.* **2004**, *6*, 222–230.
21. Kommineni, S.; Zoeckler, J.; Stocking, A.; Liang, P.S.; Flores, A.; Rodriguez, R.; Brown, T.; Per, R.; Brown, A. *3.0 Advanced Oxidation Processes*; Center for Groundwater Restoration and Protection National Water Research Institute: Los Angeles, CA, USA, 2000; pp. 110–208.
22. Hao, X.; Li, R.; Wang, J.; Yang, X. Numerical simulation of a regenerative thermal oxidizer for volatile organic compounds treatment. *Environ. Eng. Res.* **2018**, *23*, 397–405. [CrossRef]
23. Yuan, R.; Liu, D.; Wang, S.; Zhou, B.; Ma, F. Enhanced photocatalytic oxidation of humic acids using Fe^{3+}-Zn^{2+} co-doped TiO_2: The effects of ions in aqueous solutions. *Environ. Eng. Res.* **2017**, *23*, 181–188. [CrossRef]
24. Cortez, S.; Teixeira, P.; Oliveira, R.; Mota, M. Evaluation of Fenton and ozone-based advanced oxidation processes as mature landfill leachate pre-treatments. *J. Environ. Manag.* **2011**, *92*, 749–755. [CrossRef] [PubMed]
25. Li, G.; He, J.; Wang, D.; Meng, P.; Zeng, M. Optimization and interpretation of O_3 and O_3/H_2O_2 oxidation processes to pretreat hydrocortisone pharmaceutical wastewater. *Environ. Technol.* **2015**, *36*, 1026–1034. [CrossRef] [PubMed]
26. Jang, W.J.; Roh, H.S.; Jeong, D. W. An important factor for the water gas shift reaction activity of Cu-loaded cubic $Ce_{0.8}Zr_{0.2}O_2$ catalysts. *Environ. Eng. Res.* **2018**, *23*, 339–344. [CrossRef]
27. Yetilmezsoy, K.; Sakar, S. Improvement of COD and color removal from UASB treated poultry manure wastewater using Fenton's oxidation. *J. Hazard. Mater.* **2008**, *151*, 547–558. [CrossRef] [PubMed]
28. Yang, D.M.; Yuan, J.M. COD and Color removal from real dyeing wastewater by ozonation. *Water Environ. Res.* **2016**, *88*, 403–407. [PubMed]
29. Kwon, M.; Kye, H.; Jung, Y.; Yoon, Y.; Kang, J.W. Performance characterization and kinetic modeling of ozonation using a new method: ROH, O_3 concept. *Water Res.* **2017**, *122*, 172–182. [CrossRef] [PubMed]
30. Kim, K.S.; Kim, S.B.; Kim, J.B.; Jeon, Y.B.; Jang, E.S.; Lee, S.J.; Kim, S.W. Pilot Plant Study for Evaluation of High Concentration Industrial Wastewater by Ozone Oxidation System. *Korean Soc. Environ. Eng.* **2004**, 901–902.
31. Lee, E.; Lee, H.; Kim, Y.K.; Sohn, K.; Lee, K. Hydrogen peroxide interference in chemical oxygen demand during ozone based advanced oxidation of anaerobically digested livestock wastewater. *Int. J. Environ. Sci. Technol.* **2011**, *8*, 381–388. [CrossRef]
32. Albrektienė, R.; Rimeika, M.; Zalieckienė, E.; Šaulys, V.; Zagorskis, A. Determination of organic matter by UV absorption in the ground water. *J. Environ. Eng. Landsc.* **2012**, *20*, 163–167. [CrossRef]

33. Kaewsuk, J.; Seo, G. T. Verification of NOM removal in MIEX−NF system for advanced water treatment. *Sep. Purif. Technol.* **2011**, *80*, 11–19. [CrossRef]
34. Hassan, A.R.; Rozali, S.; Safari, N.H.M.; Besar, B.H. The roles of polyethersulfone and polyethylene glycol additive on nanofiltration of dyes and membrane morphologies. *Environ. Eng. Res.* **2018**, *23*, 316–322. [CrossRef]

Article

Shifts in the Microbial Community of Activated Sludge with Different COD/N Ratios or Dissolved Oxygen Levels in Tibet, China

Jin Xu [1,2], Peifang Wang [1,*], Yi Li [1,*], Lihua Niu [1] and Zhen Xing [2]

1 Ministry of Education Key Laboratory of Integrated Regulation and Resource Development on Shallow Lakes, College of Environment, Hohai University, Nanjing 210098, China; xurin123@163.com (J.X.); nlhnq55@163.com (L.N.)

2 College of Resources and Environment, Tibet Agricultural and Animal Husbandry University, Tibet Nyingtri 860000, China; xztibetan@163.com

* Correspondence: pfwang2004@hotmail.com (P.W.); envly@hhu.edu.cn (Y.L.)

Received: 29 March 2019; Accepted: 12 April 2019; Published: 16 April 2019

Abstract: In this study, we examined the influence of the organic carbon-to-nitrogen ratio (chemical oxygen demand (COD/N)) and dissolved oxygen (DO) levels on the removal efficiency of pollutants and on the change in total microflora in the cyclic activated sludge system (CASS) in the Nyingchi prefecture in Tibet. The results demonstrated that the treatment performance was the best when the COD/N ratio was 7:1 or the DO levels were 2–2.5 mg/L in comparison with four different tested COD/N ratios (4:1, 5:1, 7:1, and 10:1) and DO concentrations (0.5–1, 1–2, 2–2.5, and 2.5–3.5 mg/L). The treatment performance can be explained by the relative operational taxonomic unit richness and evenness of the microbial communities in activated sludge. Evident microbial variance was observed, especially different COD/N ratios and DO concentrations, which were conducive to the disposal of urban sewage in plateaus. The results help to understand sewage treatment under different COD/N ratios or DO concentrations on plateaus. This work provides practical guidance for the operation of any wastewater treatment plant on a plateau.

Keywords: plateau; sewage treatment; COD/N ratio; dissolved oxygen; microbial community

---

1. Introduction

Sewage treatment has contributed to the improvement in the water-environment in high-altitude regions [1]. The activated sludge method uses activated sludge to transform organic and inorganic pollutants, which has been applied to improve aquatic environments around the world. Due to its low operational cost and good performance (chemical oxygen demand (COD/N) removal efficiency can reach 85% or higher), it is still the most widely used and common water treatment measured in urban effluent treatment [2]. The wastewater treatment plants (WWTPs) at high-altitude regions face some operational barriers, mainly because of extreme climate conditions, such as thin air and low pressure [3]. The stable operation of WWTPs with widely-used sewage treatment technology, based on the activated sludge or biological process in plateau regions, is crucial [1].

The main factors influencing urban effluent treatment processes include the COD/N ratio and the dissolved oxygen (DO) concentration. If the COD/N ratio in influent concentration is too high, heterotrophic organisms consume more available ammonia-nitrogen than nitrifying organisms [4]. Yun [5] stated that nitration activity gradually fails at the reactors with organic additions. Sharma [6] reported that the nitrification rate increases with deduced COD/TKN (Kjeldahl nitrogen) in the influent. Luo et al. [7] studied the effect of the COD/N ratio in influent on the decomposition of aerobic granule sludge, and the results showed that when the COD/N ratio in the influent is 2:1 or 1:1, it strongly influenced the stability of aerobic granule sludge with regard to physical properties and nitrification efficiency, and when the COD/N ratio is reduced to 1:1, the aerobic granule sludge decomposes. Since an increased COD/N could result in a lowered nitrification efficiency and enhanced denitrification efficiency, the COD/N ratio in influent influences the quantity of microorganisms [8]. Zielinska [9] studied nitrogen removal from wastewater and the bacterial diversity of activated sludge at different COD/N ratios and DO concentrations. The results showed that the total bacteria diversity was similar in all experimental series and the diversity and abundance of ammonia-oxidizing bacteria (AOB) were higher in the reactors with a COD/N ratio of 0.7:1 in comparison with the reactors with a COD/N of 6.8:1 [9].

In any urban WWTP, the DO concentration determines the degradation rate of organic matter in microflora, aerobic growth, and the operational cost [10,11]. DO concentrations must be controlled because it determines the dominant reaction process in the reactor. High DO concentrations suppress denitrification, and low DO concentrations lead to restricted ammoxidation [12]. Dissolved oxygen can be controlled by air supply, so the cost of treating wastewater with oxygen is affected. A low COD/N ratio and DO concentration facilitate a nitrification effect to some extent and decrease oxygen demand by 25% and COD by 40% in the denitrification process [13,14]. According to the report by Ma [15], in the continuous flow system of domestic urban wastewater treatment, 0.4–0.7 mg/L DO was conducive to nitrite accumulation. Yadav [16] evaluated the microbial community in response to different DO concentrations (1, 2, and 4 mg/L) in activated sludge and the results showed that the diversity of bacteria was the greatest when the DO was 2 mg/L in the reactor. The pressure increases in the gas phase and the concentration of saturated DO increases in the water, which is conducive to the oxygen transfer to the water, whereas the pressure decreases when the sewage is aerated in the biological treatment tank. The oxygen pressure at 4000 m altitude is only 60% of that at sea level [3]. Oxygen transfer varies due to variation in pressure and its influence on microbial communities at high latitudes. Therefore, we need more information on the microbial community of activated sludge with different DO levels at high latitudes.

Many studies have provided information about low-elevation WWTPs. To provide some practical guidance on domestic wastewater treatment in plateau regions, the objectives of this study were: (1) To investigate the influence of different COD/N ratios or DO concentrations on urban wastewater treatment on plateaus, (2) to investigate microbial community structure and its diversity in activated sludge under different COD/N ratios or DO concentrations in plateaus, and (3) to determine the relationships amongst domestic wastewater treatment performance, microflora structure, and species diversity. We think these results provide practical guidance for treating domestic wastewater in plateau regions.

2. Materials and Methods

2.1. Reactor and Operation

A total of four lab-scale cyclic activated sludge systems (CASSs) were used to treat synthetic sewage in Nyingtri, Tibet. The experimental site, the Environmental Science and Engineering Laboratory of the Tibet Agricultural and Animal Husbandry University, was located at an altitude of about 3000 m above sea level. The atmospheric pressure of the experimental site was 706 hPa (measured value). A schematic diagram of the CASS reactor and the reactor volume of 0.045 m^3 are shown in Figure 1. Activated sludge (AS) originated from the Lhasa sewage treatment plant with an initial sludge concentration of 2.5 g/L. The synthetic influent sewage feed and the aeration system of the device were adjusted by a creep pump. The pH in the system was maintained between 7.2 and 8.4. The pH was adjusted with anhydrous sodium carbonate. The reactor was run under water temperature (11 ± 1 °C). The composition of the synthetic sewage is shown in Table 1 at different COD/N ratios test stages. CN-1, CN-2, CN-3, and CN-4 refer to the COD/N ratios of 4:1, 5:1, 7:1, and 10:1, respectively. The DO was kept constant, or more than 4 mg/L, to provide adequate oxygen during the different

COD/N ratios test stages. The composition of the synthetic sewage is shown in Table 2 at different DO concentration test stages. DO-1, DO-2, DO-3, and DO-4 refer to the DO concentrations of 0.5–1, 1–2, 2–2.5, and 2.5–3.5 mg/L, respectively. The COD/N ratio was controlled at 7:1 to provide adequate nutrients during the different DO concentration test stages. Under batch feeding conditions, there were four daily CASS cycles. One CASS operational cycle required 6 h, of which 4 h were the aeration stage, and 2 h were the sludge setting and drainage stage.

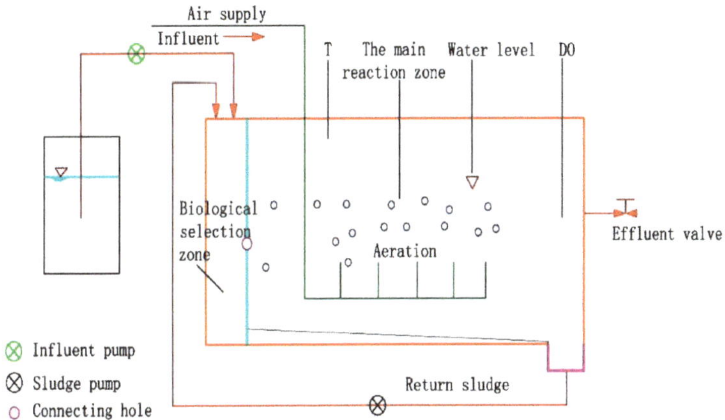

Figure 1. The cyclic activated sludge system (CASS) process unit.

Table 1. The composition of synthetic sewage (chemical oxygen demand (COD/N), H_4^+–N = 4/1, 5/1, 7/1 and 10/1).

Species	Composition	Concentration (mg/L)	Confect
Organics	COD	120, 150, 210, 300	glucose
Nitrogen compounds	NH_4^+–N	30	NH_4CI
Phosphorus compounds	TP	5	K_2HPO_4
Other additions and trace elements (mg/L)	$MgSO_4{\cdot}7H_2O$(12); $FeSO_4{\cdot}7H_2O$(10); $CaCl_2$(30); $NaHCO_3$(50); pH (7.2–8.4); H_3BO_3(0.15); $CoCl_2{\cdot}6H_2O$(0.15); $CuSO_4{\cdot}6H_2O$(0.03); $FeCl_3{\cdot}6H_2O$(1.5); KI(0.03); $MnCl_2{\cdot}2H_2O$(0.12); $(NH_4)_2Mo_7O_{24}{\cdot}2H_2O$(0.06); $ZnSO_4{\cdot}7H_2O$(0.12)		

Table 2. The composition of synthetic sewage (dissolved oxygen (DO) = 0.5–1, 1–2, 2–2.5 and 2.5–3.5 mg/L).

Species	Composition	Concentration (mg/L)	Confect
Organics	COD	210	glucose
Nitrogen compounds	NH_4^+–N	30	NH_4CI
Phosphorus compounds	TP	5	K_2HPO_4
Other additions and trace elements (mg/L)	$MgSO_4{\cdot}7H_2O$(12); $FeSO_4{\cdot}7H_2O$(10); $CaCl_2$(30); $NaHCO_3$(50); pH (7.2–8.4); H_3BO_3(0.15); $CoCl_2{\cdot}6H_2O$(0.15); $CuSO_4{\cdot}6H_2O$(0.03); $FeCl_3{\cdot}6H_2O$(1.5); KI(0.03); $MnCl_2{\cdot}2H_2O$(0.12); $(NH_4)_2Mo_7O_{24}{\cdot}2H_2O$(0.06); $ZnSO_4{\cdot}7H_2O$(0.12)		

To describe "cyclical", a decanting factor (F_d) can be obtained: $F_d = V_e/V_d$,

Where V_e is 10 L wastewater that was exchanged in each cycle, V_d is the reactor volume, 0.045 m^3. The reactors ran continuously for about 25 days, including 10 days of adjustment and 15 days of stabilization.

2.2. Collection of Samples and Analytical Methods

Daily measurements of the effluent of the reactors included: COD, NH_4–N^+, NO_2^-–N, NO_3^-–N, and total phosphorus (TP), according to Chinese standard methods during the stable stage [17]. COD was determined using the potassium dichromate method. Nitrogen content was measured with a flow analyzer. TP was determined by ammonium molybdate spectrophotometry. A portable dissolved oxygen meter (JPBJ-608, Shanghai Precision and Scientific Instrument Corporation, Shanghai, China) was used to determine the water temperature and DO. We collected 10 mL of the mixed liquor-suspended solids (MLSS) reaction mixture from each reactor during the stable stage for microbiological analysis.

2.3. Pyrosequencing Analysis

The microbial community structure and its diversity of activated sludge samples were investigated by high-throughput 16S ribosomal ribonucleic acid (rRNA) gene sequencing. The detailed operational steps of the pyrosequencing methods have been described by Xu [18]. The rarefaction analysis was conducted to reveal the diversity indices. The detailed data processing and analysis have also been described by Xu [19].

3. Results and Discussion

3.1. Richness and Evenness of Microbial Communities of Activated Sludge

The relative operational taxonomic unit (OTU) richness and evenness of microbial communities of activated sludge with different COD/N ratios or DO concentrations are shown in Figures 2–4. The relative OTU richness of the species is reflected in the length of the curve on the horizontal axis. The wider the curve, the richer the relative OTU richness of the species. The relative OTU evenness of species composition is reflected in the shape of the curve. The flatter the curve, the more relative the OTU evenness of species composition. The relative OTU richness and evenness of the bacterial communities occurred when the COD/N ratio was 7:1 or the DO was 2–2.5 mg/L (Figure 2). The relative OTU richness and evenness of the fungal communities of activated sludge peaked at a COD/N ratio of 5:1 or DO of 2–2.5 mg/L (Figure 3). The relative OTU richness and evenness of archaeal communities were the highest, at the COD/N ratio of 7:1 or the DO of 2–2.5 mg/L (Figure 4). The Shannon index values for different COD/N ratios and DO concentrations are shown in Table 3. The Shannon index values represent the diversity and evenness of a sample. Table 3 shows that the diversity and evenness of fungal communities of activated sludge were far lower at COD/N of 4:1, 7:1, and 10:1 than that at 5:1. The diversity and evenness of fungal communities of activated sludge were far lower at the DO of 0.5–1, 1–2, and 2.5–3.5 mg/L than at 2–2.5 mg/L. As shown in Figure 3, the broken curve ends at the lower horizontal axis values. The highest species diversity and evenness was observed when the COD/N ratio was 7:1 or the DO was 2–2.5 mg/L.

Table 3. The Shannon index values with different COD/N ratios or DO concentrations.

Shannon Index	CN-1	CN-2	CN-3	CN-4	DO-1	DO-2	DO-3	DO-4
bacterial	2.27	1.93	3.29	2.39	2.62	3.23	3.37	2.61
fungal	1.26	2.53	1.10	1.52	0.13	0.38	1.13	0.38
archaeal	2.21	2.22	3.38	2.17	3.05	3.06	3.37	2.87

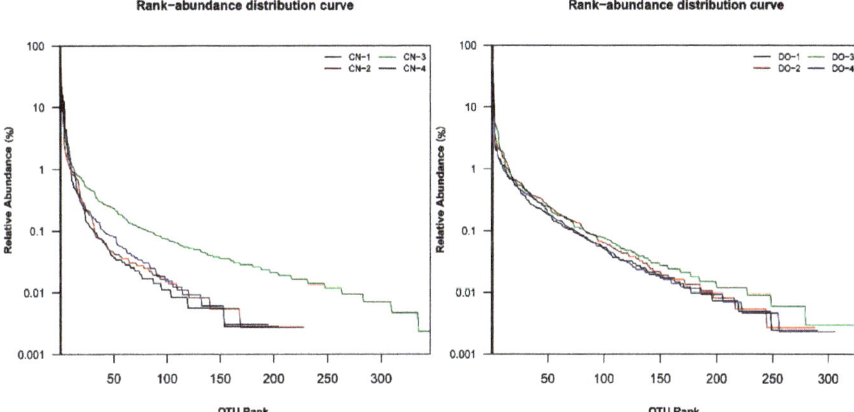

Figure 2. Relative abundance of the operational taxonomic unit () richness and evenness of bacterial communities of activated sludge.

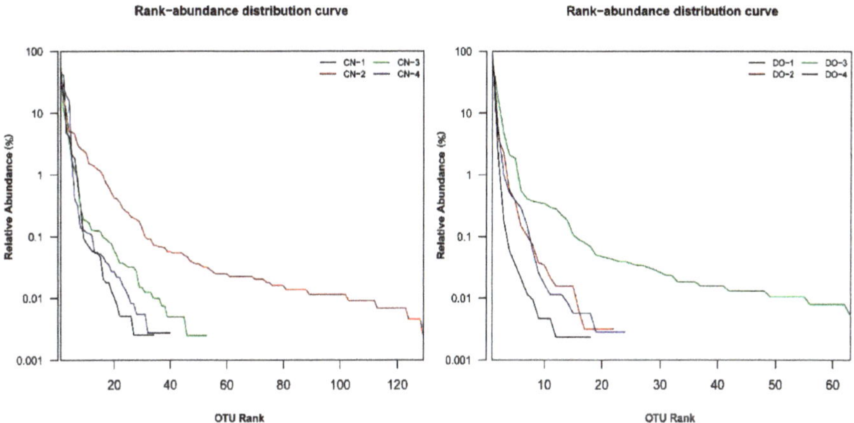

Figure 3. Relative abundance of the OTU richness and evenness of fungal communities of activated sludge.

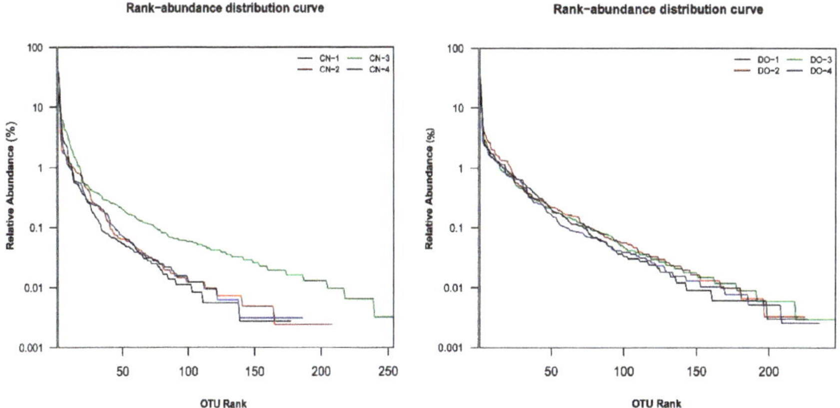

Figure 4. Relative abundance of the OTU richness and evenness of archaeal communities of activated sludge.

3.2. Microbial Community Structures

Figure 5 depicts the relative abundance of the bacterial community under different COD/N ratios or DO concentrations in the activated sludge. The results indicate that *Proteobacteria* (>80%) was the main group in the activated sludge and *Bacteroidetes* (<10%) was secondary in activated sludge when the COD/N ratio was 4:1 and 5:1. When the COD/N ratio was 7:1, *Proteobacteria* (43.51%) was the main group and *Bacteroidetes* (37.88%) was the secondary group in activated sludge. When the COD/N ratio was 10:1, *Proteobacteria* (68.38%) was the main group and *Bacteroidetes* (14.19%) was the secondary group in activated sludge. These figures demonstrate that *Proteobacteria* was sensitive to the COD/N ratio. As shown in Figure 5, under different COD/N ratios, *Proteobacteria* was the most abundant phylum and *Bacteroidetes* was the second most abundant phylum in activated sludge. This result is similar to reported research results for urban WWTPs [20,21]. The previous research revealed that some *Actinobacteria*, *Bacteroidetes*, and *Firmicutes* members can decompose complex polysaccharides [22,23]. When the COD/N ratio was 10:1, the proportion of *Actinobacteria* increased to 15.39%. As shown in Figure 5, the relative abundance of the *Actinobacteria* was lower at the COD/N ratios of 4:1, 5:1, and 7:1 than at 10:1. *Actinobacteria* play a necessary role in nutrient removal during sewage treatment [24]. The microbial abundance at the phylum level was lower at COD/N ratios of 4:1, 5:1, and 10:1 than at 7, indicating that the proportions of different phyla in the bio-community in the sludge vary with the feeding ratio.

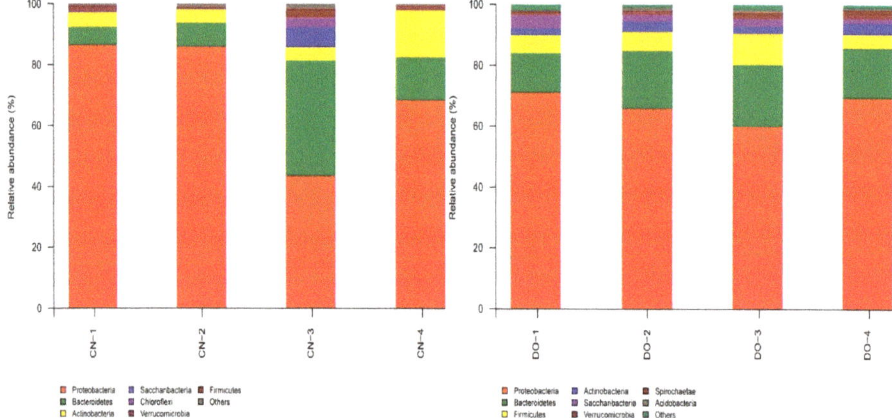

Figure 5. Relative abundance of bacterial phyla of activated sludge.

As shown in Figure 5, when the DO concentrations were 0.5–1, 1–2, 2–2.5, and 2.5–3.5 mg/L, *Proteobacteria* (>60%) was the main group in activated sludge. Under different DO concentrations, *Proteobacteria* was the most abundant phylum. When the DO concentration was 2–2.5 mg/L, *Firmicutes* (10.29%) was significantly augmented. The relative abundance of *Firmicutes* was lower when the DO concentration was 0.5–1, 1–2, and 2.5–3.5 mg/L (5.89%, 6.23%, and 4.60%, respectively) than that at 2–2.5 mg/L (10.29%). In this research, the majority of sequences were labelled as unclassified, and the most abundant phylum observed in all reactors and the previously activated biomass, such as *Actinobacteria*, *Proteobacteria*, *Firmicutes*, *Bacteroidetes*, *Acidobacteria*, *Nitrospira*, and *Chloroflexi*, were previously reported [25,26]. This result is similar to previous research results [16]. The difference in our reported richness and evenness of microbial community to those reported by Chandra Yadav et al. [16] may be related to elevational gradient. As reported by Niu [27], a significant change in the evenness of the bacterial community was observed at around 1200 m of elevation with a linear decline as elevation increased to 3660 m. The microbial abundance at the phylum level was clearly lower when the DO concentration was 0.5–1, 1–2, and 2.5–3.5 mg/L than when the DO was 2–2.5 mg/L, indicating that the bio-community proportions in the sludge vary with the DO concentration.

Figure 6 summarizes the relative abundances of the fungi community on the phylum level of activated sludge with different COD/N ratios or DO concentrations. The figure shows that *Ascomycota* was the most abundant phylum. *Basidiomycota* was the second most abundant phylum, followed by unclassified fungi when the COD/N ratio was at 4:1, 5:1, 7:1, or 10:1. *Ascomycota* (81.38%) was predominant and *Basidiomycota* (17.50%) was also a major group when the COD/N ratio was at 7. *Ascomycota* also played a critical role in nutrient removal during sewage treatment. As shown in Figure 6, *Basidiomycota* (>70%) was predominant, followed by *Ascomycota* (<30%) in activated sludge, when the DO was at 0.5–1, 1–2, 2–2.5, or 2.5–3.5 mg/L. In Figure 6, the relative abundances of *Ascomycota* at the DO of 0.5–1, 1–2, and 2.5–3.5 mg/L (2.04%, 4.58%, and 7.21%, respectively) were lower than when the DO was 2–2.5 mg/L (23.81%).

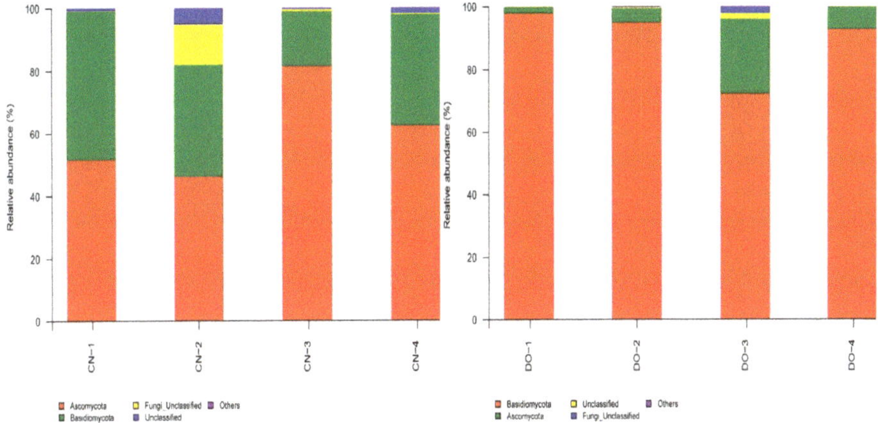

Figure 6. Relative abundance of fungal phyla of activated sludge.

Figure 7 demonstrates the relative abundances of the archaeal community on the phylum level of activated sludge at different COD/N ratios or DO concentrations. Figure 7 show that *Thaumarchaeota* represented the most abundant phylum. *Euryarchaeota* was the second most abundant phylum, followed by *Bacteroidetes* when the COD/N ratio was 4:1 and 5:1. *Bacteroidetes* was the most abundant phylum, *Euryarchaeota* was the second most abundant phylum, followed by *Thaumarchaeota* when the COD/N ratio was 7:1. *Euryarchaeota* was the most abundant phylum, *Thaumarchaeota* was the second most abundant phylum, followed by *Bacteroidetes* when the COD/N ratio was 10:1. Figure 7 suggests that *Euryarchaeota* was sensitive to the COD/N ratio. *Thaumarchaeota* (>50%) was predominant, and *Euryarchaeota* (<40%) was also a major group in activated sludge when the DO was 0.5–1, 1–2, 2–2.5, and 2.5–3.5 mg/L. The relative abundances of *Euryarchaeota* when the DO was 0.5–1, 1–2, and 2.5–3.5 mg/L (18.89%, 24.77%, and 26.22%, respectively) were lower than when the DO was 2–2.5 mg/L (39.82%). These results demonstrate that shifts in microbial community occur with different COD/N ratios or DO concentrations.

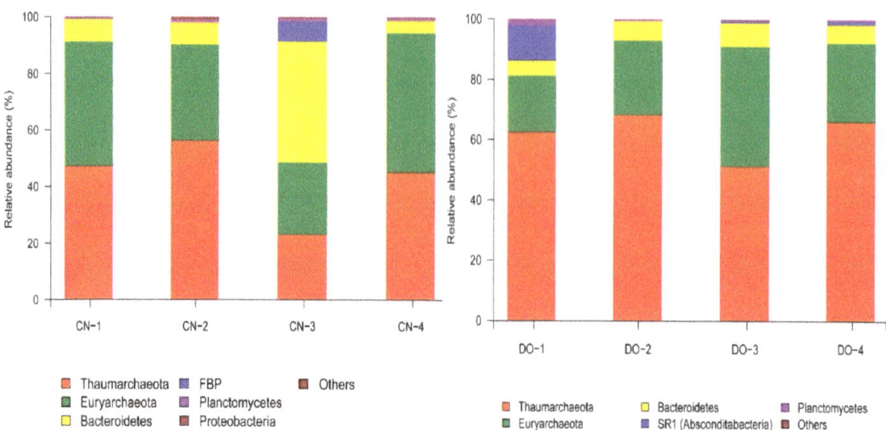

Figure 7. Relative abundance of archaeal phyla of activated sludge.

3.3. Process Performance

The COD/N ratio in domestic sanitary wastewater is normally lower than these specified values, where the removal of nitrogen was restricted by the lack of organic carbon sources [28]. When the COD/TN (Total Nitrogen) ratio in influent concentration was less than four, removing the nitrogen was difficult [29,30]. Carbon shortage was detrimental to denitrification [31]. Figure 8 demonstrates the removal result of carbon, nitrogen, and phosphorus under different COD/N ratios. Figure 8 shows that the removal rate of COD and TP was the highest when the COD/N ratio was 7:1. The average removal rate of COD and TP was about 81% and 66%, respectively, when the COD/N ratio was 7:1. A lack of nutrients in the aerobic tank or excessive sludge load could lead to sludge disintegration.

Figure 8 shows that, in every reactor, the ammonia–nitrogen concentration in effluent did not exceed 4 mg/L, and nitrites and nitrates were the main nitrogen oxide form. The NO_3^-–N and NO_2^-–N concentrations in the effluent gradually increased in the stable stage. In the stable stage, the accumulated NO_3^-–N and NO_2^-–N concentrations were 19.28, 19.19, 19.36, and 19.04 mg/L when the COD/N ratio was 4:1, 5:1, 7:1, and 10:1, respectively. As shown in Figure 8, the effluent NH_4–N^+ (average 2.38 mg/L) at a COD/N ratio of 7:1 were lower than when the COD/N ratio was 4:1 (2.59 mg/L), 5:1 (2.67 mg/L), and 10:1 (2.64 mg/L). In this batch of experiments, the ammonia oxidation efficiency exceeded 89%, with nitrites and nitrates as the main products. The treatment performance was the best when the COD/N ratio was 7:1, compared with all tested COD/N ratios.

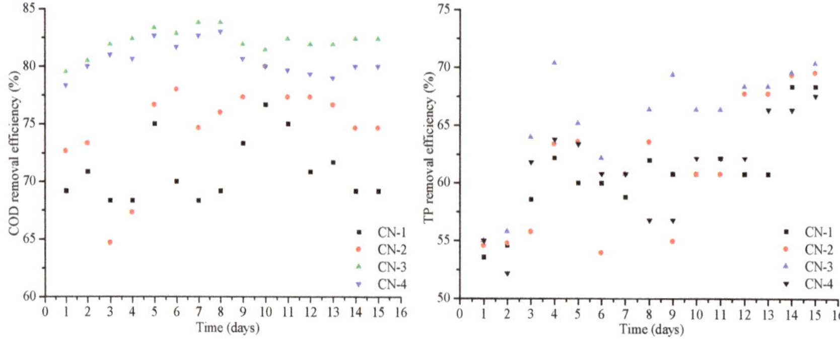

Figure 8. *Cont.*

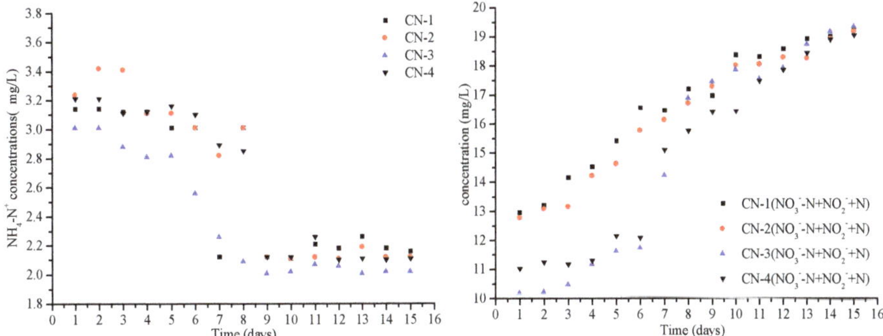

Figure 8. The effluent parameter NH_4–N^+, $NO_3{}^-$–N, and $NO_2{}^-$–N, COD and total phosphorus (TP) removal efficiency under different COD/N ratios.

Figure 9 demonstrates the removal result of carbon, nitrogen, and phosphorus under different DO concentrations. The water quality blackened for a long time in the aerobic pond due to lack of oxygen. The figure shows that the COD degradative efficiency was the lowest when operating at 0.5–1 mg/L. Operation at 2–2.5 mg/L DO produced the highest COD removal rates throughout the 15-day stabilization period of the study. The average removal rates of COD and TP were about 83% and 75%, respectively, when the DO was 2–2.5 mg/L. High DO concentration lead to the oxidation and disintegration of sludge. It was easier to sediment active sludge as a result of degassing.

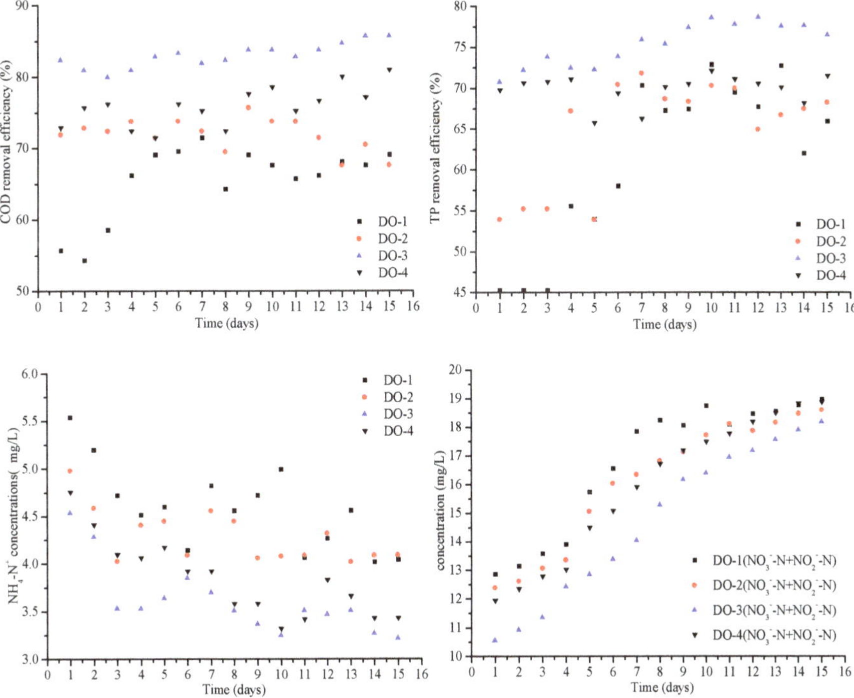

Figure 9. The effluent parameter, NH_4–N^+, $NO_3{}^-$–N, and $NO_2{}^-$–N, COD, and TP removal efficiency under different DO concentrations.

The nitrogen content in wastewater was also important as it not only caused eutrophication of a water body, but excessive ammonia in wastewater affected the degradation capacity and microbial growth [32–34]. In every reactor, the ammonia–nitrogen concentration in effluent did not exceed 6 mg/L (Figure 9), and nitrites and nitrates were the main nitrogen oxide form. The effluent NO_3^-–N and NO_2^-–N concentrations gradually increased. In the stable stage, the accumulated NO_3^-–N and NO_2^-–N concentrations increased up to 18.97, 18.60, 18.19, and 18.88 mg/L at DO concentrations of 0.5–1, 1–2, 2–2.5, and 2.5–3.5 mg/L, respectively. As can be seen in Figure 9, the effluent NH_4–N^+ (average 3.61 mg/L) concentration at a DO of 2–2.5 mg/L was lower than when the DO was 0.5–1 (4.58 mg/L), 1–2 (4.29 mg/L), and 2.5–3.5 mg/L (3.84 mg/L). In this experiment, the ammonia oxidation efficiency exceeded 81% with nitrites and nitrates as the main products. The treatment performance was the best when the DO was 2–2.5 mg/L.

As the test site was located in Nyingtri in Tibet, the atmospheric pressure of Nyingtri was 706 hPa (measured value). The oxygen content at 3000 m above sea level (ASL) was only 79% of that at sea level. Therefore, the calculation method should be revised according to the actual situation when the sewage treatment plant calculates the aeration amount [35,36].

3.4. The Relationship among the Changes of Pollutants, Microbial Community Structure, and Species Diversity

The treatment performance can be explained by the relative OTU richness and evenness of microbial communities of activated sludge [19,27]. The community richness and evenness were two important factors that influenced the functional stability and general performance of WWTPs [37]. As Wittebolle [38] reported, communities with higher evenness have more functional resistance to environmental stress. According to Werner [39], in full-scale bioenergy systems, the methanogenic activity and substrate removal efficiency are correlated with community evenness. As shown in Figures 2–4, the highest OTU richness and evenness of microbial communities was observed when the COD/N ratio was 7:1 or the DO was 2–2.5 mg/L. As shown in Table 3, the highest species diversity and evenness were also observed when the COD/N ratio was 7:1 or the DO was 2–2.5 mg/L. The treatment performance was the best of the four conditions studied when the COD/N ratio was 7:1 or the DO was 2–2.5 mg/L (Figures 8 and 9). The shifts in microbial community with different operational conditions could be correlated with differential treatment performances. The treatment performance was the best when the COD/N ratio was 7:1 or the DO was 2–2.5 mg/L because the relative OTU richness and evenness were the highest at these values. The richness and evenness of the community may be important for determining the general performance of WWTPs.

4. Conclusions

In this study, we proved that the wastewater treatment performance was the most efficient when the COD/N ratio was 7:1 (average 81% COD and 66% TP removal efficiency) or the DO was 2–2.5 mg/L (average 83% COD and 75% TP removal efficiency) of the four different COD/N ratios (4:1, 5:1, 7:1, and 10:1) and DO concentrations (0.5–1, 1–2, 2–2.5, and 2.5–3.5 mg/L) tested in a plateau region. The ammonia oxidation efficiency exceeded 89% for different COD/N ratios or 81% for different DO concentrations with nitrites and nitrates as the main products. We also demonstrated that the relative abundance species evenness of the microbial community structure of activated sludge was the greatest when the COD/N ratio was 7:1 or the DO was 2–2.5 mg/L. The microbial community structure was shown to vary with different COD/N ratios or DO concentrations. Due to the difference in altitude between plateaus and plains, the calculation method of aeration in the sewage treatment plant should be adapted according to the actual situation. The analyses provided a better understanding of sewage treatment with different COD/N ratios or DO concentrations in plateau areas.

Author Contributions: Conceptualization, J.X.; data curation, J.X.; formal analysis, J.X.; investigation, J.X. and L.N.; methodology, J.X.; project administration, P.W. and Y.L.; resources, P.W. and Y.L.; supervision, P.W. and Y.L.; validation, J.X.; visualization, J.X.; writing—original draft, J.X.; writing—review and editing, J.X. and Z.X.

Funding: This research was funded by the National Natural Science Foundation of China (no. 51568059).

Conflicts of Interest: The authors declare no conflict of interest. The founding sponsors had no role in the design of the study; in the collection, analyses, or interpretation of data; in the writing of the manuscript, and in the decision to publish the results.

References

1. Xu, J.; Wang, P.F.; Li, Y.; Niu, L.H. Performance and characterization of the microbial community structures in the activated sludge from wastewater treatment plant at high altitudes in Tibet of China. *Desalin. Water. Treat.* **2018**, *106*, 108–115. [CrossRef]
2. Yang, M.; Liu, X.C.; Zhang, Y. Analysis of bacterial community structures in two sewage treatment plants with different sludge properties and treatment performance by nested PCR-DGGE method. *J. Environ. Sci.* **2007**, *19*, 60–66.
3. Juanico, M.; Weinberg, H.; Soto, N. Process design of waste stabilization ponds at high altitude in Bolivia. *Water Sci. Technol.* **2000**, *42*, 307–313. [CrossRef]
4. Strauss, E.A.; Lamberti, G.A. Regulation of nitrification in aquatic sediments by organic carbon. *Limnol. Oceanogr.* **2000**, *45*, 1854–1859. [CrossRef]
5. Yun, Z.; Jung, Y.H.; Lim, B.R. The stability of nitrite nitrification with strong nitrogenous wastewater: Effects of organic concentration and microbial diversity. *Water Sci. Technol.* **2004**, *49*, 89–96. [CrossRef]
6. Sharma, R.; Gupta, S.K. Influence of chemical oxygen demand to total Kjeldahl nitrogen ratio and sludge age on nitrification of nitrogenous wastewater. *Water Environ. Res.* **2004**, *76*, 155–161. [CrossRef]
7. Luo, J.H.; Hao, T.W.; Wei, L. Impact of Influent COD/N Ratio on Disintegration of Aerobic Granular Sludge. *Water Res.* **2014**, *62*, 127–135. [CrossRef]
8. Gieseke, A.; Purkhold, U.; Wagner, M.; Amann, R.; Schramm, A. Community structure and activity dynamics of nitrifying bacteria in a phosphate-removing biofilm. *Appl. Environ. Microbiol.* **2001**, *67*, 1351–1362. [CrossRef]
9. Zielinska, M.; Bernat, K.; Cydzik-Kwiatkowsa, A.; Sobolewska, J.; Wojnowska-Baryla, I. Nitrogen removal from wastewater and bacterial diversity in activated sludge at different COD/N ratios or dissolved oxygen concentrations. *J. Environ. Sci.* **2012**, *24*, 990–998. [CrossRef]
10. Kapley, A.; Tolmare, A.; Purohit, H.J. Role of oxygen in the utilization of phenol by Pseudomonas CF600 in continuous culture. *J. Microbiol. Biotechnol.* **2001**, *17*, 801–804. [CrossRef]
11. Wells, G.F.; Park, H.D.; Yeung, C.H.; Eggleston, B.; Francis, C.A.; Criddle, C.S. Ammonia-oxidizing communities in a highly aerated full-scale activated sludge bioreactor: Betaproteobacterial dynamics and low relative abundance of Crenarchaea. *Environ. Microbiol.* **2009**, *11*, 2310–2328. [CrossRef]
12. Garrido, J.M.; van Benthum, W.A.J.; van Loosdrecht, M.C.M.; Heijnen, J.J. Influence of dissolved oxygen concentration on nitrite accumulation in a biofilm airlift suspension reactor. *Biotechnol Bioeng.* **1997**, *53*, 168–178. [CrossRef]
13. Ganigué, R.; López, H.; Balaguer, M.D.; Colprim, J. Partial ammonium oxidation to nitrite of high ammonium content urban landfill leachates. *Water Res.* **2007**, *41*, 3317–3326. [CrossRef] [PubMed]
14. Pollice, A.; Tandoi, V.; Lestingi, C. Influence of aeration and sludge retention time on ammonium oxidation to nitrite and nitrate. *Water Res.* **2002**, *36*, 2541–2546. [CrossRef]
15. Ma, Y.; Peng, Y.Z.; Wang, S.Y.; Yuan, Z.G.; Wang, X.L. Achieving nitrogen removal via nitrite in a pilot-scale continuous pre-denitrification plant. *Water Res.* **2009**, *43*, 563–572. [CrossRef]
16. Chandra Yadav, T.; Khardenavis, A.A.; Kapley, A. Shifts in microbial community in response to dissolved oxygen Levels in activated sludge. *Bioresour. Technol.* **2014**, *165*, 257–264. [CrossRef] [PubMed]
17. The State Environmental Protection Administration. *Water and Wastewater Monitoring Analysis Method*, 4th ed.; China Environmental Science Press: Beijing, China, 2002. (In Chinese)
18. Xu, J.; Li, S.W. The Changes of Pollutants and Microbial Community Structures of the Activated Sludge in Response to Different Temperature Levels in Tibet of China. *IOP Conf. Ser. Earth Environ. Sci.* **2018**, *189*, 052080. [CrossRef]
19. Xu, J.; Wang, P.F.; Li, Y.; Niu, L.H. Activated sludge culture domestication at high altitudes in Tibet of China. *Desalin. Water. Treat.* **2019**, *142*, 98–103. [CrossRef]
20. Zhang, T.; Shao, M.F.; Ye, L. 454 pyrosequencing reveals bacterial diversity of activated sludge from 14 sewage treatment plants. *ISME J.* **2012**, *6*, 1137–1147. [CrossRef]

21. Xia, S.; Duan, L.; Song, Y.; Li, J.; Piceno, Y.M.; Andersen, G.L.; Alvarez-Cohen, L.; Moreno-Andrade, I.; Huang, C.L.; Hermanowicz, S.W. Bacterial community structure in geographically distributed biological wastewater treatment reactors. *Environ Sci. Technol.* **2010**, *44*, 7391–7396. [CrossRef] [PubMed]
22. Flint, H.J.; Scott, K.P.; Duncan, S.H.; Louis, P.; Forano, E. Microbial degradation of complex carbohydrates in the gut. *Gut Microbes* **2010**, *3*, 289–306. [CrossRef]
23. Větrovský, T.; Steffen, K.T.; Baldrian, P. Potential of cometabolic transformation of polysaccharides and lignin in lignocellulose by soil actinobacteria. *PLoS ONE* **2014**, *9*, 1–9.
24. Gao, F.; Nan, J.; Zhang, X.H.; Wu, T.H. A dynamic modelling of nutrient metabolism in a cyclic activated sludge technology (CAST) for treating low carbon source wastewater. *Environ. Sci. Pollut. Res.* **2017**, *24*, 17016–17030. [CrossRef] [PubMed]
25. LeCleir, G.R.; Buchan, A.; Hollibaugh, J.T. Chitinase gene sequences retrieved from diverse aquatic habitats reveal environment-specific distributions. *Appl. Environ. Microbiol.* **2004**, *70*, 6977–6983. [CrossRef]
26. Zang, K.; Kurisu, F.; Kasuga, I.; Furumai, H.; Yagi, O. Analysis of the phylogenetic diversity of estrone-degrading bacteria in activated sewage sludge using microautoradiography–fluorescence in situ hybridization. *Syst. Appl. Microbiol.* **2008**, *31*, 206–214. [CrossRef] [PubMed]
27. Niu, L.H.; Li, Y.; Wang, P.F. Understanding the linkage between elevation and activated sludge bacterial community along a 3600 m elevational gradient in China. *Appl. Environ. Microbiol.* **2015**, *81*, 6567–6576. [CrossRef] [PubMed]
28. Ryu, H.D.; Lee, S.I. Comparison of 4-stage biological aerated filter (BAF) with MLE process in nitrogen removal from low carbon-to-nitrogen wastewater. *Environ. Eng. Sci.* **2009**, *26*, 163–170. [CrossRef]
29. Her, J.J.; Huang, J.S. Influences of carbon source and C/N ratio on nitrate/nitrite denitrification and carbon breakthrough. *Bioresour. Technol.* **1995**, *54*, 45–51. [CrossRef]
30. Liu, H.B.; Yang, C.Z.; Pu, W.H. Removal of nitrogen from wastewater for reusing to boiler feed-water by an anaerobic/aerobic/membrane bioreactor. *Chem. Eng. J.* **2008**, *140*, 122–129. [CrossRef]
31. Sheng, P.S.; Carles, P.N.; Brian, M. Effective Biological Nitrogen Removal Treatment Processes for Domestic Wastewaters with Low C/N Ratios: A Review. *Environ. Eng. Sci.* **2010**, *27*, 111–126.
32. Puigagut, J.; Salvadó, H.; García, J. Short-term harmful effects of ammonia nitrogen on activated sludge microfauna. *Water Res.* **2005**, *39*, 4397–4404. [CrossRef]
33. Khardenavis, A.A.; Kapley, A.; Purohit, H.J. Simultaneous nitrification and denitrification by diverse *Diaphorobacter* sp. *Appl. Microbiol. Biotechnol.* **2007**, *77*, 403–409. [CrossRef] [PubMed]
34. Rajagopal, R.; Massé, D.I.; Singh, G. A critical review on inhibition of anaerobic digestion process by excess ammonia. *Bioresour. Technol.* **2013**, *143*, 632–641. [CrossRef]
35. Shen, H.Y.; Guo, Y.; Wang, Y.Y.; Shen, Y.T. Case study on engineering projects of sewage treatment plants in plateau region. *Water. Pure Technol.* **2014**, *33*, 9–12. (In Chinese)
36. Cui, J. Discussion on the calculation of air supply for sewage treatment plant in plateau region. *Water Wastewater Eng.* **2012**, *38*, 41–45. (In Chinese)
37. Johnson, D.R.; Lee, T.K.; Park, J.; Fenner, K.; Helbling, D.E. The functional and taxonomic richness of wastewater treatment plant microbial communities are associated with each other and with ambient nitrogen and carbon availability. *Environ. Microbiol.* **2015**, *17*, 4851–4860. [CrossRef]
38. Wittebolle, L.; Marzorati, M.; Clement, L.; Balloi, A.; Daffonchio, D.; Heylen, K.; De Vos, P.; Verstraete, W.; Boon, N. Initial community evenness favours functionality under selective stress. *Nature* **2009**, *458*, 623–626. [CrossRef] [PubMed]
39. Werner, J.J.; Knights, D.; Garcia, M.L.; Scalfone, N.B.; Smith, S.; Yarasheski, K.; Cummings, T.A.; Beers, A.R.; Knight, R.; Angenent, L.T. Bacterial community structures are unique and resilient in full-scale bioenergy systems. *Proc. Natl. Acad. Sci. USA* **2011**, *108*, 4158–4163. [CrossRef]

Review

Sustainable Management and Successful Application of Constructed Wetlands: A Critical Review

Angela Gorgoglione [1,*] and Vincenzo Torretta [2]

1 Department of Fluid Mechanics and Environmental Engineering (IMFIA), College of Engineering, Universidad de la República, Julio Herrera y Reissig 565, 11300 Montevideo, Uruguay

2 Department of Theoretical and Applied Sciences, Universitá degli Studi dell'Insubria, G.B. Vico 46, 21100 Varese, Italy; vincenzo.torretta@uninsubria.it

* Correspondence: agorgoglione@fing.edu.uy

Received: 24 September 2018; Accepted: 25 October 2018; Published: 27 October 2018

Abstract: Constructed wetlands (CWs) are affordable and reliable green technologies for the treatment of various types of wastewater. Compared to conventional treatment systems, CWs offer an environmental-friendly approach, are low cost, have fewer operational and maintenance requirements, and have a high potential for being applied in developing countries; particularly in small rural communities. However, the sustainable management and successful application of these systems remain a challenge. Therefore, after briefly giving basic information on wetlands and summarizing the classification and use of current CWs, this study aims to provide sustainable solutions for the performance and applications of CWs. To accomplish this objective, design and management parameters of CWs, including macrophyte species, media types, water level, hydraulic retention time (HRT), and hydraulic loading rate (HLR), are discussed. The current study collects and presents results of more than 120 case studies from around the world. This work provides a tool for researchers and decision-makers for using CWs to treat wastewater in a particular area. This study presents an aid for informed analysis, decision-making, and communication.

Keywords: constructed wetlands; design and operation; macrophyte; substrate; hydraulic conditions; sustainability; treatment system; artificial wetland

---

1. Introduction

In the last few decades, constructed wetlands (CWs) have been widely used to treat several types of wastewater such as domestic sewage, industrial effluent, agricultural wastewater, landfill leachate, polluted river water, and urban runoff [1–5]. The studies conducted by Seidel in the 1960s [6–10] and by Kickuth in the 1970s [11–13] in Germany, are considered the kick-off research on CWs. Since then, much research has been done, and the technology has evolved, which has made CWs more feasible from the operational point of view.

Several studies have focused on the design, development, and the performance of CWs [14,15], and on the ability of these engineered systems to remove specific pollutants (organic compounds, suspended solids, nutrients, heavy metals, benzene, toluene, ethylene, xylene—BTEX), pharmaceutical contaminants, pathogens, etc.) from wastewater [3,16–18]. However, there are fewer studies on the sustainable operation and successful application of these systems. Therefore, this topic remains a challenge. CWs' sustainability is influenced by several factors including vegetation, media types, and hydraulics/hydrology. In particular, plant species and substrate material are critical influencing factors to the pollutant removal ability of CWs. They are considered the primary biological components of CWs and, as such, directly or indirectly modify the processes of primary pollutant removal over time [19–21]. Moreover, the treatment performance of CWs is based on optimal operating parameters (i.e., water depth, hydraulic retention time (HRT), and pollutant load). The variation of these parameters affects the efficiency of contaminant removal [22–24]. Furthermore, a variety of pollutant removal processes (e.g., sedimentation, filtration, precipitation, volatilization, adsorption, plant uptake, and various microbial processes) are directly and/or indirectly influenced by the different environmental conditions, both inside and outside the CWs. Such environmental conditions are, for instance, temperature, availability of dissolved oxygen (DO) and organic carbon source, operation strategies, pH, and redox conditions [3,25,26].

It is worth noting that previous research has almost exclusively focused on how to improve and optimize the efficiency of the treatment performances of CWs [22,24,27]. However, there is still a gap in the understanding of these technology systems, since the current research only focuses on achieving sustained levels of water quality enhancement. This gap in information may result in reduced use of CWs. Before undertaking expensive experimental studies to gather and analyze additional steps to the treatment performance, it is necessary first to understand what enhancement in operation and application of CWs would result if we develop an in-depth knowledge of these systems. This thorough understanding should allow for converting these "man-made" wetlands into sustainable solutions for wastewater treatment made for each particular area. Therefore, to promote and develop sustainable operations and successful application of CWs, it is necessary to review and discuss recent information on the sustainability of these treatment technology systems.

Based on these considerations, this work aims to categorize and provide an overall review of the applications of CWs for wastewater in recent years. In addition, it analyzes the developments in CWs considering plants, substrates selection, and operational parameters in order to optimize the sustainability of wastewater treatments. The conceptual framework of this study is made for incorporating future research considerations aimed to improve the sustainability of CWs. By collecting the results of more than 120 case studies from around the world, this review will allow decision-makers and researchers to assess and quantify the most sustainable solution for CWs wastewater treatment in a particular area. This study presents a useful tool for decision-making, informed analysis, and communication.

With the aim of avoiding repetition throughout the manuscript, the following synonyms will be used to indicate CW: artificial wetland, treatment system, constructed shallow water system, artificial shallow water system.

The remainder of this review is organized as follows. In Section 2, the well-known definition and classification of CWs based on hydrologic factors are summarized. In Section 3, the importance of the knowledge of the wastewater type that a specific area may produce is highlighted. In Section 4, a thorough description and in-depth analysis of the parameters that contribute to sustainable design and maintenance of CWs is reported. Consideration on the sustainability of CWs and main conclusions are presented in Section 5.

2. Research Methodology

Considering the objective of this study, research was focused on published articles where an experimental-scale or field-scale artificial wetland was design, built and studied.

Search parameters included the results of multiple web-based libraries [28–32] using the same keywords reported after the abstract section. The studies selected and cited in this manuscript are between the years 1960 and 2018. Non-peer-reviewed articles were excluded from the selection.

An initial extensive comparison of field and experimental treatment systems was performed. Not all data acquired were found in sufficient quantity or quality to be reported. Therefore, it was chosen to address the following factors only: system type, wastewater type, substrate used, plant species, and HRT.

3. Definition and Classifications of Constructed Wetlands

CWs may be classified according to several design criteria. The three most important parameters are hydrology (open water surface flow and subsurface flow); type of macrophytic growth (emergent, submerged, floating-leaved, and free-floating); and flow path in sub-surface wetlands (horizontal and vertical). It is possible to combine different types of CWs, creating hybrid systems, which utilize the particular advantages of the individual systems [33,34]. Descriptions of several types of CWs are reported in the following paragraphs.

3.1. Constructed Wetlands with Horizontal Subsurface Flow

The horizontal subsurface flow (HSSF) systems are characterized by tanks waterproofed with plastic membranes, filled with inert material of appropriate particle size (e.g., gravel), in which emergent macrophytes develop their roots (*Phragmites australis* is commonly used, even though it is considered an invasive weed in some countries), as schematically represented in Figure 1a.

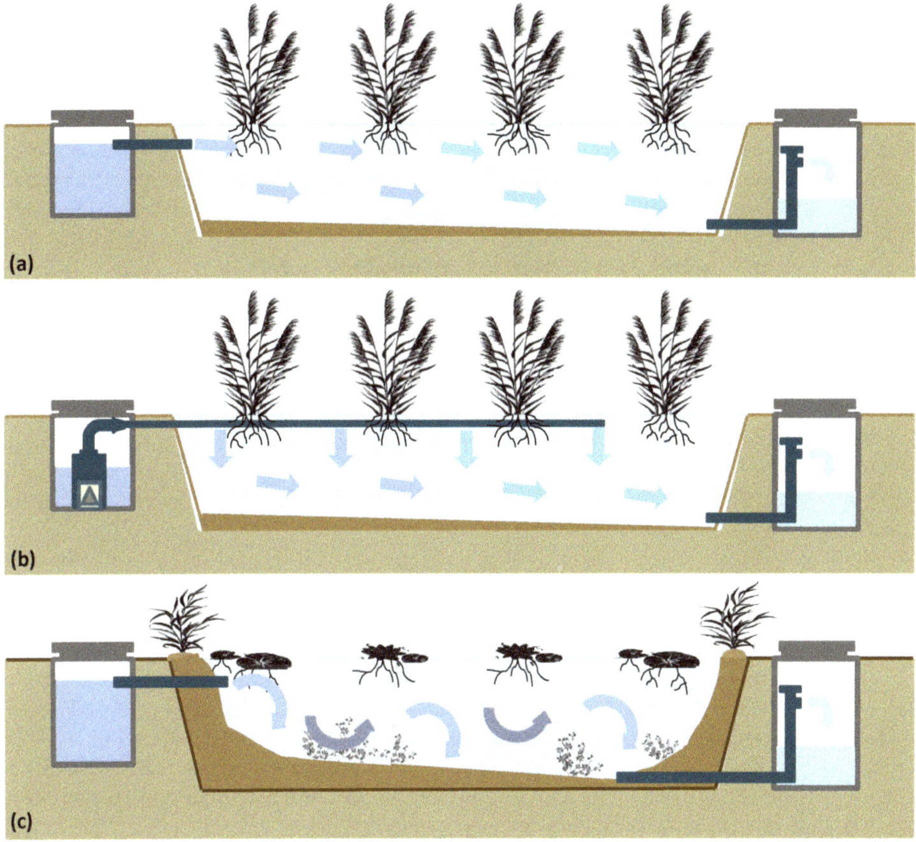

Figure 1. Schematic representation of a CW with (**a**) HSSF; (**b**) VSSF; and (**c**) FWS (arrows indicate the general flow pattern).

The water flow is continuously maintained below the surface of the inert material. This creates a predominantly anoxic environment, rich in aerobic micro-sites in close proximity to plant roots, which operate as oxygen transfer systems from the atmosphere to the inside of the filter bed. The redox conditions of this system allow it to be highly elastic, versatile and efficient with the various types of wastewater that need to be treated, and with the variations of the pollutant content. In these systems, the wastewater passes through the inert material and is in contact with the macrophytes' rhizosphere. The organic and nitrogen matter is degraded by the microbial action, while phosphorus and heavy metals are adsorbed by the inert material.

The plant species contribute to the purification process, firstly, by favoring the development of an active aerobic microbial population in the rhizosphere and, secondly, through the action of atmospheric oxygen pumping from the emerged part of the root system to the surrounding ground portion. This creates better oxidation of the wastewater and creation of alternating aerobic, anoxic, and anaerobic zones. Such conditions allow the development of several families of specific microorganisms and the almost complete disappearance of pathogens, as they are particularly sensitive to the rapid changes in dissolved oxygen content.

HSSF systems are not very tolerant of cold climates. In fact, their performances are always reduced under these weather conditions. These systems keep the septic influent warm with insulation to maximize the functioning of microorganisms and, therefore, maintain treatment performance constant during the season. Dead plant material is also used as a natural insulation layer and protects the filter bed during winter. For systems implemented in areas with unusually cold weather, it is good practice to lower the water level in the tank to prevent freezing.

High organic removal efficiencies can be reached with traditional HSSF CWs. With respect to nitrogen, the limited oxygen availability in some zones decreases nitrification rates and, in turn, the nitrogen removal performance despite rapid denitrification. Considering phosphorous, its removal mechanisms are mainly physical (e.g., precipitation with Ca^{2+}, Al^{3+} or Fe^{3+} that may be present in the soil material), so they are not influenced by oxygen concentration.

The plants most used in HSSF systems are emergent macrophytes like *Phragmites* sp., *Typha* sp., *Scirpus (Schoenoplectus)* sp., *Phalaris arundinacea*, and *Iris* sp. It was found that, besides the macrophyte species mentioned above, local species, which are easily available and grow well under local climatic conditions are regularly used in HSSF CWs.

3.2. Constructed Wetlands with Vertical Subsurface Flow

The geometrical configuration and physical layout of the vertical subsurface flow (VSSF) systems is very similar to the one of HSSF (Figure 1b).

The main difference between HSSF and VSSF systems is how the wastewater flows through the inert medium. While in HSSF systems there is a continuous inlet and a flow in the horizontal direction, in VSSF systems the effluent is introduced into the tanks in a discontinuous way and flows in the vertical direction. The intermittent inlet, with the filling and emptying cycles, creates the conditions of a "batch" reactor. It often requires at least two tanks in parallel, which operate with an alternating flow, so it is possible to adjust the timing of re-oxygenation of the bed by varying the frequency and quantity of the hydraulic load of the wastewater input.

The filling medium of this type of systems is made of inert particles finer than the HSSF system to allow a slow water percolation and, thus, a distribution as homogeneous as possible on the entire surface of the bed. The coarse sands used in VSSF systems have suitable hydraulic conductivity for the slow vertical filtration, and they offer a ratio between volume and surface area higher than gravel used in HSSF systems, to facilitate the biomass attachment.

The intermittent supply of the wastewater, associated with a substrate with various particle sizes, facilitates the drainage in the medium; which is alternately in conditions of deficient or excess oxygen. Therefore, the higher aeration of the substrate increases the aerobic processes such as the removal of organic matter and nitrification.

Since the traditional VSSF CWs provide an ideal environment for aerobic bacterial respiration, it shows better organic removal treatment performance from pre-treated domestic wastewater than traditional HSSF CWs. With respect to nitrogen, the excess of oxygen increases the nitrification process eliminating more nitrogen or, at least, converting the main part in ammoniacal nitrogen. Considering phosphorous, the performance is very similar to one of the HSSF CW.

Depending on the climate, *Phragmites australis* (*reed*), *Typha* sp. (*cattails*), and *Echinochloa pyramidalis* are common plant options for VSSF systems.

3.3. Constructed Wetlands with Free Water Surface

The free water surface (FWS) systems are organized with tanks or channels that are naturally or artificially waterproofed, in which the water level is constantly maintained above the surface of the medium (Figure 1c), with a water depth that typically ranges between 0.3 and 0.6 m.

The flow follows a path that includes the inlet area and all areas of the system until reaching one or more outlets. The regions characterized by low water depth, with low flow velocity and the presence of plant bodies, standardize the flow through the formation of a multitude of small channels that simulate the behavior of a plug flow reactor. The primary design goals of an FWS system are to ensure the contact of the wastewater with the active biologic surface of the system, to allow an efficient HRT of the wastewater in the system, and to prevent the formation of hydraulic short-circuits [35].

In these systems, the mechanisms of pollutant removal attempt to reproduce those processes that characterize natural wetlands for pathogenic organisms, Biochemical Oxygen Demand (BOD), Chemical Oxygen Demand (COD), suspended solids, nutrients, heavy metals, and other micropollutant removal. Organic and nitrogenous substances are mainly removed by biological processes under oxygenated conditions (at the surface) or anoxic conditions (in depth). While the suspended solids can on one hand be removed (by sedimentation and/or filtration through plants), on the other hand, they can be created (for example for the presence of microalgae, fragmentation of plant tissues, production of phytoplankton, formation of chemical precipitates). Phosphorus removal is small and occurs through adsorption, complexation, and precipitation processes. The FWS systems show very high performance in pathogenic microorganism removal. However, this efficiency has extreme variability mainly due to the complex combination of physical, chemical and biological processes that affect the removal mechanisms, such as the attachment of microorganisms on the sediment, UV radiation in the deeper areas not occupied by vegetation, and the presence of colonies of birds that can cause the contribution of feces [36]. Finally, heavy metals may be removed through processes like the uptake by the plants, physical–chemical interactions with the ground, or the formation of complexes and resulting precipitation [37].

The plants most used in FWS systems are common marsh species such as *Scirpus* sp., *Eleocharis* sp., *Cyperus* sp., *Glyceria maxima*, *Juncus* sp., *Phragmites australis*, *Phalaris arundinacea*, and *Typha* sp. Most FWS systems use a single species or a surface species in combination with submerged species.

A comparison of some advantages and disadvantages of the three different types of CWs are listed in Table 1.

Table 1. Advantages and disadvantages of HSSF, VSSF, and FWS CWs.

Advantages	Disadvantages
HSSF CWs	
Long flowing distances possible; nutrient gradients can establish	Higher area demand
Nitrification and denitrification possible	Careful calculation of hydraulics necessary for optimal O_2-supply
Formation of humic acids for N and P removal	Equal wastewater supply is complicated
Longer life cycle	
VSSF CWs	
Smaller area demand	Short flow distances
Good oxygen supply - good nitrification	Poor denitrification
Simple hydraulics	Higher technical demands
High purification performance from the beginning	Loss of performance esp. in P-removal (saturation)
FWS CWs	
Addition to the "green space" in a community	Higher area demand
BOD, TSS, COD, metals, and organic material removal in a reasonable detention time	Anoxic environment—poor nitrification
N and P removal in a significantly longer detention time	Mosquito production
Minimization of mechanical equipment, energy, and skilled operator requirements	

4. Types of Wastewater Treated

For an efficient and sustainable selection of CWs for a particular area, a thorough understanding of the type of wastewater being treated is necessary. CWs have long been used primarily for the treatment of municipal or domestic wastewaters. However, at present, they are utilized for other wastewaters including agricultural, industrial, and several runoff waters.

4.1. Municipal Wastewater

HSSF CWs are commonly used to treat domestic (single house or households) and municipal (clusters of houses or community) sewage as both secondary and tertiary treatment stages. The "typical" composition of municipal wastewaters is reported in the study of Kadlec and Knight of 1996: BOD_5 = 220 mg L^{-1}, COD = 500 mg L^{-1}, TSS = 220 mg L^{-1}, NH_4–N = 25 mg L^{-1}, NO_x–N = 0 mg L^{-1}, N_{org} = 15 mg L^{-1}, TKN = 40 mg L^{-1}, TP = 8 mg L^{-1} [38]. It is worth noting that HSSF CWs can successfully treat wastewaters with low concentrations of organics. For instance, conventional treatment systems such as activated sludge cannot treat wastewater with low organic concentrations (usually less than 50–80 mg/L BOD_5). By considering all the case studies included in this work, it is possible to calculate an average of the treatment performance in terms of removed load (RL) (kg/ha d) of HSSF CWs treating municipal and domestic wastewaters: $BOD_{5\,RL}$ = 77.6, COD_{RL} = 149, TSS_{RL} = 83, TN_{RL} = 10, NH_4–N_{RL} = 5.3, TP_{RL} = 1.9 [39,40]. Besides pollutants usually detected in municipal wastewaters, HSSF CWs were also used for removal of linear alkylbenzene sulfonates [41–43] and pharmaceuticals [44] from the sewage.

4.2. Industrial Wastewaters

A variety of industrial wastewaters have been treated in CWs. Vymazal [39] classified them on the basis of the industrial processes: petrochemical and chemical industries; pulp and paper, textile and tannery industries; abattoir and meat processing effluents; food processing; wineries and distilleries. Treatment of contaminated waters from the petrochemical industry is aimed at removal of various hydrocarbons including diesel range organics, BTEX [17,45,46]. One of the most extensive horizontal flow (HF) CWs in Europe (total area of 49,000 m^2) was built at the Air Products chemical works at Billingham, Teeside, United Kingdom [47]. The use of HSSF CWs for the treatment of tannery wastewaters is relatively new, and experiments were carried out in Turkey, Portugal, Greece, and the USA [48–51]. The use HSSF CWs for the treatment of textile wastewaters was carried out as early as the late 1980s and early 1990s in Germany [52] and Australia [53]. The first experiments to treat abattoir wastewaters were reported by Finlayson et al. [54] from Australia. The most recent ones were conducted in Lithuania by Gasiunas and Strusevičius [55], and Gasiunas et al. [56], who presented the results from an 1880 m^2 HF CW designed to treat meat-processing wastewaters. One of the first reports on the use of HF CW for food-processing sewage was by White [57] on seafood processor wastewater. Recently, HSSF CWs have been used to treat cheese-processing wastewaters. Mantovi et al. [58] described the use of HF CWs to treat wastewaters from the production of Italian cheese "Parmigiano-Reggiano" (400 m^2, 10.5 m^3 d^{-1}) and "Grana Padano" (2700 m^2, 70 m^3 d^{-1}). The treatment efficiency in both systems was very high and amounted to 94%, 96%, 98%, 62% and 45% for TSS, COD, BOD_5, TKN, and TP, respectively. In addition, the reduction of vegetable fats and oils was very high; the inflow concentrations of 59 mg L^{-1} (Parmigiano) and 167 mg L^{-1} (Grana Padano) were reduced to 1 and 2 mg L^{-1}, respectively. Winery wastewaters are characterized by a high content of organics (up to 45,000 mg L^{-1} BOD_5) and solids, high acidity, and significant variations in seasonal flow production [59,60]. In addition, the winery wastewaters are characterized by low N/C and P/C ratios. Cork boiling wastewater is known for its high content of phenolic compounds and toxic nature. Gomes et al. [61] evaluated the total phenolic compounds (TPh) removal over a 2.5-year monitoring period, in an HSSF CW planted with *Phragmites australis*. Average TPh removal reached 69.1%, with respective mass removal rates up to 0.5 $g/m^2/d$.

4.3. Agricultural Wastewaters

Wastewater from feedlot operations is commonly treated with FWS CWs with a series of lagoons as a pretreatment step (separation of settleable solids, digestion of solids, and treatment of the liquid portion) [40,62]. HSSF CWs are used to a lesser extent, but many excellent examples could be found in the scientific literature [39,63–66]. In Table 2, an average treatment performance for HSSF CWs treating wastewaters from agro-industrial operations is presented. This average was calculated by considering all the case studies included in this work. The inflow concentrations are much lower as compared to raw wastewaters because of intensive pretreatment. On the contrary, Rozena et al. [67], declared that there is no one CW design (HSSF, VSSF, FWS) that is the most effective for agricultural wastewater, so hybrid designs may prove to be the most efficient and practical.

Table 2. Average treatment performance of HSSF CW treating agricultural wastewater. (Adapted from [39]).

	Concentration (mg/L)		Eff. (%)	n *	Loading (kg/ha d)			n *
	In	Out			In	Out	Rem	
BOD_5	464	183	68.2	43(19)	541	294	246	43(18)
COD	871	327	63	38(17)	1239	602	637	37(17)
TSS	516	180	76.9	56(26)	1430	779	651	54(23)
TN	116	57.5	51.3	31(13)	68	42	26	31(13)
NH_4-N	71.5	39.6	33.8	45(18)	74.6	19	55.6	45(18)
TP	19.8	8.5	54.3	44(18)	13.7	7	6.7	44(18)

In = inflow to a vegetated bed. Out = final outflow. Rem = removed load. * The number denotes the number of annual means with a number of systems in parentheses.

4.4. Stormwater Runoff

Agricultural and urban runoff represent the two most common polluted storm waters that threaten the quality of surface waters [68–71]. For this reason, many studies have classified stormwater runoff from urban regions as a contribution to non-point source pollution to surface water [72,73]. An example of HSSF CWs used to treat agriculture stormwater runoff was reported by Zhou et al. [74]. The average total nitrogen (TN) inflow concentration was approximately 22 mg L^{-1} in which about

80% was nitrate, 10% ammonia and 10% organic nitrogen. The removal varied between 27% and 80% depending on the HRT.

For treatment of urban stormwater runoff, FWS CWs are mostly used. Walker et al. [75], in their two-year field study, used a FWS design to treat stormwater runoff from an existing urban area in Queensland, Australia. However, there are some examples of the utilization of HSSF and VSSF designs as well. For instance, Geary et al. [76] reported on the use of an HSSF CW to treat urban runoff from a 21 ha urban catchment at Blue Haven, Australia. Scholz et al. [77], assessed over two years the treatment efficiencies of VF wetland filters containing macrophytes and granular media of different adsorption capacities.

Stormwater monitoring campaigns usually provide records of pollutants by measuring the event mean concentration (EMC), defined as [78]:

$$EMC = \frac{\sum_{i=1}^{n} C_i V_i}{V} \tag{1}$$

where V is the total runoff volume per event (L), V_i is the runoff volume during time period i (L), C_i is the pollutant concentration during time period i (mg/L), and n is the total number of samples during a single storm event.

Knowing EMC, it is possible to evaluate the pollutant concentration removal efficiency (CRE) and the efficiency ratio (ER). CRE calculates the reduction in pollutant concentration for a given stormwater treatment device, while ER is described in terms of the average pollutant EMCs calculated over the duration of the analyzed storm events for a given stormwater treatment device. They are respectively defined as:

$$CRE = \frac{\sum\left(\frac{EMC_{in} - EMC_{out}}{EMC_{in}}\right)}{no\ of\ events} \tag{2}$$

$$ER = 1 - \left(\frac{\mu_{EMC_{out}}}{\mu_{EMC_{in}}}\right) \tag{3}$$

where EMC_{in} and EMC_{out} are respectively the pollutant EMCs measured at the inlet and outlet of the system, μ_{EMCin} and μ_{EMCout} represent the mean of EMCs measured respectively at the inlets and at the outlets of the CW.

Walker et al. [75] calculated these two efficiencies by taking into account some of the common pollutants detected in urban runoff (TSS, TN, TP, NH_3–N, NO_3–N, NO_x–N), measured over two years (Table 3).

Table 3. CREs and ERs of FWS system treating urban runoff (Adapted from [75]).

	TSS	TN	TP	NH_3–N	NO_3–N	NO_x–N
CRE ± std. dev. [%]	58 ± 29	7 ± 48	33 ± 33	45 ± 140	50 ± 33	49 ± 33
ER [%]	81	17	52	8	47	47

5. Sustainability of the Design and Management of Constructed Wetlands

After understanding the type of wastewater, it is then possible to consider the factors that contribute to improving the sustainability of CWs. Factors to be considered include the site, plant and substrate selection, water level, wastewater type, HRT, hydraulic loading rate (HLR), installation, operation and maintenance procedures. These parameters have been investigated experimentally [51], by using modeling [79–81] and Artificial Neural Networks [82]. In particular, factors such as plant selection, substrate selection, and hydraulic conditions (water level, HLR, and HRT) are critical in creating a viable CW system and achieve the sustainable treatment performance. In the following paragraphs, the importance of these factors is described.

5.1. Plant Selection

Macrophytes are common in wetlands and are considered a significant design element in natural and constructed systems [83,84]. The plant species used in natural treatment systems are plants that commonly live in wetlands (aquatic and hydrophilic plants), adapted to grow in soils that are moderately or constantly saturated. The presence or absence of these macrophytes often delineates CWs as green technology [85]. Wetland plants can adsorb pollutants from the wastewater and accumulate them in their tissue in addition to providing microorganisms in the system with a complimentary growing environment [86]. Furthermore, these plants are able to transfer oxygen from their roots to the rhizosphere, creating aerobic conditions that enhance the contaminant degradation in the system.

The selection of plant species needs to take into account factors such as the climatic conditions of the site, the characteristics of the wastewater to be treated, and the effluent quality required. On the basis of these considerations, Rozena et al. [67], in their review, found that *Typha spp.* tends to be most commonly used in Northeastern–North America. The most suited vegetation to the proposed CW system should be selected by taking into account adaptability to the saturation conditions of the terrain, the growth potential of roots and their oxygen carrying capacity, the high capacity of photosynthetic activity, the tolerance to high pollutant concentrations, disease resistance, and management requirements. On the basis of these considerations, only a few plant species have been widely used in CWs [87]. Macrophytes frequently employed in CW treatments include emergent plants, submerged plants, floating-leaved plants, and free-floating plants. The most common and used emergent species we found included *Typha* spp. (Typhaceae), *Phragmites* spp. (Poaceae), *Iris* spp. (Iridaceae), *Scirpus* spp. (Cyperaceae), *Juncus* spp. (Juncaceae), and *Eleocharis* spp. (Spikerush). The most frequently used submerged plants are *Hydrilla verticillata*, *Vallisneria natans*, *Ceratophyllum demersum*, *Myriophyllum verticillatum*, and *Potamogeton crispus*. The main floating-leaved plants are *Nymphoides peltata*, *Nymphaea tetragona*, *Trapa bispinosa*, and *Marsilea quadrifolia*. The free-floating plants are *Eichhornia crassipes*, *Salvinia natans*, *Hydrocharis dubia* and *Lemna minor* [21]. Considering that plants are one of the leading factors influencing water quality in wetlands, numerous studies were performed on the uptake capacity of plants in CWs. Plant uptake ability may differ according to different technical parameters, such as system configurations, retention times, loading rates, wastewater types and climatic conditions [3]. The impact of plants regarding nitrogen and phosphorus removals is considered high, accounting for 15–80% N and 24–80% P [88,89]. However, several researchers found that it was lower and within the range of 14.29–51.89% of total nitrogen removal, and 10.76–34.17% of total phosphorus removal, respectively [90,91]. Regarding heavy metal removal, Ha et al. [92] evaluated the accumulating capability of *Eleocharis acicularis* in different concentrations of In, Ag, Pb, Cu, Cd, and Zn, and the results showed that *E. acicularis* had an excellent ability to accumulate metals from water. In addition, Ranieri et al. [17] reported the removal of more than 60% of BTEX from wastewater at an HRT higher than 100 h by *Phragmites australis* and *Typha latifolia*.

A summary of the plant contributions to the CW systems is presented in Table 4.

Table 4. Summary of mycrophite roles in CWs.

Role of Mycrophytes	Source
Roots: physical effects	
Filtering effect	[93]
Improved hydraulic conductivity	[94,95]
Reduced velocity	[93]
Prevention from clogging	[96]
Roots: microorganisms	
Surface for attachment	[93,94]
Oxygen	[93,94,97–100]
Uptake function	
Nutrients	[88–91]
Metals	[92,101]
BTEX	[17,102]
Evapotranspiration	
Increased water loss	[103,104]

5.2. Substrate Selection

Substrate materials have a strong influence on the movement of water through CW (hydraulic conductivity) and on plant growth. These materials provide a vast surface area for microorganisms to attach additionally to plant biomass (roots, stems, and leaves) and also act either as a filtration and/or adsorption medium for pollutants [105]. Both the chemical soil composition and physical parameters—such as particle size distributions, pore spaces, degree of irregularity—and the coefficient of permeability, are the key criteria influencing treatment performance [106]. For this reason, the selection of optimal substrates is determined in terms of the hydraulic permeability and capacity for adsorbing pollutants. Poor hydraulic conductivity would lead to the clogging of systems, with the consequent severe reduction in the effectiveness of the system. Low adsorption by substrates could also negatively influence the long-term removal performance of CWs [107]. Wu et al. [21] summarized several studies carried out on the selection of wetland substrates, in particular for sustainable phosphorus removal from wastewater. Ionized ammonia can be removed from wastewater through exchange with soil strata, detritus, humic substances, and organic and inorganic sediments or else fixed within the clay lattice in CWs [106]. On the other hand, adsorbed ammonium binds loosely to the materials and can be released effortlessly in response to changes in water chemistry [108]. Numerous studies have been carried out to assess the impact of different substrates used to enhance pollutant adsorption capacity. Meng et al. [26] confirmed the results obtained from previous research [3,4,109], which evaluated the use of different media substrates such as rice husk and organic mulch on system efficiency. The results showed that these media improved nitrogen removal due to organic carbon content. However, these results contradicted those of others regarding the use of expensive substrate materials to improve the CW performance. For example, the use of granular activated carbon did not enhance the adsorption capacity of CW media [14]. Furthermore, the use of zeolite and bauxite media did not provide a significant improvement in CW system efficiency as reported by Stefanakis and Tsihrintzis [110].

Frequently used substrates include natural material, artificial media and industrial by-products (Table 5). Outcomes from these studies suggest that substrates such as sand, gravel, and rock are poor for long-term phosphorus storage, while synthetic and industrial products with high phosphorus sorption capacity and hydraulic conductivity may be more effective alternative substrates in CWs.

Table 5. Substrates commonly selected for CW wastewater treatment (Adapted from [21]).

Type of Substrate	Source
Natural material	
Sand	[4]
Gavel	[111]
Clay	[111]
Calcite	[112]
Marble	[19]
Vermiculite	[19]
Bentonite	[113]
Dolomite	[112]
Limestone	[114]
Shell	[115]
Shale	[4]
Peat	[4]
Wollastonite	[116]
Maerl	[4]
Zeolite	[117]
Industrial by-product	
Slag	[16]
Fly ash	[113]
Coal cinder	[118]
Alum sludge	[119]
Hollow brick crumbs	[118]
Moleanos limestone	[120]
Wollastonite tailings	[121]
Oil palm shell	[122]
Synthetic products	
Activated carbon	[118]
Lightweight aggregates	[4]
Compost	[4]
Calcium silicate hydrate	[20]
Ceramsite	[20]

5.3. Hydraulic Conditions

The water level is an essential element in determining which plant types will become established [123], and it also affects the biochemical reactions responsible for removing contaminants by changing the redox status and dissolved oxygen level in CWs [21]. By comparing 0.27 m deep wetland beds with 0.5 m deep wetland beds, García et al. [124] showed that differences occur in the transformations of pollutants within systems of different depths. In addition, experiments conducted by Aguirre et al. [125], with the aim of investigating the effect of water depth on organic matter removal efficiency in HF CWs, concluded that the relative contribution of different metabolic pathways varied with water depth. Hydrology is another primary factor in controlling CW efficiency. Flow rate should be monitored to accomplish a satisfactory treatment performance [126].

HRT is one of the few operational factors that can be controlled in CWs. For example, a critical BOD removal efficiency can be obtained at an HRT shorter than one day, while the system efficiency will be enhanced at an HRT of about seven days, as shown by Reed and Brown [127]. Based on this, HRT is an important factor that affects the efficiency of the CW treatment, which is usually decided by designers. Despite the advantage of enhancing the treatment efficiency, when increasing the HRT, this can also be considered as the main disadvantage for large wetland areas, particularly when land availability is restricted [128]. The optimal design of HLR and HRT plays a significant role in the removal efficiency of CWs. Greater HLR promotes quicker passage of wastewater through the media, thus reducing the optimum contact time. A proper microbial community may be established in CWs and have suitable contact time to remove pollutants at a longer HRT [3,21,129]. Toet et al. [130] found positive nitrogen removal in CWs with an HRT of 0.8 days compared results with a 0.3 day HRT. Furthermore, the effect of HRT may differ between CWs depending on the dominant plant species and temperature, as those factors can affect the hydraulic efficiency of wetlands. In a long-term experiment, Cui et al. [16] observed a minor decrease of ammonium and TN removal from domestic wastewater in vertical flow (VF) CWs when HLR changed from 7 to 21 cm/d. Mean ammonium removal decreased from 65% to 60%, and TN reduced from 30% to 20%.

Proper organization and maintenance of a wetland system are essential to ensure the following objectives: (i) achieve and maintain the pollutant removal efficiency established in the design phase; (ii) minimize malfunctions and maximize environmental protection and economic savings; (iii) maximize species efficiency and longevity.

6. Useful Considerations and Conclusions

After years of studies and implementations, the scientific community has widely recognized that CWs are a reliable treatment technology. This review demonstrates that the advances in the design and operation of CWs accomplished over the years have significantly increased pollutant removal efficiency, and the sustainable application of this treatment system has also been significantly improved. In Table 6, recommendations on the design and operation parameters of CWs are presented.

Table 6. Recommendations on the design and operation parameters.

Parameter	Recommendation
Plant selection	Native plant species
Substrate selection	Synthetic and industrial products
Water level (m)	0.27–0.5
HRT (day)	1–7

Considering the increasingly stringent water quality standards for wastewater treatments and water quality worldwide, CWs have some limitations, and future studies and development work are necessary. In particular, as it is represented in Figure 2:

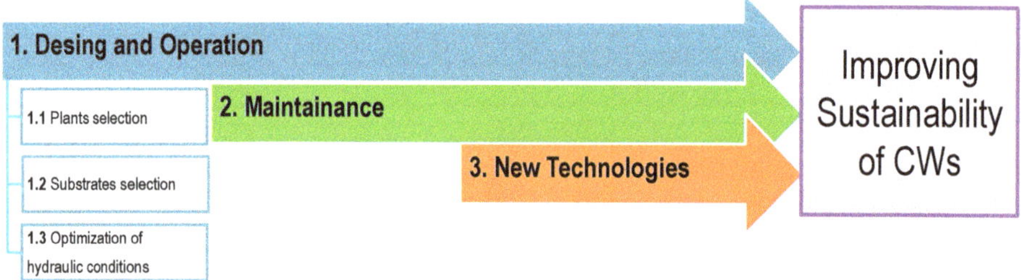

Figure 2. Considerations for improving the sustainability of CWs.

Wetland macrophytes (1.1 in Figure 2) and substrates (1.2 in Figure 2) represent two factors that influence the efficiency of pollutant removal in CWs. More attention should be given to appropriate plant species selection studied for CWs in temperate and cold climates. An intensive evaluation of differences between species and season is also needed. In addition, some non-conventional wetland media, characterized by high sorption capacity, should be studied and used for CWs. Moreover, the review of the design and operating parameters shows that the optimal treatment performance is vitally dependent on environmental, hydraulic and operating conditions (1.3 in Figure 2). Therefore, understanding how to manage and optimize these conditions warrants more investigation.

Additional research on the critical pathways and mechanisms corresponding to higher pollutant removal should be taken into consideration. The review of design and operation parameters (plant and substrate selection, and hydraulic conditions) shows that the optimal treatment performance is crucially dependent on hydraulic, environmental, and operating conditions. Therefore, if an optimization of the design and management of these systems wants to be accomplished, further studies on the aspects above mentioned would be needed.

For the sake of clarification, it is worth mentioning that as well as studies on design and operation parameters, additional research on maintenance processes (2. in Figure 2) and new strategies and technologies (3. in Figure 2) is necessary for sustainable CW systems and water quality improvement.

Taking into account the efficient and sustainable implementation of full-scale CWs, future studies should focus on a comprehensive assessment of plants and substrates in field trials under real conditions, optimization of environmental and operational parameters, exploration of novel enhancement technologies and maintenance strategies.

New studies will provide information that will increase the successful application and sustainability of CWs.

Funding: This research received no external funding.

Acknowledgments: We thank our colleagues Bruno J.L. Pitton and Bridget Giffei from Department of Plant Sciences, University of California, Davis (UCD) for their thorough Use of English review.

Conflicts of Interest: The authors declare no conflict of interest.

References

1. Yalcuk, A.; Ugurlu, A. Comparison of horizontal and vertical constructed wetland systems for landfill leachate treatment. *Bioresour. Technol.* **2009**, *100*, 2521–2526. [CrossRef] [PubMed]
2. Harrington, C.; Villa, M. Assessment of pre-digested piggery wastewater treatment operations with surface flow integrated constructed wetland systems. *Bioresour. Technol.* **2010**, *101*, 7713–7723. [CrossRef] [PubMed]
3. Saeed, T.; Sun, G. A review on nitrogen and organics removal mechanisms in subsurface flow constructed wetlands: Dependency on environmental parameters, operating conditions and supporting media. *J. Environ. Manag.* **2012**, *112*, 429–448. [CrossRef] [PubMed]
4. Saeed, T.; Sun, G. A lab-scale study of constructed wetlands with sugarcane bagasse and sand media for the treatment of textile wastewater. *Bioresour. Technol.* **2013**, *128*, 438–447. [CrossRef] [PubMed]
5. Badhe, N.; Saha, S.; Biswas, R.; Nandy, T. Role of algal biofilm in improving the performance of free surface, up-flow constructed wetland. *Bioresour. Technol.* **2014**, *169*, 596–604. [CrossRef] [PubMed]
6. Seidel, K. Zur Problematik der Keim- und Pflanzengewasser. *Verh. Intern. Verein. Limnol.* **1961**, *14*, 1035–1039. (In German)
7. Seidel, K. Abau von Bacterium coli durch höhere Pflanzen. *Naturwissenschaften* **1964**, *51*, 395. (In German) [CrossRef]
8. Seidel, K. Phenol-Abbau in Wasser durch Scirpus lacustris L. wehrend einer versuchsdauer von 31 Monaten. *Naturwissenschaften* **1965**, *52*, 398–406. (In German) [CrossRef]
9. Seidel, K. Neue Wege zur Grundwasseranreicherung in Krefeld, Vol. II. Hydrobotanische Reinigungsmethode. *GWF Wasser/Abwasser* **1965**, *30*, 831–833. (In German)
10. Seidel, K. Reinigung von Gewässern durch höhere Pflanzen. *Naturwissenschaften* **1966**, *53*, 289–297. [CrossRef] [PubMed]
11. Kickuth, R. Degradation and incorporation of nutrients from rural wastewaters by plant hydrosphere under limnic conditions. In *Utilization of Manure by Land Spreading*; EUR 5672e; Comm. Europ. Commun.: London, UK, 1977; pp. 335–343.
12. Kickuth, R. Elimination gelöster Laststoffe durch Röhrichtbestände. *Arbeiten Deutschen Fischereiverbandes* **1978**, *25*, 57–70. (In German)
13. Kickuth, R. Abwasserreinigung in Mosaikmatritzen aus aeroben and anaeroben Teilbezirken. In *Grundlagen der Abwasserreinigung*; Moser, F., Ed.; Verlag Oldenburg: Munchen, Wien, 1981; pp. 639–665. (In German)
14. Scholz, M.; Xu, J. Performance comparison of experimental constructed wetlands with different filter media and macrophytes treating industrial wastewater contaminated with lead and copper. *Bioresour. Technol.* **2002**, *83*, 71–79. [CrossRef]

15. Carty, A.; Scholz, M.; Heal, K.; Gouriveau, F.; Mustafa, A. The universal design, operation and maintenance guidelines for farm constructed wetlands (FCW) in temperate climates. *Bioresour. Technol.* **2008**, *99*, 6780–6792. [CrossRef] [PubMed]
16. Cui, L.; Ouyang, Y.; Lou, Q.; Yang, F.; Chen, Y.; Zhu, W.; Luo, S. Removal of nutrients from wastewater with Canna indica L. under different vertical-flow constructed wetland conditions. *Ecol. Eng.* **2010**, *36*, 1083–1088. [CrossRef]
17. Ranieri, E.; Gorgoglione, A.; Montanaro, C.; Iacovelli, A.; Gikas, P. Removal capacity of BTEX and metals of constructed wetlands under the influence of hydraulic conductivity. *Desalin. Water Treat.* **2014**. [CrossRef]
18. Gorgoglione, A. Control and Modeling Non-Point Source Pollution in Mediterranean Urban Basins. Ph.D. Thesis, Doctoral Program in Environmental and Territorial Safety and Control; Scuola Interpolitecnica di Dottorato—Politecnico di Bari, Bari, Italy, 2016. [CrossRef]
19. Arias, C.A.; Del Bubba, M.; Brix, H. Phosphorus removal by sands for use as media in subsurface flow constructed reed beds. *Water Res.* **2001**, *35*, 1159–1168. [CrossRef]
20. Li, C.J.; Wan, M.H.; Dong, Y.; Men, Z.Y.; Lin, Y.; Wu de, Y.; Kong, H.N. Treating surface water with low nutrients concentration by mixed substrates constructed wetlands. *J. Environ. Sci. Health* **2011**, *46*, 771–776. [CrossRef] [PubMed]
21. Wu, H.; Zhang, J.; Ngo, H.H.; Guo, W.; Hu, Z.; Liang, S.; Fan, J.; Liu, H. A review on the sustainability of constructed wetlands for wastewater treatment: Design and operation. *Bioresour. Technol.* **2015**, *175*, 594–601. [CrossRef] [PubMed]
22. Persson, J.; Wittgren, H.B. How hydrological and hydraulic conditions affect performance of ponds. *Ecol. Eng.* **2003**, *21*, 259–269. [CrossRef]
23. Kadlec, R.H.; Wallace, S.D. *Treatment Wetlands*, 2nd ed.; CRC Press/Taylor & Francis Group: Boca Raton, FL, USA, 2009.
24. Wu, S.; Kuschk, P.; Brix, H.; Vymazal, J.; Dong, R. Development of constructed wetlands in performance intensifications for wastewater treatment: A nitrogen and organic matter targeted review. *Water Res.* **2014**, *57C*, 40–55. [CrossRef] [PubMed]
25. Chen, Y.; Wen, Y.; Cheng, J.; Xue, C.H.; Yang, D.H.; Zhou, Q. Effects of dissolved oxygen on extracellular enzymes activities and transformation of carbon sources from plant biomass: Implications for denitrification in constructed wetlands. *Bioresour. Technol.* **2011**, *102*, 2433–2440. [CrossRef] [PubMed]
26. Meng, P.; Pei, H.; Hu, W.; Shao, Y.; Li, Z. How to increase microbial degradation in constructed wetlands: Influencing factors and improvement measures. *Bioresour. Technol.* **2014**, *157*, 316–326. [CrossRef] [PubMed]
27. Matamoros, V.; Rodríguez, Y.; Bayona, J.M. Mitigation of emerging contaminants by full-scale horizontal flow constructed wetlands fed with secondary treated wastewater. *Ecol. Eng.* **2014**, *99*, 222–227. [CrossRef]
28. Springer. Available online: https://www.springer.com/ (accessed on 22 October 2018).
29. Scopus. Available online: https://www.scopus.com/ (accessed on 22 October 2018).
30. Mdpi. Available online: https://www.mdpi.com/ (accessed on 22 October 2018).
31. Taylor & Francis Online. Available online: https://www.tandfonline.com/ (accessed on 22 October 2018).
32. ScienceDirect. Available online: https://www.sciencedirect.com/ (accessed on 22 October 2018).
33. Vymazal, J. Constructed wetlands for treatment of industrial wastewaters: A review. *Ecol. Eng.* **2014**, *73*, 724–751. [CrossRef]
34. Salvati, S.; Bianco, A.; Bardasi, G.; Cecilia, M.; De Mattia, M.C.; Giovannelli, L.; Canepel, R. *Guida Tecnica per la Progettazione e Gestione dei Sistemi di Fitodepurazione per il Trattamento delle Acque Reflue Urbane*; ISPRA, Manuali e Linee Guida 81/2012: Roma, Italy, 2012; ISBN 978-88-448-0548-7. (In Italy)
35. Vymazal, J. Constructed wetlands, surface flow. In *Encyclopedia of Ecology*; Elsevier: Amsterdam, The Netherlands, 2008; pp. 765–776. [CrossRef]
36. Ghermandi, A.; Bixio, D.; Thoeye, C. The role of free water surface constructed wetlands as polishing step in municipal wastewater reclamation and reuse. *Sci. Total Environ.* **2007**, *380*, 247–258. [CrossRef] [PubMed]
37. Kleinmann, M.A. Girts, Acid mine water treatment in wetlands: An overview of an emergent technology. In *Acquatic Plants for Water Treatment and Resource Recovery*; Reddy, K.R., Smith, W.H., Eds.; Magnolia Publications: Orlando, FL, USA, 1987.
38. Kadlec, R.H.; Knight, R.L. *Treatment Wetlands*; CRC Press/Lewis Publishers: Boca Raton, FL, USA, 1996.
39. Vymazal, J. The use constructed wetlands with horizontal sub-surface flow for various types of wastewater. *Ecol. Eng.* **2009**, *35*, 1–17. [CrossRef]

40. Vymazal, J.; Kröpfelová, L. *Wastewater Treatment in Constructed Wetlands with Horizontal Sub-Surface Flow*; Springer: Dordrecht, The Netherlands, 2008.
41. Thomas, R.; Freeman, C.; Rehman, N.; Fox, K. Removal of linear alkylbenzene sulphonate (LAS) in constructed wetlands. In *Wetlands-Nutrients, Metals and Mass Cycling*; Vymazal, J., Ed.; Backhuys Publishers: Leiden, The Netherlands, 2003; pp. 35–47.
42. Huang, Y.; Latorre, A.; Barcelo, D.; Garcia, J.; Aqguirre, P.; Mujeriego, R.; Bayona, J.M. Factors affecting linear alkylbenzene sulfonates removal in subsurface flow constructed wetlands. *Environ. Sci. Technol.* **2004**, *38*, 2657–2663. [CrossRef] [PubMed]
43. Kantawanichkul, S.; Wara-Aswapati, S. LAS removal by a horizontal flow constructed wetland in tropical climate. In *Natural and Constructed Wetlands. Nutrients, Metals and Management*; Vymazal, J., Ed.; Backhuys Publishers: Leiden, The Netherlands, 2005; pp. 261–270.
44. Matamoros, V.; García, J.; Bayona, J.M. Elimination of PPCPs in subsurface and surface flow constructed wetlands. In *Abstracts of the International Symposium on Wetland Pollutant Dynamics and Control*; Ghent University: Ghent, Belgium, 2005; pp. 107–108.
45. Ranieri, E.; Gorgoglione, A.; Petrella, A.; Petruzzelli, V.; Gikas, P. Benzene removal in horizontal subsurface flow constructed wetlands treatment. *Intern. J. Appl. Eng. Res.* **2015**, *10*, 14603–14614.
46. Ranieri, E.; Gorgoglione, A.; Ionescu, G. A Sustainable solution for Ethylbenzene removal: Horizontal subsurface flow constructed wetlands treatment. *Fresenius Environ. Bull.* **2016**, *25*, 1.
47. Sands, Z.; Gill, L.S.; Rust, R. Effluent treatment reed beds: Results after ten years of operation. In *Wetlands and Remediation*; Means, J.F., Hinchee, R.E., Eds.; Battelle Press: Columbus, OH, USA, 2000; pp. 273–279.
48. Küçük, O.S.; Sengul, F.; Kapdan, I.K. Removal of ammonia from tannery effluents in a reed bed constructed wetland. *Water Sci. Technol.* **2003**, *48*, 179–186. [CrossRef] [PubMed]
49. Dotro, G.; Fitch, M.; Larsen, D.; Palazolo, P. Treatment of chromium-bearing wastewaters from tannery operations with constructed wetlands. In Proceedings of the 10th International Conference on Wetland Systems for Water Pollution Control, Lisbon, Portugal, 23–29 September 2006; pp. 1725–1733.
50. Akratos, C.S.; Tsihrintzis, V.A. Effect of temperature, HRT, vegetation and porous media on removal efficiency of pilot-scale horizontal subsurface flow constructed wetlands. *Ecol. Eng.* **2007**, *29*, 173–191. [CrossRef]
51. Calheiros, C.S.C.; Rangel, A.O.S.S.; Castro, P.K.L. Constructed wetland systems vegetated with different plants applied to the treatment of tannery wastewater. *Water Res.* **2007**, *41*, 1790–1798. [CrossRef] [PubMed]
52. Winter, M.; Kickuth, R. Elimination of sulphur compounds from wastewater by the root zone process. I. Performance of a large-scale purification plant at a textile finishing industry. *Water Res.* **1989**, *23*, 535–546. [CrossRef]
53. Davies, T.H.; Cottingham, P.D. The use of constructed wetlands for treating industrial effluent. In Proceedings of the 3rd International Conference on Wetland Systems in Water Pollution Control, Sydney, Australia, 11–16 November 1992; pp. 53.1–53.5.
54. Finlayson, M.; von Oertzen, I.; Chick, A.J. Treating poultry abattoir and piggery effluents in gravel trenches. In *Constructed Wetlands in Water Pollution Control*; Cooper, P.F., Findlater, B.C., Eds.; Pergamon Press: Oxford, UK, 1990; pp. 559–562.
55. Gasiunas, V.; Strusevičius, Z. The experience of wastewater treatment using constructed wetlands with horizontal subsurface flow in Lithuania. In *Proceedings of the International Conference on Constructed and Riverine Wetlands for Optimal Control of Wastewater at Catchment Scale*; Mander, Ü., Vohla, C., Poom, A., Eds.; University of Tartu, Institute of Geography: Tartu, Estonia, 2003; Volume 94, pp. 242–249.
56. Gasiunas, V.; Strusevičius, Z.; Strusevičiéne, M.-S. Pollutant removal by horizontal subsurface flow constructed wetlands in Lithuania. *J. Environ. Sci. Health* **2005**, *40A*, 1467–1478. [CrossRef]
57. White, K.D. Enhancement of nitrogen removal in subsurface-flow constructed wetlands by employing a 2-stage configuration, an unsaturated zone, and recirculation. In Proceedings of the 4th International Conference on Wetland Systems for Water Pollution Control, ICWS'94 Secretariat, Guangzhou, China, 6–10 November 1994; pp. 219–229.
58. Mantovi, P.; Piccinni, S.; Lina, F.; Marmiroli, M.; Marmiroli, N. Treating wastewaters from cheese productions in H-SSF constructed wetlands. In Proceedings of the International Conference on Multi Functions of Wetland Systems, Padova, Italy, 26–29 June 2007; pp. 72–73.

59. Shepherd, H.L.; Tchobanoglouis, G.; Grismer, M.E. Time-dependant retardation model for chemical oxygen demand removal in a sub-surface flow constructed wetland for winery wastewater treatment. *Water Environ. Res.* **2001**, *73*, 597–606. [CrossRef] [PubMed]
60. Masi, F.; Conte, G.; Martinuzzi, N.; Pucci, B. Winery high organic content wastewaters treated by constructed wetlands in Mediterranean climate. In Proceedings of the 8th International Conference on Wetland Systems for Water Pollution Control, University of Dar-es-Salaam, Dar-es-Salaam, Dar-es-Salaam, Tanzania, 16–19 September 2002; pp. 274–282.
61. Gomes, A.C.; Silva, L.; Albuquerque, A.; Simões, R.; Stefanakis, A.I. Investigation of lab-scale horizontal subsurface flow constructed wetlands treating industrial cork boiling wastewater. *Chemosphere* **2018**, *207*, 430–439. [CrossRef] [PubMed]
62. Kadlec, R.H.; Knight, R.L.; Vymazal, J.; Brix, H.; Cooper, P.; Haberl, R. Constructed wetlands for pollution control. In *Processes, Performance, Design and Operation. IWA Specialist Group on the Use of Macrophytes in Water Pollution Control*; IWA Scientific and Technical Report No. 8; IWA Publishing: London, UK, 2000.
63. Mantovi, P.; Piccinini, S.; Marmiroli, N.; Maestri, E. Treating dairy parlor wastewater using subsurface-flow constructed wetlands. In *Wetlands and Remediation II*; Nehring, K.W., Brauning, S.E., Eds.; Battelle Press: Columbus, OH, USA, 2002; pp. 205–212.
64. Mantovi, P.; Marmiroli, M.; Maestri, E.; Tagliavini, S.; Piccinini, S.; Marmiroli, N. Application of a horizontal subsurface flow constructed wetland on treatment of dairy parlor wastewater. *Bioresour. Technol.* **2003**, *88*, 85–94. [CrossRef]
65. Drizo, A.; Twohig, E.; Weber, D.; Bird, S.; Ross, D. Constructed wetlands for dairy effluent treatment in Vermont: Two years of operation. In Proceedings of the 10th International Conference on Wetland Systems for Water Pollution Control, MAOTDR 2006, Lisbon, Portugal, 23–29 September 2006; pp. 1611–1621.
66. Chazarenc, F.; Boumecied, A.; Brisson, J.; Boulanger, Y.; Comeau, Y. Phosphorus removal in a freshwater fish farm using constructed wetlands and slag filters. In *Proceedings of the International Conference on Multi-Functions of Wetland Systems*; Borin, M., Bacelle, S., Eds.; P.A.N. s.r.l.: Padova, Italy, 2007; pp. 50–51.
67. Rozena, E.R.; VandeerZaag, A.C.; Wood, J.D.; Drizo, A.; Zheng, Y.; Madani, A.; Gordon, R.J. Constructed wetlands for agricultural wastewater treatment in Northeastern North America: A review. *Water* **2016**, *8*, 173. [CrossRef]
68. Di Modugno, M.; Gioia, A.; Gorgoglione, A.; Iacobellis, V.; la Forgia, G.; Piccinni, A.F.; Ranieri, E. Build-up/Wash-Off monitoring and assessment for sustainable management of first flush in an urban area. *Sustainability* **2015**, *7*, 5050–5070. [CrossRef]
69. Gorgoglione, A.; Gioia, A.; Iacobellis, V.; Piccinni, A.F.; Ranieri, E. A predictive model for pollutant concentrations in ungauged urban basins. In Proceedings of the XXXV Convegno Nazionale di Idraulica e Costruzioni Idrauliche, Bologna, Italy, 14–16 September 2016.
70. Gorgoglione, A.; Gioia, A.; Iacobellis, V.; Piccinni, A.F.; Ranieri, E. A rationale for pollutograph evaluation in ungauged areas, using daily rainfall patterns: Case studies of the Apulian region in Southern Italy. *Appl. Environ. Soil Sci.* **2016**, *2016*, 9327614. [CrossRef]
71. Gorgoglione, A.; Bombardelli, F.; Young, T.M. Influence of rainfall event characteristics on urban pesticide runoff. In Proceedings of the 253rd American Chemical Society National Meeting & Exposition "Advanced Materials, Technologies, Systems & Processes", San Francisco, CA, USA, 2–6 April 2017.
72. Liao, C.; Richards, J.; Taylor, A.R.; Gan, J. Development of polyurethane-based passive samplers for ambient monitoring of urban-use insecticides in water. *Environ. Pollut.* **2017**, *231*, 1412–1420. [CrossRef] [PubMed]
73. Gorgoglione, A.; Bombardelli, F.A.; Pitton, B.J.L.; Oki, L.R.; Haver, D.L.; Young, T.M. Role of sediments in insecticide runoff from urban surfaces: Analysis and modeling. *Int. J. Environ. Res. Public Health* **2018**, *15*, 1464. [CrossRef] [PubMed]
74. Zhou, Q.; Zhang, R.; Shi, Y.; Li, Y.; Paing, J.; Picot, B. Nitrogen and phosphorus removal in subsurface constructed wetland treating agriculture stormwater runoff. In Proceedings of the 9th International Conference Wetland Systems for Water Pollution Control, Lyon, France, 26–30 September 2004; pp. 75–82.
75. Walker, C.; Tondera, K.; Lucke, T. Stormwater treatment evaluation of a constructed floating wetland after two years operation in an urban catchment. *Sustainability* **2017**, *9*, 1687. [CrossRef]
76. Geary, P.M.; Mendez, H.; Dunstan, R.H. Design considerations in the performance of stormwater devices incorporating constructed wetlands. In Proceedings of the 10th International Conference on Wetland Systems for Water Pollution Control, Lisbon, Portugal, 23–29 September 2006; pp. 1833–1842.

77. Scholz, M.; Hohn, P.; Minall, R. Mature experimental constructed wetlands treating urban water receiving high metal loads. *Biotechnol. Prog.* **2002**, *18*, 1257–1264. [CrossRef] [PubMed]
78. Gorgoglione, A.; Bombardelli, F.A.; Pitton, B.J.L.; Oki, L.R.; Haver, D.L.; Young, T.M. Uncertainty in the parameterization of sediment build-up and wash-off processes in the simulation of sediment transport in urban areas. *Environ. Model. Soft.* **2018**. [CrossRef]
79. Langergraber, G.; Simunek, J. Reactive transport modeling of subsurface flow constructed wetlands using the HYDRUS wetland module. *Vadose Zone J.* **2012**, *11*. [CrossRef]
80. Liolios, K.A.; Moutsopoulos, K.N.; Tsihrintzis, V.A. Modeling of flow and BOD fate in horizontal subsurface flow constructed wetlands. *Chem. Eng. J.* **2012**, *200–202*, 681–693. [CrossRef]
81. Llorens, E.; Saaltink, M.W.; Poch, M.; Garcia, J. Bacterial transformation and biodegradation processes simulation in horizontal subsurface flow constructed wetlands using CWM1-RETRASO. *Bioresour. Technol.* **2011**, *102*, 928–936. [CrossRef] [PubMed]
82. Akratos, C.S.; Papaspyros, J.N.; Tsihrintzis, V.A. Total nitrogen and ammonia removal prediction in horizontal subsurface flow constructed wetlands: Use of artificial neural networks and development of a design equation. *Bioresour. Technol.* **2009**, *100*, 586–596. [CrossRef] [PubMed]
83. Scholz, M. Classification methodology for sustainable flood retention basins. *Landsc. Urbran Plan.* **2007**, *81*, 246–256. [CrossRef]
84. Villa, J.A.; Mitsch, W.J.; Song, K.; Miao, S. Contribution of different wetland plant species to the DOC exported from a mesocosm experiment in the Florida Everglades. *Ecol. Eng.* **2014**, *71*, 118–125. [CrossRef]
85. Stefanakis, A.; Akratos, C.S.; Tsihrintzis, V.A. *Vertical Flow Constructed Wetlands: Eco-Engineering Systems for Wastewater and Sludge Treatment*; Newnes: Oxford, UK, 2014.
86. Vymazal, J. The use of sub-surface constructed wetlands for wastewater treatment in the Czech Republic: 10 years experience. *Ecol. Eng.* **2002**, *18*, 633–646. [CrossRef]
87. Vymazal, J. Emergent plants used in free water surface constructed wetlands: A review. *Ecol. Eng.* **2013**, *61*, 582–592. [CrossRef]
88. Greenway, M.; Woolley, A. Changes in plant biomass and nutrient removal over 3 years in a constructed wetland, Cairns, Australia. *Water Sci. Technol.* **2001**, *44*, 303–310. [CrossRef] [PubMed]
89. Vymazal, J. Removal of nutrients in various types of constructed wetlands. *Sci. Total Environ.* **2007**, *380*, 48–65. [CrossRef] [PubMed]
90. Wu, H.; Zhang, J.; Li, C.; Fan, J.; Zou, Y. Mass balance study on phosphorus removal in constructed wetland microcosms treating polluted river water. *CLEAN Soil Air Water* **2013**, *41*, 844–850. [CrossRef]
91. Wu, H.; Zhang, J.; Wei, R.; Liang, S.; Li, C.; Xie, H. Nitrogen transformations and balance in constructed wetlands for slightly polluted river water treatment using different macrophytes. *Environ. Sci. Pollut. Res.* **2013**, *20*, 443–451. [CrossRef] [PubMed]
92. Ha, N.T.; Sakakibara, M.; Sano, S. Accumulation of Indium and other heavy metals by Eleocharis acicularis: An option for phytoremediation and phytomining. *Bioresour. Technol.* **2011**, *102*, 2228–2234. [CrossRef] [PubMed]
93. Vymazal, J. Plants used in constructed wetlands with horizontal subsurface flow: A review. *Hydrobiologia* **2011**, *674*, 133–156. [CrossRef]
94. Brix, H. Do macrophytes play a role in constructed treatment wetlands? *Water Sci. Technol.* **1997**, *35*, 11–17. [CrossRef]
95. Petticrew, E.L.; Kalff, J. Water-flow and clay retention in submerged macrophyte beds. *Can. J. Fish. Aquat. Sci.* **1992**, *49*, 2483–2489. [CrossRef]
96. Brix, H. Functions of macrophytes in constructed wetlands. *Water Sci. Technol.* **1994**, *29*, 71–78. [CrossRef]
97. Luederitz, V.; Eckert, E.; Lange-Weber, M.; Lange, A.; Gersberg, R.M. Nutrient removal efficiency and resource economics of vertical flow and horizontal flow constructed wetlands. *Ecol. Eng.* **2001**, *18*, 157–171. [CrossRef]
98. Yang, L.; Chang, H.T.; Huang, M.N.L. Nutrient removal in gravel- and soil-based wetland microcosms with and without vegetation. *Ecol. Eng.* **2001**, *18*, 91–105. [CrossRef]
99. Munch, C.; Kuschk, P.; Roske, I. Root stimulated nitrogen removal: Only a local effect or important for water treatment? *Water Sci. Technol.* **2005**, *51*, 185–192. [CrossRef] [PubMed]

100. Ruiz-Rueda, O.; Hallin, S.; Baneras, L. Structure and function of denitrifying and nitrifying bacterial communities in relation to the plant species in a constructed wetland. *FEMS Microbiol. Ecol.* **2009**, *67*, 308–319. [CrossRef] [PubMed]
101. Lee, B.H.; Scholz, M. What is the role of phragmites australis in experimental constructed wetland filters treating urban runoff? *Ecol. Eng.* **2007**, *29*, 87–95. [CrossRef]
102. Rakoczy, J.; Remy, B.; Vogt, C.; Richnow, H.H. A bench-scale constructed wetland as a model to characterize benzene biodegradation processes in freshwater wetlands. *Environ. Sci. Technol.* **2011**, *45*, 10036–10044. [CrossRef] [PubMed]
103. Headley, T.R.; Davison, L.; Huett, D.O.; Muller, R. Evapotranspiration from subsurface horizontal flow wetlands planted with phragmites australis in sub-tropical Australia. *Water Res.* **2012**, *46*, 345–354. [CrossRef] [PubMed]
104. Borin, M.; Milani, M.; Salvato, M.; Toscano, A. Evaluation of Phragmites australis (cav.) trin. Evapotranspiration in northern and southern Italy. *Ecol. Eng.* **2011**, *37*, 721–728. [CrossRef]
105. Taleno, V.C. Comparison of Two Constructed Wetland with Different Soil Depth in Relation to Their Nitrogen Removal. Ph.D. Thesis, Universidad Autónoma De San Luis Potosí, San Luis Potosí, Mexico, 2012.
106. Valipour, A.; Ahn, Y.-H. Constructed wetlands as sustainable ecotechnologies in decentralization practices: A review. *Environ. Sci. Pollut. Res.* **2016**, *23*, 180–197. [CrossRef] [PubMed]
107. Wang, R.; Korboulewsky, N.; Prudent, P.; Domeizel, M.; Rolando, C.; Bonin, G. Feasibility of using an organic substrate in a wetland system treating sewage sludge: Impact of plant species. *Bioresour. Technol.* **2010**, *101*, 51–57. [CrossRef] [PubMed]
108. Kadlec, R.H. Wastewater treatment at the Houghton Lake wetland: Soils and sediments. *Ecol. Eng.* **2009**, *35*, 1333–1348. [CrossRef]
109. Tee, H.C.; Lim, P.E.; Seng, C.E. Newly developed baffled subsurface flow constructed wetland for the enhancement of nitrogen removal. *Bioresour. Technol.* **2012**, *104*, 235–242. [CrossRef] [PubMed]
110. Stefanakis, A.I.; Tsihrintzis, V.A. Effects of loading, resting period, temperature, porous media, vegetation and aeration on performance of pilot-scale vertical flow constructed wetlands. *Chem. Eng. J.* **2012**, *181*, 416–430. [CrossRef]
111. Calheiros, C.S.; Rangel, A.O.; Castro, P.M. Evaluation of different substrates to support the growth of Typha latifolia in constructed wetlands treating tannery wastewater over long-term operation. *Bioresour. Technol.* **2008**, *99*, 6866–6877. [CrossRef] [PubMed]
112. Ann, Y.; Reddy, K.R.; Delfino, J.J. Influence of chemical amendments on phosphorus immobilization in soils from a constructed wetland. *Ecol. Eng.* **1999**, *14*, 157–167. [CrossRef]
113. Xu, D.; Xu, J.; Wu, J.; Muhammad, A. Studies on the phosphorus sorption capacity of substrates used in constructed wetland systems. *Chemosphere* **2006**, *63*, 344–352. [CrossRef] [PubMed]
114. Tao, W.; Wang, J. Effects of vegetation, limestone and aeration on nitritation, anammox and denitrification in wetland treatment systems. *Ecol. Eng.* **2009**, *35*, 836–842. [CrossRef]
115. Seo, D.C.; Cho, J.S.; Lee, H.J.; Heo, J.S. Phosphorus retention capacity of filter media for estimating the longevity of constructed wetland. *Water Res.* **2005**, *39*, 2445–2457. [CrossRef] [PubMed]
116. Brooks, A.S.; Rozenwald, M.N.; Geohring, L.D.; Lion, L.W.; Steenhuis, T.S. Phosphorus removal by wollastonite: A constructed wetland substrate. *Ecol. Eng.* **2000**, *15*, 121–132. [CrossRef]
117. Bruch, I.; Fritsche, J.; Bänninger, D.; Alewella, U.; Sendelov, M.; Hürlimann, H.; Hasselbach, R.; Alewell, C. Improving the treatment efficiency of constructed wetlands with zeolite-containing filter sands. *Bioresour. Technol.* **2011**, *102*, 937–941. [CrossRef] [PubMed]
118. Ren, Y.; Zhang, B.; Liu, Z.; Wang, J. Optimization of four kinds of constructed wetlands substrate combination treating domestic sewage. Wuhan Univ. *J. Nat. Sci.* **2007**, *12*, 1136–1142.
119. Babatunde, A.O.; Zhao, Y.Q.; Zhao, X.H. Alum sludge-based constructed wetland system for enhanced removal of P and OM from wastewater: Concept, design and performance analysis. *Bioresour. Technol.* **2010**, *101*, 6576–6579. [CrossRef] [PubMed]
120. Mateus, D.M.R.; Vaz, M.M.N.; Pinho, H.J.O. Fragmented limestone wastes as a constructed wetland substrate for phosphorus removal. *Ecol. Eng.* **2012**, *41*, 65–69. [CrossRef]
121. Hill, D.T.; Payne, V.W.E.; Rogers, J.W.; Kown, S.R. Ammonia effects on the biomass production of five constructed wetland plant species. *Bioresour. Technol.* **1997**, *62*, 109–113. [CrossRef]

122. Chong, H.L.; Chia, P.S.; Ahmad, M.N. The adsorption of heavy metal by Bornean oil palm shell and its potential application as constructed wetland media. *Bioresour. Technol.* **2013**, *130*, 181–186. [CrossRef] [PubMed]
123. Ranieri, E.; Gorgoglione, A.; Solimeno, A. A comparison between model and experimental hydraulic performances in a pilot-scale horizontal subsurface flow constructed wetland. *Ecol. Eng.* **2013**, *60*, 45–49. [CrossRef]
124. García, J.; Aguirre, P.; Mujeriego, R.; Huang, Y.; Ortiz, L.; Bayona, J.M. Initial contaminant removal performance factors in horizontal flow reed beds used for treating urban wastewater. *Water Res.* **2004**, *38*, 1669–1678. [CrossRef] [PubMed]
125. Aguirre, P.; Ojeda, E.; García, J.; Barragán, J.; Mujeriego, R. Effect of water depth on the removal of organic matter in horizontal subsurface flow constructed wetlands. *J. Environ. Sci. Health* **2005**, *40*, 1457–1466. [CrossRef]
126. Lee, C.; Fletcher, T.D.; Sun, G. Nitrogen removal in constructed wetland systems. *Eng. Life Sci.* **2009**, *9*, 11–22. [CrossRef]
127. Reed, S.C.; Brown, D. Surface flow wetlands—A performance evaluation. *Water Environ. Res.* **1995**, *67*, 244–248. [CrossRef]
128. Deblina, G.; Brij, G. Effect of hydraulic retention time on the treatment of secondary effluent in a subsurface flow constructed wetland. *Ecol. Eng.* **2010**, *36*, 1044–1051.
129. Yan, Y.; Xu, J. Improving winter performance of constructed wetlands for wastewater treatment in Northern China: A review. *Wetlands* **2014**, *34*, 243–253. [CrossRef]
130. Toet, S.; Logtestijn, R.S.P.V.; Kampf, R.; Schreijer, M.; Verhoeven, J.T.A. The effect of hydraulic retention time on the removal of pollutants from sewage treatment plant effluent in a surface-flow wetland system. *Wetlands* **2005**, *25*, 375–391. [CrossRef]

Article

The Effects of Plants on Pollutant Removal, Clogging, and Bacterial Community Structure in Palm Mulch-Based Vertical Flow Constructed Wetlands

Marina Carrasco-Acosta [1], Pilar Garcia-Jimenez [1], José Alberto Herrera-Melián [2,*], Néstor Peñate-Castellano [2] and Argimiro Rivero-Rosales [3]

1 Department of Biology, Instituto Universitario de Investigación en Estudios Ambientales y Recursos Naturales i-UNAT, Universidad de Las Palmas de Gran Canaria, Campus de Tafira, Las Palmas de Gran Canaria 35017, Las Palmas, Spain; marina.carrasco@ulpgc.es (M.C.-A.); pilar.garcia@ulpgc.es (P.G.-J.)

2 Department of Chemistry, Instituto Universitario de Investigación en Estudios Ambientales y Recursos Naturales i-UNAT, Universidad de Las Palmas de Gran Canaria, Campus de Tafira, Las Palmas de Gran Canaria 35017, Las Palmas, Spain; nestorp1990@gmail.com

3 Department of Chemistry, Instituto de Oceanografía y Cambio Global IOCAG, Universidad de Las Palmas de Gran Canaria, Campus de Tafira, Las Palmas de Gran Canaria 35017, Las Palmas, Spain; argimiro.rivero@ulpgc.es

* Correspondence: josealberto.herrera@ulpgc.es

Received: 27 December 2018; Accepted: 23 January 2019; Published: 25 January 2019

Abstract: In this study, the effects of plants on the performance and bacterial community structure of palm mulch-based vertical flow constructed wetlands was studied. The wetlands were built in August 2013; one of them was planted with *Canna indica* and *Xanthosoma* sp., and the other one was not planted and used as a control. The experimental period started in September 2014 and finished in June 2015. The influent was domestic wastewater, and the average hydraulic surface loading was 208 L/m^{2}d, and those of COD, BOD, and TSS were 77, 57, and 19 g/m^{2}d, respectively. Although the bed without plants initially performed better, the first symptoms of clogging appeared in December 2014, and then, its performance started to fail. Afterwards, the wetland with plants provided better removals. The terminal restriction fragment length polymorphism (T-RFLP) analysis of *Enterococci* and *Escherichia coli* in the effluents suggests that a reduction in their biodiversity was caused by the presence of the plants. Thus, it can be concluded that the plants helped achieve better removals, delay clogging, and reduce *Enterococci* and *E. coli* biodiversity in the effluents.

Keywords: Constructed wetland; mulch-based substrate; clogging; plants; T-RFLP

1. Introduction

Constructed wetlands (CWs) are particularly well suited for wastewater treatment in small communities, i.e., those with a <2.000 p.e. (person equivalent) because of their low cost, easy maintenance, high treatment efficiency, and visual appeal [1]. Similarly, CWs are environmentally friendly and a real alternative for the treatment of different types of wastewaters [2]. CWs are open, shallow reactors composed basically of an impervious layer, a mineral substrate (usually gravel or sand), helophytes, water, and the associated microbes. According to the water flow, CWs can be classified as horizontal flow (HF) or vertical flow (VF) and surface or subsurface flow. One of the main constraints of CWs is their high surface area, usually 3–5 m^2/p.e. VF CWs demand less surface and are more efficient than HFs because of the higher aeration of the substrates [3]. However, VF CWs usually include smaller-sized substrates, such as sand, to achieve a good surface influent distribution and appropriate hydraulic retention times. One of the main problems associated with the use of sand in VF CWs is the risk of clogging [4]. A remarkable exception is the so-called "French system", which can treat raw domestic wastewater without primary settling [5].

CWs are considered to be more sustainable than conventional wastewater treatments, such as activated sludge [6,7]. However, a relevant proportion of the environmental impact of CWs is associated with the construction phase and, particularly, the use of gravel and/or sand as porous media. A conventional VF for 200 p.e. (3 m^2/p.e. with depth: 80 cm) would require 480 m^3 of gravel and/or sand Fuchs et al. [8] claimed that the construction phase impact could be significantly reduced by using local materials to minimize transport. Sand extraction poses serious threats to the environment in terms of erosion, loss of land and biological diversity, and the increase of poverty among people [9]. Additionally, some regions face supply constraints due to the overexploitation of natural aggregates in construction [10]. In the particular case of the Canary Islands, most of the sand used in the construction sector and beach regeneration has serious legal, environmental, economic, and political implications, because it is imported from the near Western Sahara in the African coast. This region, a former Spanish colonial territory, is still in disputes with Morocco and the Polisario Front [11]. The volume of sand and gravel devoted to the construction of CWs is a small proportion of the total amount extracted, but the point is to understand how sustainable CWs should be, ideally.

Over the last decade, many studies have been devoted to finding cost-effective substrates to increase the treatment efficiency of and/or minimize clogging in CWs [12]. The environmental impact of CWs can be drastically reduced by using wastes as substrates. Different materials, such as construction wastes, dewatered alum sludge [13], bamboo rings [14], palm tree mulch [15,16], and rice straw [17], have been tested.

Plants are considered to be an essential part of CWs. They help stabilize the bed surface, prevent clogging and channeled flow [18], and remove pollutants (N, P, heavy metals, etc.). Another important role of plants in CWs is their ornamental value [19] and the improvement of the CW's ecological significance, including wildlife habitat creation, cooling by means of evapotranspiration, recreation, and landscaping [20]. Additionally, plants can accelerate the development of microbial communities by promoting alternate aerobic and anaerobic micro-environments in roots [21]. Thus, not only the presence of plants, but also plant biodiversity plays a key role in the microbial communities and the CW's ability to remove pollutants.

The study of the bacterial communities allows us to analyze the contribution of these organisms to CWs and, to a large extent, to understand CWs work appropriately. Similarly, the study of bacterial communities could determine how different substrates, in combination with the presence of plants, affect the efficiency of CWs. From myriad of bacteria, fecal indicator bacteria (FIB) are the most commonly analyzed as a consequence of their involvement in European regulations and their impact on human wellness. FIB are represented by *Escherichia coli* and *Enterococci* [22,23]. The understanding of the whole bacterial community requires a detailed screening to unveil their diversity and distribution. Unlike bacteria culture methods, culture-independent methods based on molecular markers, such as 16s rRNA, and analysis of terminal restriction fragment length polymorphisms (T-RFLP) are an excellent approach to studying the bacterial community structure of CWs [24–27]. The analysis of T-RFLP patterns from different CWs and the comparison among them allows us to unveil to what extent a bacterial community can change and how far experimental conditions (flow, substrate type, presence of plants) would increase the performance of CWs.

The main goal of this work was to study the effect of plants on the performance of palm mulch-based VF CWs treating domestic wastewater. The role of plants was analyzed through both chemical and microbiological parameters. Furthermore, we aimed to compare the evolution over time of the structure of bacterial communities in the effluents of planted and unplanted mulch-based VF CWs.

2. Material and Methods

2.1. Location of Constructed Wetland and Sampling

The studied CW was designed to treat raw wastewater from a group of 4 households located in a rural zone of the Island of Gran Canaria, Spain. The households have a permanent population of 5 people, which is increased to 15 people during holidays and weekends.

Wastewater was discharged directly into a partially clogged, non-waterproofed, 36 m^3 seepage pit (Figure 1) with an approximate hydraulic retention time of 1.5 days. Two submergible, 700-W pumps, which were timer-controlled to function for 1 min each hour, were installed. Influent pulses were of approximately 15–20 L. The pit was supposed to provide partial sedimentation and homogenization. Thus, the influent can be considered to be partially decanted domestic wastewater.

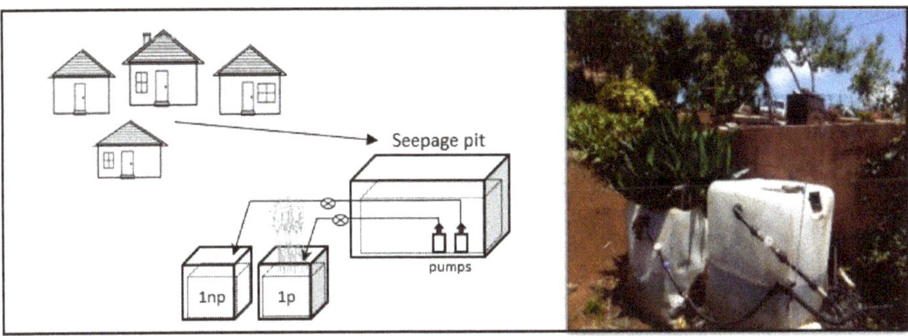

Figure 1. Layout and photograph of the constructed wetlands (CWs). 1p: planted vertical flow (VF) CW; and 1np: unplanted VF CW.

The CW was placed at 650 m above sea level. The average temperature is 18 °C, with January being the coldest month with an average temperature of 12 °C and August the warmest month with an average temperature of 23 °C. The average relative air moisture is 80% and the predominant wind direction is NNE. The average pluviosity is 485 mm/year, with the rainy season being between October and March.

The two VF CWs were constructed with 1.080-L plastic, cubic-like recipients. A layer of washed gravel with a height of 10–15 cm and an average diameter of 1 cm was deposited at the bottom of the recipients. The organic substrate was placed directly above the gravel and consisted of approximately 1-m-high mulch made of dry branches of Canarian palm tree (*Phoenix canariensis*). Additionally, two flow meters and valves were placed at the inlet of each reactor to measure and control the inflow. The influent was distributed on the surface of the VF CWs by means of 5 perforated tubes with a 12-mm diameter.

In order to value the contribution of the plants, one VF CW was planted with *Canna indica* and *Xanthosoma* sp. and compared with another VF CW (the control), which was not planted. These species of plant are characterized by large roots and big tubers, respectively. The analyses were started one year after the construction of the VF CWs, because the mulch substrate had been reduced by degradation and compaction.

Chemical and microbiological analyses were completed in unfiltered and homogenized samples of the influent and the effluents of the planted (1p) and unplanted (1np) VF CWs.

2.2. Analytical Methods

2.2.1. Water Analysis

Sampling was performed in the morning at 08:00 h, coinciding with when the pumps started a working period. The influent sample was taken during pumping from the influent outlet. The volume sampled was 1 L for each sampling point. The sampling recipients were 1-L plastic bottles for standard parameters, sterilized 50-mL glass bottles for *E. coli* and total *E. coli* enumeration, and sterilized 1-L amber glass bottles for DNA studies.

The water quality parameters were measured according to standard methods (APHA, 2005). BOD_5 (Biological Oxygen Demand, henceforth BOD) and Chemical Oxygen Demand (COD) were measured in homogenized, unfiltered samples. Ammonia-N and Na ions were determined with selective electrodes from Crison (Spain). Ca^{2+} and Mg^{2+} ions were determined by the EDTA titrimetric method. Anions (PO_4^{3-}, Cl^-, F^-, NO_2^-, NO_3^-, and SO_4^{2-}) were analyzed with a 792 Basic IC ionic chromatograph from Metrohm, equipped with a Metrosep A sup 4 column. The mobile phase and suppression solutions were 1.8 mM Na_2CO_3/1.7 mM $NaHCO_3$ buffer and 20 mM H_2SO_4, respectively.

The concentrations of *Escherichia coli* and total coliforms (TC) were determined by the membrane filter method with incubation at 37 °C for 24 h with a chromogenic agar (Panreac, Spain). *E. coli* displayed dark blue to violet colonies, while TC were assumed to be the sum of the *E. coli* colonies plus the salmon-red colonies.

2.2.2. Molecular Method

In order to analyze molecularly the bacterial community structure, a terminal restriction fragment length polymorphism (T-RFLP) analysis was carried out in the effluents of both CWs. The analyses were performed in December 2014, when the initial clogging symptoms were observed in 1np (light clogging), and in May 2015, when the clogging in 1 np was severe (severe clogging).

DNA Isolation

First, 1 liter of water of each sample was centrifuged, and pellets pooled to extract genomic DNA. The DNA isolation was performed according to the CTAB method described by Murray and Thompson [28]. In detail, CTAB was made of CTAB 0.1% (*w*/*v*), PVPP 0.1% (*w*/*v*), TRIS-HCl pH 8.6, SDS 10, EDTA 0.5M pH 8, NaCl 4M, and β-Mercaptoethanol 2% (*v*/*v*). To obtain bacterial samples, 800-μg pellets were introduced into microtubes. Then, an extraction buffer (800 μL) was added to a microtube containing the source of the DNA template. The sample was held for 1 h in a bath at 65 °C, and mixed gently by inversion approximately every 20 min. Later, 800 μL of CIA (chloroform: isoamyl alcohol, 24:1 *v*/*v*) was added and centrifuged for 20 min at 3000 rpm in a Beckman centrifuge.

Successive washes in CIA and centrifugations were carried out until the supernatant became whitish. To continue, 2/3 of isopropyl alcohol at −20 °C was added, and a centrifugation for 30 min at 15,000× *g* was carried out. Afterwards, the isopropyl alcohol was removed, and 20 μL of ethanol (80%) was added. Finally, the samples were centrifuged for 5 min at 15,000× *g*. The supernatant was then discarded, and the pellets were resuspended in 15 μL of H_2O and stored at 4 °C until used.

The yield and purity of genomic DNA were calculated from the A260/A280 ratio measured using a NanoDrop spectrophotometer (ThermoFisher Scientific, Waltham, MA, USA). For each VF CW type and date, an assay was repeated three times with two replicates of each one.

T-RFLP Analysis

The bacterial 16s rRNA gene of genus-specific *Enterococcus* sp. [29] was amplified by using a HEX fluorescent-labeled forward primer E1 (5′-TCA ACC GGG GAG GGT-3′) and an unlabeled reverse primer E2 (5′-ATT ACT AGC GAT TCC GG-3′). The sequence encoding *lacZ* for *E. coli* [30] was amplified using a fluorescent-labeled 6 FAM forward primer ZL-1675 (5′-ATG AAA GCT GGC TAC AGG AAG GCC-3′). The corresponding unlabeled reverse primer sequence was ZR-2548

(5′-CAC CAT GCC GTG GGT TTC AAT ATT-3′). DNA (5 μL) was thus amplified in a PCR reaction mix (50 μL of total volume) containing 10 μL of the 5X PrimeSTAR™ buffer (Takara Bio Inc), 0.5 μL of PrimeSTAR™ HS DNA Polymerase (2.5 U/μL) (Takara Bio Inc), 4 μL of dNTPs mix (2.5 mM each one, Takara Bio Inc), and 2 μL of each oligonucleotide (10 μM).

The amplifications were performed in a thermocycler MyCyclerTM (Biorad, Hercules, CA, USA), with an initial denaturation at 96 °C for 4 min followed by 40 cycles at 94 °C for 1 min, 45 °C for 1 min, and 72 °C for 5 min. The final elongation was carried out at 72 °C for 10 min. The PCR-amplification products were checked by agarose (1%) electrophoresis at 75 V, resulting in gene fragments of 733 pb and 876 pb for *Enterococcus* sp. and *E. coli*, respectively. The amplifications were carried out by triplicate.

PCR-amplification products were separately purified using a Wizard®SV Gel and PCR Clean-Up System (Promega, Madison WI, USA). Purified products from *E. coli* and *Enterococcus* sp. were checked by electrophoresis on a 2% agarose gel. The concentrations were estimated as previously mentioned.

T-RFLP analysis was carried out based on the previous enzyme digestion assay. The NEBcutter v2.0 (New England BioLabs) software program was used to estimate a range of terminal restriction fragments (T-RFs) of varying sizes in silico. The restriction enzyme MseI (New England BioLabs) was selected in silico according to the yielded T-RFs of the respective PCR amplicons for *E. coli* and *Enterococcus* sp.

Enzyme digestion was performed with 100 ng of purified PCR amplicon for 3 h at 37 °C and inactivated at 65 °C for 20 min. Amplicon was then cut with 0.5 U of the restriction enzyme MseI (recognition site: T'TAA) according to the manufacturer's instructions (BioLabs, New England BioLabs). Restriction digestion was carried out 3 times for the *E. coli* and *Enterococcus* sp., respectively.

Sequencing was carried out on an ABI 3730 capillary sequencer (Life Technologies) by Secugen (Madrid, Spain), and the resulting sequences were analyzed using GeneMarker v1.85 (SoftGenetics). Optimizing was performed with different dilutions according to Secugen.

2.3. Statistics

2.3.1. Chemical and Microbiological Analysis

Average values of concentrations, surface loadings, and removals were compared by means of ANOVA if the data were homoscedastic (Bartlett test) and normally distributed (Shapiro–Wilk test). If these conditions were not met, the Kruskal–Wallis non-parametric test was used. In all cases, a significance level of 95% (p-value > 0.05) was utilized.

2.3.2. T-RFLP Analysis

The T-RFLP fragment sequences were obtained in .fsa format and GeneMarker software (SoftGenetics) was used for data scoring. The analyses of the polymorphic bands were only scored when they were highly reproducible and according to the peak height. Thus, small T-RFs ranging from 0 to 100 bp were excluded from the analysis, as they were considered to be PCR artifacts. Moreover, those T-RFs with intensities lower than 1% florescence, regarded as background interference, were also excluded from the matrices. Next, relative peak areas were recalculated according to the removal of artifactual peaks. The relative abundance of T-RFs was determined by calculating the ratio between the peak height of each peak and the total peak height of all the peaks within 1 sample and assuming that unambiguous T-RFs ranged in size from 100 to 850 bp.

For a comparative analysis of T-RFLP profiles, alignment was carried out through T-Align software with a default value threshold of 0.5 [31]. This software allowed for the compiling of peak profiles of different electropherograms, which corresponded to separate sampling days and sampling sites. In addition, the resulting consensus profile (henceforth overall consensus electropherogram) valued the presence or absence of T-RFs through a binary matrix by comparing with the other consensus profile of each of the other sites. For the purposes of qualitative valuation, the number of peaks in the consensus electropherograms was interpreted to be the operational taxonomic units (OTUs) [27], although non-sequencing was carried out. Statistical comparisons of a number of T-RFs were performed using R software [32]. A one-way ANOVA followed by post hoc Tukey HSD and Dunnett T3 tests was used to detect significant differences ($p \leq 0.1$) between 1p and 1np under light and severe clogging. Additionally, Venn diagrams were performed with Venn Diagram Plotter (Integrative Omics, Pacific Northwest National Laboratory) [33].

3. Results and Discussion

3.1. Features of the Influent

As mentioned above, wastewater from the households was discharged directly into the seepage pit. The pit acted as a primary settler, making the influent more stable and reducing the concentration of the total suspended solids (TSS) as compared with other systems without a primary settler [16]. Table 1 shows the chemical and microbiological parameters of the influent. The number of data points changes for each parameter because only analytically valid data and non-outliers are included. Other parameters such as Na, electrical conductivity, and Sodium Absorption Ratio (SAR)were analyzed less frequently with the goal to obtain the agronomic quality of the effluents. As can be observed, the biodegradability (BOD/COD) ratio was 71% and turbidity was 521 NTU. With these traits, the influent can be considered to range from medium to high strength for organic matter (COD and BOD), low strength for TSS, and medium strength for nutrients N and P [34].

Table 1. Chemical and microbiological parameters, expressed as average value ± SD, associated with influent from September 2014 to May 2015.

Parameter	Average ± Std. Dev.	Max–Min, Number of Data Points	Units
COD	390 ± 114	783–260, 26	mg/L
BOD	279 ± 89	470–140, 22	mg/L
Total Suspended Solids	94 ± 27	137–48, 26	mg/L
Turbidity	521 ± 132	723–266, 26	NTU
N-ammonia	31.1 ± 6.2	42.8–20.2, 25	mg/L
Escherichia coli	$1.5 (\pm 0.92) \times 10^6$	3.0×10^6–3.6×10^5, 23	CFU/100 mL
Total coliforms	$3.1 (\pm 2.1) \times 10^6$	8.0×10^6–8.0×10^5, 23	CFU/100 mL
Na^+	114 ± 39	157–41, 9	mg/L
Hardness ($Ca^{2+}Mg^{2+}$)	3.0 ± 1.5	5.4–1.6, 10	meq/L
Sodium Absorption Ratio	6.2 ± 2.9	10.7–2.5, 9	
Electrical Conductivity	1945 ± 169	2190–1304, 25	µS/cm
pH	7,8 ± 0	8.0–7.5, 25	
N-Nitrites	0.09 ± 0.4	1.6–0.0, 17	mg/L
N-Nitrates	1.1 ± 0.83	2.6–0.0, 17	mg/L
P-Phosphates	10.4 ± 3.9	14.7–2.0, 17	mg/L
Sulfates	14 ± 17	76–4, 16	mg/L
Cl^-	99 ± 11	122–83, 17	mg/L
F^-	7.9 ± 1.2	10.23–6.27, 17	mg/L

3.2. Average Surface Loading Rates and Removals

The hydraulic loading rate (HLR) and organic loading rates (OLR) are represented in Table 2. The average BOD-LRs applied in the present study were 51 and 70 g/m^{2}d, for 1np and 1p, respectively. These values fall in the upper range of those found in the literature. For instance, the OLR applied to the French System's first stage ranged between 40 and 50 g BOD/m^{2}d [5].

The average HLR of 1p (233 L/m^{2}d) was significantly higher (p: 0.026) than that of 1np (189 L/m^{2}d). The same result was obtained for the BOD-LR (1p: 70 g/m^{2}d, 1np: 51 g/m^{2}d, p: 0.0233) and ammonia-N (1p: 9 g/m^{2}d, 1np: 7 g/m^{2}d, p: 0.0212). Non-significant differences were reported for the COD-LR (1p: 90 g/m^{2}d and 1np: 70 g/m^{2}d, p: 0.0965) and TSS-LR (1p: 22 g/m^{2}d and 1np: 18 g/m^{2}d, p: 0.129). Considering that in the cases of dissimilarity, the p-values were close to 0.05

(HLR, COD, and ammonia-N) and that there was no difference for COD and TSS, it can be concluded that both VF CWs received almost similar surface loadings and their differences in removal can be mainly ascribed to the presence of plants.

Table 2. Hydraulic loading rate (HLR, L/m^{2}d) and organic loading rates (OLR, g/m^{2}d) for 1p and 1np.

Parameters	Average ± Std. Dev. Max–Min	
	1p	1np
HLR	233 ± 54 323–146	189 ± 78 349–44
COD	90 ± 29 137–46	74 ± 40 171–18
BOD	70 ± 29 126–24	51 ± 23 98–6
TSS	22 ± 9 40–8	18 ± 10 41–4
Ammonia-N	9 ± 3 18–5	7 ± 3 16–1

Although for some parameters, 1p received higher surface loadings, its performance was equal to or better than 1np (Table 3). The removals were slightly higher for 1p, although not significantly ($p > 0.1$) for COD (1p: 58%, 1np: 50%), BOD (1p: 62%, 1np: 59%), TSS (1p: 77%, 1np: 72%), *E. coli* (1p: 79%, 1np: 73%), and TC (1p: 82%, 1np: 74%). The exception was turbidity removal (1p: 90%, 1np: 73%, p: 0.005) for which the presence of plants improved clearly. This result is of interest, because according to Spanish legislation, turbidity is a basic parameter for the reuse of treated wastewater [35].

Table 3. Average removals ± standard deviation, expressed as %, and range of maximum and minimum values.

Parameters	Average ± Std. Dev. Max–Min	
	1p	1np
COD	58 ± 17 77–8	50 ± 17 86–27
BOD	62 ± 28 97–(−3)	59 ± 30 92–(−34)
Turbidity	90 ± 10 98–62	73 ± 19 99–46
TSS	77 ± 20 99–20	72 ± 15 99–39
Ammonia-N	−39 ± 50 55–(−123)	−53 ± 59 48–(−192)
Phosphate-P	−38 ± 169 39–(−666)	−46 ± 163 36–(−675)
Sulfate	−297 ± 282 86–(−1071)	−175 ± 239 61–(−777)
E. coli	79 ± 26 99.9–11	73 ± 28 99.8–(−8)
Total coliforms	82 ± 26 99.6–(−20)	74 ± 25 99.9–5.5

The removals of ammonia-N, phosphate-P, and sulfate were negative in all cases. The increased concentrations of the ions can only come from the substrate, because it is an organic material and from the accumulated sludge. As mentioned above, high surface loadings were applied, and no rest periods were allowed. Additionally, the reactors were very efficient in removing TSS, which were retained as sludge, mainly on the surface of the CWs. The negative removals of ammonia-N, phosphate, and sulfate ions are more likely to come from the leaching or degradation of the accumulated sludge, rather than from the mulch, which had been in operation for 1 year. In other works (e.g., [16]) with palm mulch, we have observed good physical and chemical stability of the substrate, both in vertical and horizontal CWs, once it became stabilized, after 2–3 months in operation. However, under the stringent conditions imposed to the reactors, a further degradation of the substrate cannot be rejected.

It can be thought that because the substrate was mainly of an organic nature, the release of these nutrients can be caused by its decomposition. In fact, Saeed and Sun [36] observed that HF CWs, which employed wood mulch and gravel-mulch media, released organic matter, phosphorus, and TSS. This behavior was not observed in the VF CWs tested in our study. Additionally, in the present research, the substrate height was reduced by 40% by compaction and degradation during the first few months of operation. However, the analyses were initiated 1 year after construction. Thus, it seems more probable that the released ions came from the fast mineralization of the particulate matter retained on the surface of the VF CWs under the appropriate conditions of temperature, moisture, and aeration.

3.3. Performance Evolution of the Wetlands

The evolution of parameters such as COD, BOD, turbidity, TSS, ammonia-N, *E. coli*, and TC are represented for the influent and the effluents of 1p and 1np (Figure 2). Missing points have been removed after the statistical analysis of outliers.

(a)

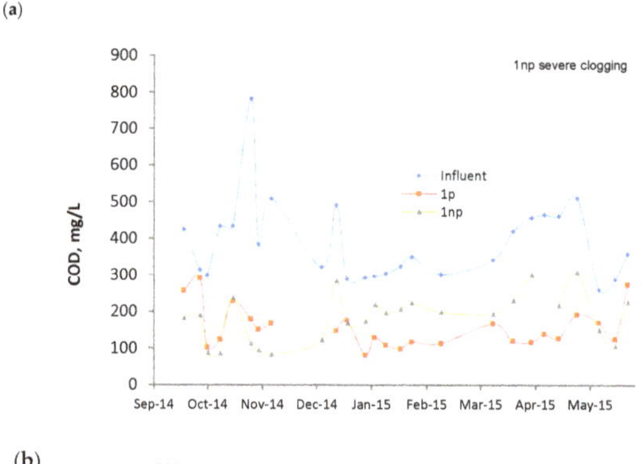

(b)

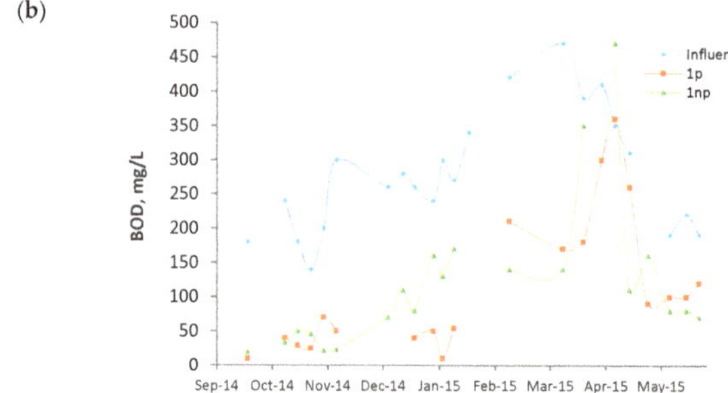

Figure 2. *Cont.*

(**c**)

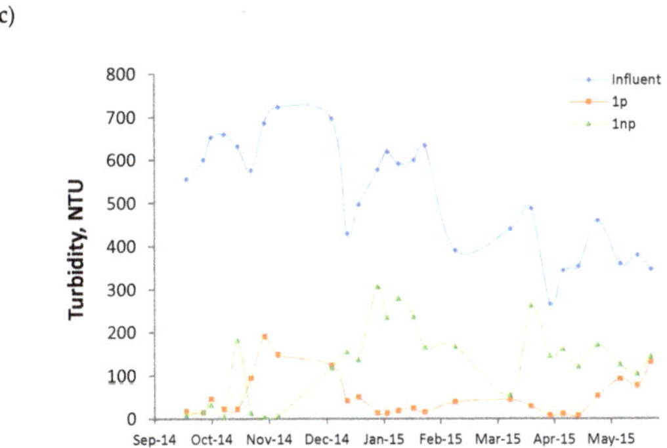

(**d**)

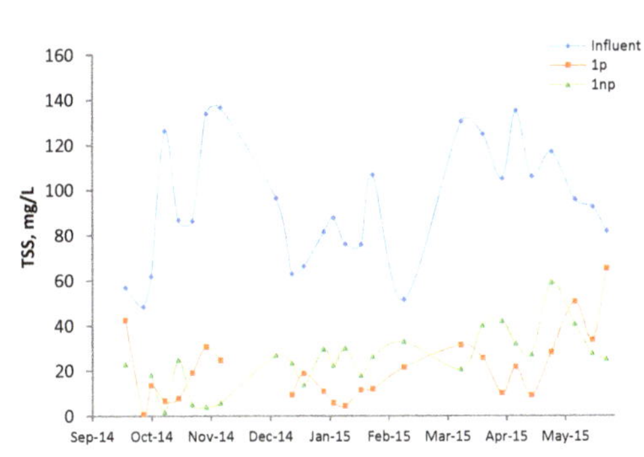

(**e**)

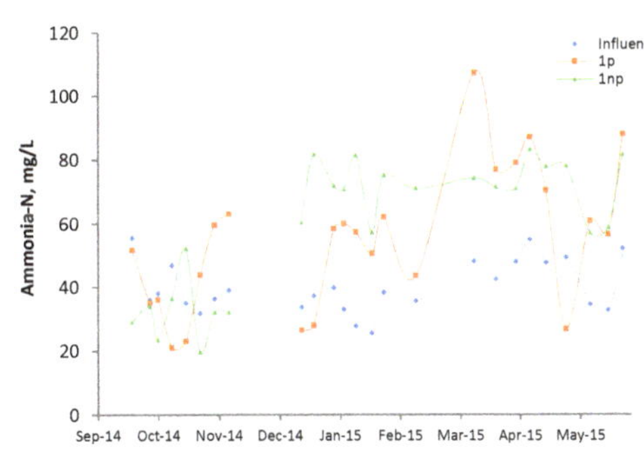

(**f**)

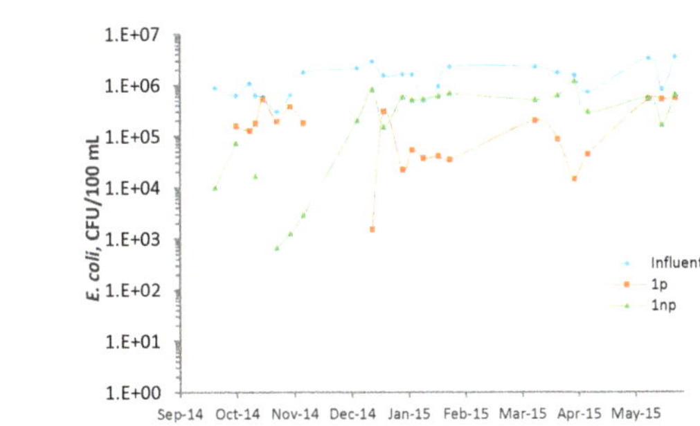

Figure 2. *Cont.*

(g)

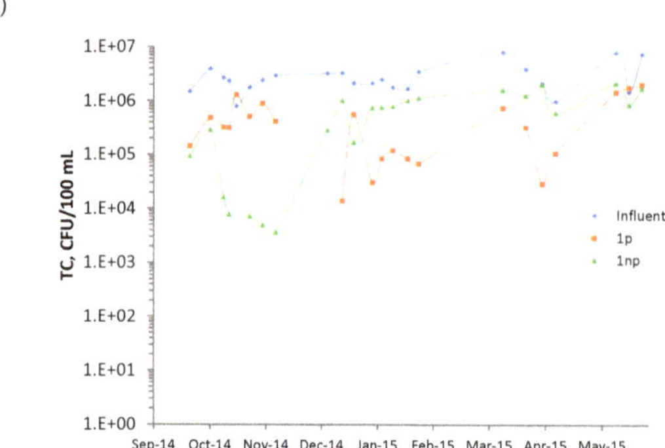

Figure 2. Concentrations in the influent (-◆-) and effluents of 1p (-■-) and 1np (-▲-) of (**a**) COD, (**b**) BOD, (**c**) turbidity, (**d**) TSS, (**e**) ammonia-N, (**f**) *E. coli*, and (**g**) TC.

Figure 3 shows the timeline of removals. During December 2014, the first clogging symptoms were observed in 1np. After that, the relative performance of the wetlands changed dramatically. Table 4 shows the average removals of chemical and microbiological parameters before and after clogging. Before clogging, 1np performed better than 1p with respect to turbidity (1np: 95% and 1p: 90%; p: 0.028), *E. coli* (1p: 59%, 1np: 97%, p: 0.007), and TC (1p: 82%, 1np:98 %, p: 0.001). Nevertheless, the removals of BOD (1p: 83%, 1np: 85%), COD (1p: 64%, 1np: 62%), and TSS (1p: 84%, 1np: 80%) were similar ($p > 0.288$). After clogging, 1np performance was significantly damaged ($p < 0.006$) for COD, BOD, turbidity, *E. coli*, and TC, while that of 1p was improved or remained unaltered.

(a)

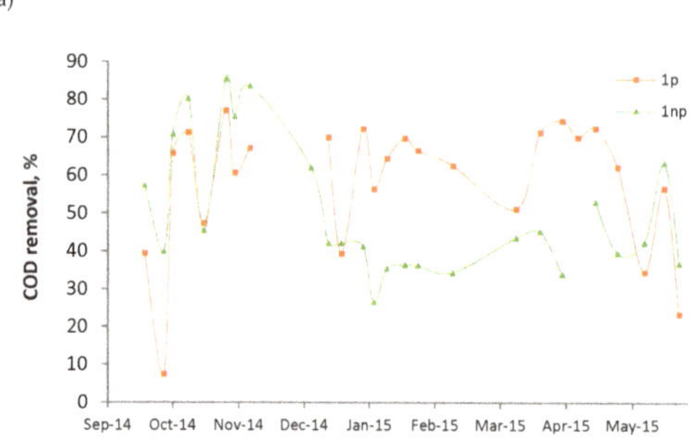

(b)

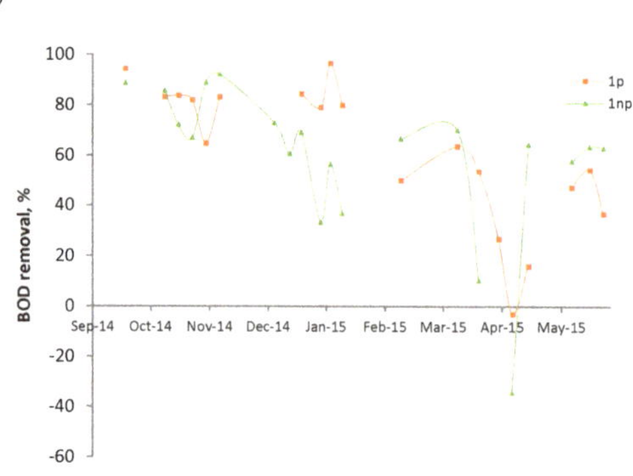

Figure 3. *Cont.*

(c)

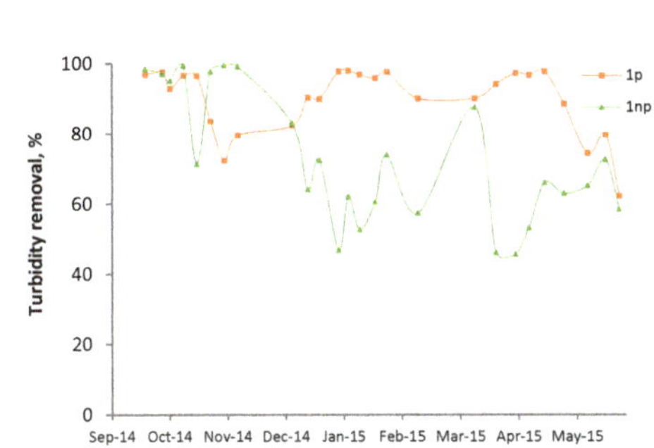

(d)

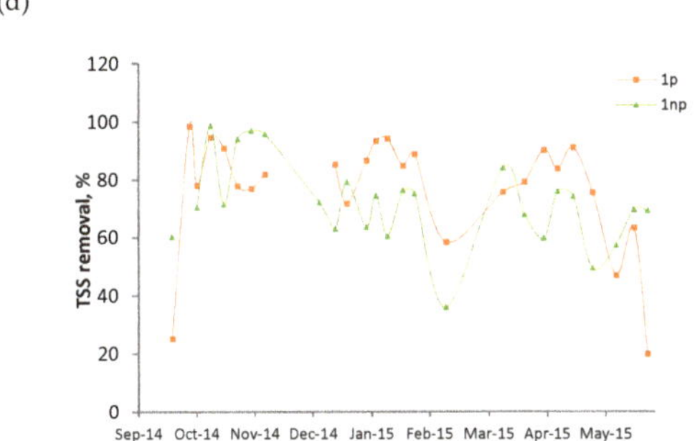

(e)

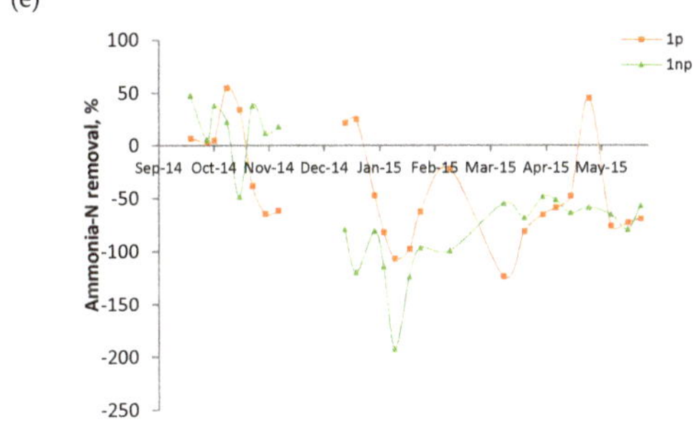

(f)

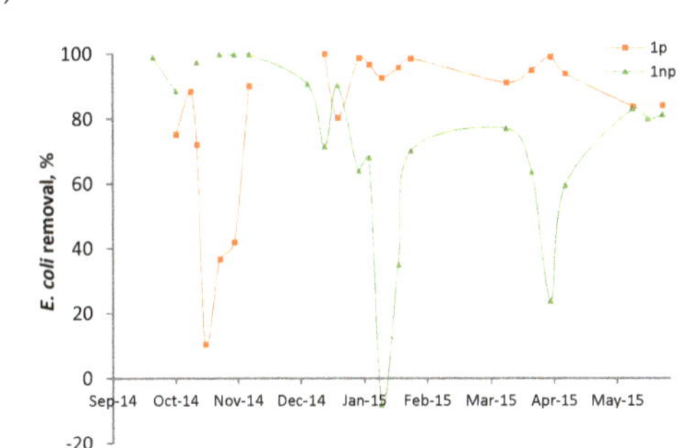

Figure 3. *Cont.*

(g)

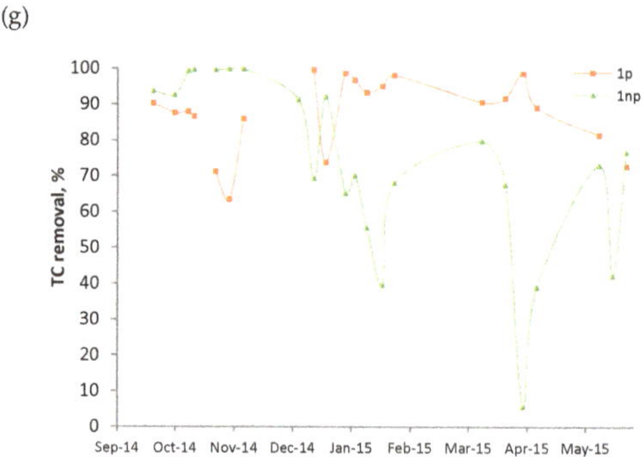

Figure 3. Evolution of the removal of (**a**) COD, (**b**) BOD, (**c**) turbidity, (**d**) TSS, (**e**) ammonia-N, (**f**) *E. coli*, and (**g**) TC for 1p (-■-) and 1np (-▲-).

The better performance of 1np before clogging can be explained by the stronger media compaction in the absence of plants and the consequent increase in the hydraulic retention time (HRT). In 1p, plant roots would be able to keep the media unclogged and, as a consequence, reduce the HRT and removal efficiency. In fact, it was observed that the effluent of 1np exited more slowly than that of 1p. Torrens et al. [37] observed a similar behavior in planted VF CWs, because the effluent from the first batches flowed through the beds more quickly than through unplanted ones. The authors claimed that the plant rhizomes could create preferential pathways. The preventing effect of plant roots regarding clogging has been discussed for a long time. Some studies have suggested that plants can make the substrate more porous [38], while others, such as Teixeira et al., [39] found a reduction in the porosity of the medium of HF CWs caused by the development of roots of Vetiver and Tifton 85 grasses. Yet, they considered that the low root volume, only 3.07% and 4.11% of the total pore space, could not have a great influence on clogging. However, Hua et al. [40] claimed that the role of plants varied throughout the clogging process. In the early stages, plant roots restricted water flow, while in the later stages, growing roots opened new pore spaces in the substrate. Our results are in agreement with those of Hua et al. [40] for the later stage but are the opposite for the early stages. This illustrates the complexity of the clogging process and the role of plant rhizomes.

Table 4. Average removals (%) achieved by 1p and 1np, before and after the light clogging of 1np.

	Wetland	Before	After
COD	1p	55	63
	1np	67	40
BOD	1p	82	85
	1np	83	55
Turbidity	1p	90	94
	1np	95	65
TSS	1p	75	87
	1np	84	71
Ammonia-N	1p	−7	−50
	1np	17	−115
E. coli	1p	59	95
	1np	97	60
Total coliforms	1p	82	94
	1np	98	69

The fact that 1np became clogged in such a short time, after about two years in operation, was unexpected. Additionally, in the case of 1p, a negative trend in performance along time can be observed, particularly for BOD (Figure 3). To understand these results, the following must be considered:

(i) The mulch employed was not fully stable. Although the analyses were made 1 year after the construction of the beds, the results indicate that the substrate was not fully stabilized. The palm branch is basically composed of a central rachis on which leaflets and spines are inserted. Once the branch is dead, the leaflets degrade relatively fast and release color, organic matter, and ions. Hence, their presence in palm mulch for CW substrate should be minimized.

(ii) The average surface loadings were relatively high (BOD: 59–62 g/m^2d, TSS: 18–22 g/m^2d, Table 2).

(iii) The reactors were in continuous operation without any rest period. In a survey of 169 full-scale VF CWs treating raw domestic wastewater, Paing et al. [5] claimed that the nominal BOD load calculated on the first stage was generally 40–50 g/m^2d, which is a similar value to that applied in this study. However, the success of the French system is partially justified by the good aeration of the substrate. This was achieved by different measures that include the application of rest periods, allowing the sludge accumulated on the surface to become mineralized.

The reduction in the performance of 1np caused by clogging underlines the importance of the substrate aeration. Regarding pathogen removal, Headley et al. [41] observed that aeration improved *E. coli* removal in horizontal flow wetlands. The better performance of 1p could be caused by the accumulation of sludge on the bed surface and the consequent HRT increase. At the same time, the presence of the roots would keep the substrate unclogged.

These results show that the effect of plants in mulch-based VF CWs was not univocal. In the early stages of the CWs, the presence of the plants reduced the efficiency of the removal of pollutants by augmenting the substrate porosity with their roots. However, in the long term, they helped to retard clogging and, as a consequence, to achieve better removals.

3.4. Analysis of T-RFLP

Terminal restriction fragment length polymorphism (T-RFLP) is a rapid, highly reproducible, and robust molecular tool for the study of bacterial community structure [27,42]. It is known that bacteria communities are affected by several factors, such as physicochemical properties, climate conditions, and the presence of plants [43,44]. Hence, bacterial communities of 1p and 1np were analyzed via T-RFLP in order to ascertain the effects of the presence of plants.

Bacterial diversity, as estimated by the number of T-RFs, showed reproducible patterns across the MseI enzyme. In addition, the size distribution of the T-RFs generated by the restriction enzyme was consistent with that of the T-RFs derived from in silico digestion.

Interestingly, shifts in the sizes of the T-RFs occurred as a result of two different factors: the presence of plants and the progressive clogging of the substrate, as can be inferred from the analysis of the operational taxonomic units (OTUs). The two VF CWs shared 31 T-RFs for *E. coli* and 19 for *Enterococcus* sp. (Figure 4). The shared T-RFs indicate the presence of a common bacteria population regardless of the presence of plants (Figure 4). Furthermore, the size of the T-RFs ranged from 117 to 810 bp for *E. coli* and from 115 to 714 bp for *Enterococcus* sp.

In addition, 1np showed a significant increase ($p < 0.1$) in T-RFs compared with that of 1p for both *E. coli* and *Enterococcus* sp. (Figure 5).

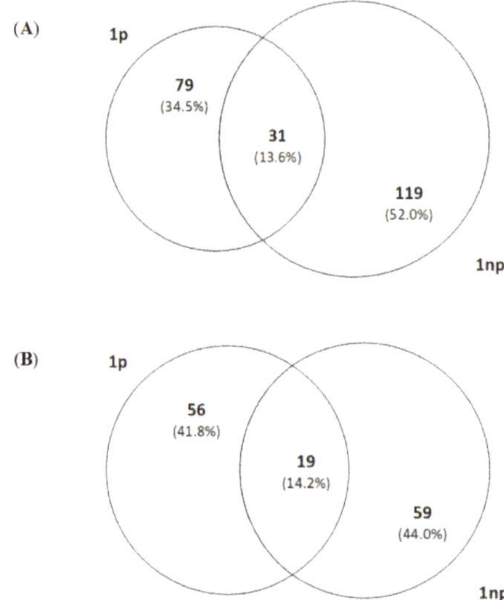

Figure 4. (**a**) Venn diagram of the size of the *Escherichia coli* terminal restriction fragments (T-RFs) shared between 1p and 1np. The number of T-RFs of 1p = 79; the number of T-RFs of 1np = 119; common number of T-RFs shared between both VF CWs = 31; (**b**) Venn diagram of the size of the *Enterococcus* sp. T-RFs shared between 1p and 1np. The number of T-RFs of 1p = 56; the number of T-RFs of 1np = 59; the number of T-RFs shared between both VF CWs = 19.

In addition, the results from the light and severe clogging periods showed that 1p reported a significant diminution ($p < 0.1$) in the T-RFs of *E. coli* over time (22.22% in light clogging versus 17.68% severe clogging). This behavior was also shown for 1np (38.89% light clogging versus 21.21% severe clogging; Figure 5a). This difference was more significant ($p < 0.01$) when relative abundance of T-RFs in 1np (38.89%) were compared to that in 1p (22.22%) at light clogging (Figure 5a). This result may infer that 1np was slightly clogged in December 2014 and continued to clog over time. In this sense, the chemical parameters sustain the T-RFs results (Table 4).

Additionally, the same trend in relative abundance of T-RFs was reported for *Enterococcus* sp., although the lower biodiversity of 1np under severe clogging was unexpected (Figure 5b). Hua et al. [26] claimed that clogging reduced bacterial diversity. Hence, an approach to explain low diversity in 1np was attempted by comparing shared T-RFs in 1p and 1np over time (Figure 6). Venn diagrams showed that 1np shared a major number of T-RFs between two sampled periods (1.8% light clogging versus 7.8% severe clogging). The presence of low shared-T-RFs, namely OTUs, in 1p would highlight the role of plant roots in combination with palm mulch substrate at retaining FIB bacteria. Kadam et al. [27] argued that the diminution of T-RFs in the planted VF CWs was committed to the diminution of genetic features, which could be correlated with the presence of root-associated microbiota. In addition, we assumed that root presence and microfauna might affect 1p performance.

The results obtained in this work indicate that plants increase the substrate porosity, but the results can be quite different depending on the level of substrate clogging. During the early stage of the experimental period, when the substrate was far from being clogged, the effect was negative as the CW with plants demonstrated worse pollutant removal. In this case, an increase in porosity led to shorter HRT. The sludge accumulated on the substrate would help to increase the HRT and thus the removals. However, later, when the sludge accumulation becomes excessive and clogging starts to develop, the consequences were very positive, because the plants were able to notably delay clogging by increasing the substrate porosity.

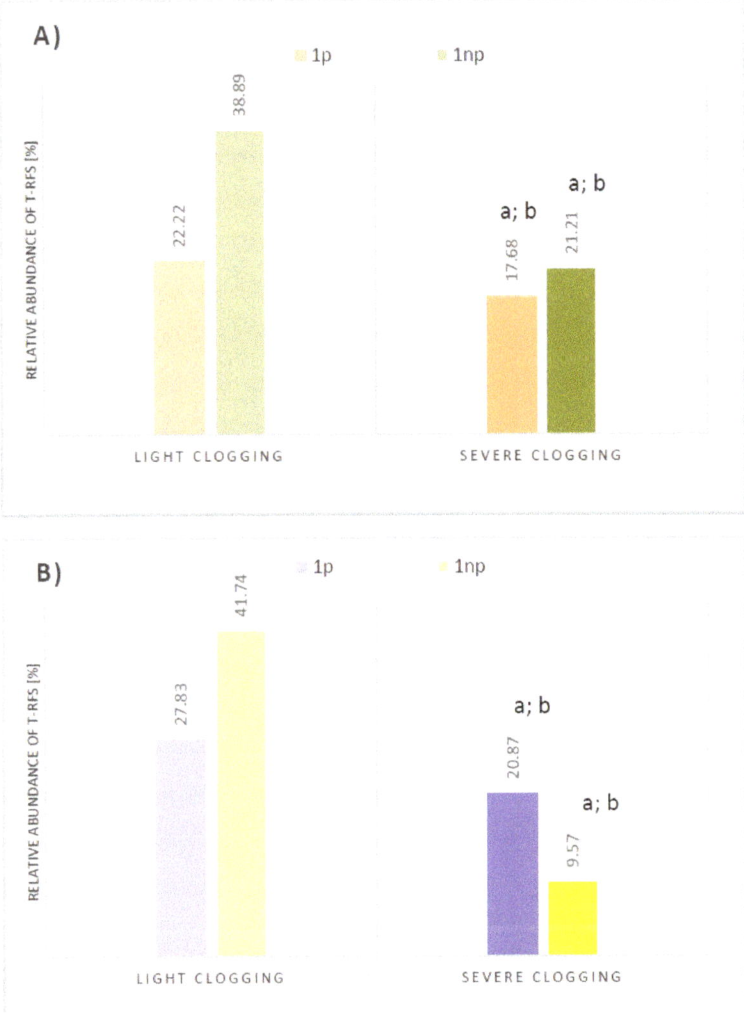

Figure 5. Relative abundance (%) of T-RFs at light and severe clogging for 1p and 1np. (**a**) Corresponds to the relative abundance of *Escherichia coli*. The 100% was assumed as the $\sum$ T-RFs (198 T-RFs) obtained for both VF CWs (1p and 1np) and both stages (light and severe clogging); (**b**) Corresponds to the relative abundance *Enterococcus* sp. The 100% was assumed as the $\sum$ T-RFs (114 T-RFs) obtained for both VF CWs (1p and 1np), and both stages (light and severe clogging); [a] indicates differences over time; [b] indicates differences between 1p and 1np.

The strong reduction of the palm mulch during the first year of operation indicates the importance of the substrate selection. The application of rest periods is highly recommended to delay clogging and improve the pollutant removals and the accumulated sludge mineralization. Similarly, the constructed wetland performance was optimized by the presence of plants as the reduction in the number of OTUs demonstrated.

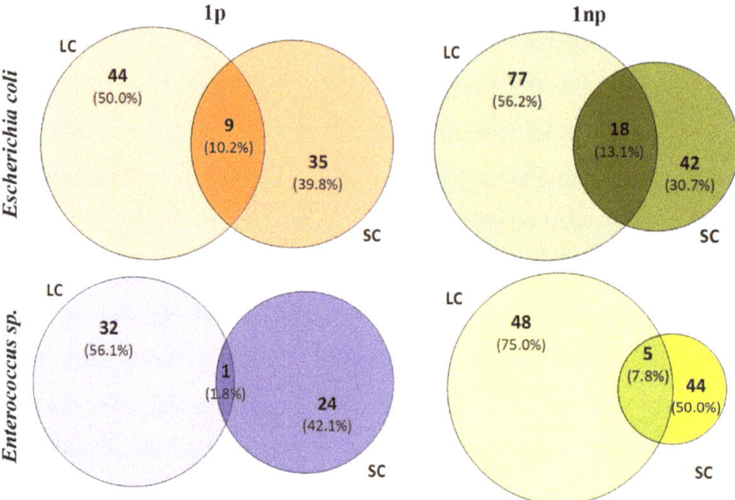

Figure 6. Venn diagrams showing the numbers of *Escherichia coli* T-RFs and *Enterococcus* sp. T-RFs in light and severe clogging and the number of T-RFs shared between both stages in each case. LC = light clogging; and SC = severe clogging.

Author Contributions: J.A.H.-M. designed the research and wrote part of the manuscript, N.P.-C. built the reactors and performed chemical and bacterial analyses, A.R.-R. performed chemical analyses, and M.C.-A. and P.G.-J. designed and analyzed the T-RFLP assays and wrote part of the manuscript. All the authors read and approved the final manuscript.

Funding: This research was funded by a grant from la Consejería de Economía, Industria, Comercio y Conocimiento del Gobierno de Canarias to P.G-J. (CGL2016-78442-C2-2-R; GOB-ESP2017-04 ULPGC). M.C.A. was supported by a predoctoral fellowship granted by la Universidad de Las Palmas de Gran Canaria (ULPGC-2016).

Conflicts of Interest: The authors declare no conflicts of interest.

References

1. Carvalho, P.N.; Arias, C.A.; Brix, H. Constructed wetlands for water treatment: New developments. *Water* **2017**, *9*, 397. [CrossRef]
2. Samsó, R.; García, J.; Molle, P.; Forquet, N. Modelling bioclogging in variably saturated porous media and the interactions between surface/subsurface flows: Application to Constructed Wetlands. *J. Environ. Manag.* **2016**, *165*, 271–279. [CrossRef]
3. Saeed, T.; Sun, G. Pollutant removals employing unsaturated and partially saturated vertical flow wetlands: A comparative study. *Chem. Eng. J.* **2017**, *325*, 332–341. [CrossRef]
4. Knowles, P.; Dotro, G.; Nivala, J.; García, J. Clogging in subsurface-flow treatment wetlands: Occurrence and contributing factors. *Ecol. Eng.* **2011**, *37*, 99–112. [CrossRef]
5. Paing, J.; Guilbert, A.; Gagnon, V.; Chazarenc, F. Effect of climate, wastewater composition, loading rates, system age and design on performances of French vertical flow constructed wetlands: A survey based on 169 full scale systems. *Ecol. Eng.* **2015**, *80*, 46–52. [CrossRef]
6. ElZein, Z.; Abdou, A.; Abd ElGawad, I. Constructed wetland as a sustainable wastewater treatment method in communities. *Procedia Environ. Sci.* **2016**, *34*, 605–617. [CrossRef]
7. Casas Ledón, Y.; Rivas, A.; López, D.; Vidal, G. Life-cycle greenhouse gas emissions assessment and extended exergy accounting of a horizontal-flow constructed wetland for municipal wastewater treatment: A case study in Chile. *Ecol. Indicators* **2017**, *74*, 130–139. [CrossRef]
8. Fuchs, V.J.; Mihelcic, J.R.; Gierke, J.S. Life cycle assessment of vertical and horizontal flow constructed wetlands for wastewater treatment considering nitrogen and carbon greenhouse gas emissions. *Water Res.* **2011**, *45*, 2073–2081. [CrossRef]

9. Farahani, H.; Bayazidi, S. Modeling the assessment of socio-economical and environmental impacts of sand mining on local communities: A case study of Villages Tatao River Bank in North-western part of Iran. *Resour. Policy* **2018**, *55*, 87–95. [CrossRef]
10. Ioannidou, D.; Meylan, G.; Sonnemann, G.; Haberte, G. Is gravel becoming scarce? Evaluating the local criticality of construction aggregates. *Resour. Conserv. Recycle* **2017**, *126*, 25–33. [CrossRef]
11. Campos-Serrano, A.; Rodríguez-Esteban, J.A. Imagined territories and histories in conflict during the struggles for Western Sahara, 1956–1979. *J. Hist. Geogr.* **2017**, *55*, 44–59. [CrossRef]
12. Yang, Y.; Zhao, Y.; Liu, R.; Morgan, D. Global development of various emerged substrates utilized in constructed wetlands. *Bioresour. Technol.* **2018**, *261*, 441–452. [CrossRef]
13. Zhao, Y.Q.; Zhao, X.H.; Babatunde, A.O. Use of dewatered alum sludge as main substrate in treatment reed bed receiving agricultural wastewater: Long-term trial. *Bioresour. Technol.* **2009**, *100*, 644–648. [CrossRef]
14. Zanella, L.; Nour, E.; Roston, D. Use of bamboo rings as substrate in subsuperficial constructed wetland system in Brazil. In Proceedings of the 12th International Conference on Wetland Systems for Water Pollution Control, Venice, Italy, 3–6 October 2010.
15. Herrera-Melián, J.A.; González-Bordón, A.; Martín-González, M.A.; García-Jiménez, P.; Carrasco, M.; Araña, J. Palm tree mulch as substrate for primary treatment wetlands processing high strength urban wastewater. *J. Environ. Manag.* **2014**, *139*, 22–31. [CrossRef]
16. Herrera-Melián, J.A.; Borreguero-Fabelo, A.; Araña, J.; Peñate-Castellano, N.; Ortega-Méndez, J.A. Effect of Substrate, Feeding Mode and Number of Stages on the Performance of Hybrid Constructed Wetland Systems. *Water* **2018**, *10*, 39. [CrossRef]
17. Cao, W.; Wang, Y.; Sun, L.; Jiang, J.; Zhang, Y. Removal of nitrogenous compounds from polluted river water by floating constructed wetlands using rice straw and ceramsite as substrates under low temperature conditions. *Ecol. Eng.* **2016**, *88*, 77–81. [CrossRef]
18. Thomas, R.; Gough, R.; Freeman, C. Linear alkylbenzene sulfonate (LAS) removal in constructed wetlands: The role of plants in the treatment of a typical pharmaceutical and personal care product. *Ecol. Eng.* **2017**, *106*, 415–422. [CrossRef]
19. Arunbabu, V.; Sruthy, S.; Antony, I.; Ramasamy, E.V. Sustainable greywater management with *Axonopus compressus* (broadleaf carpet grass) planted in sub surface flow constructed wetlands. *J. Water Process Eng.* **2015**, *7*, 153–160. [CrossRef]
20. Masi, F.; Rizzo, A.; Regelsberger, M. The role of constructed wetlands in a new circular economy, resource oriented, and ecosystem services paradigm. *J. Environ. Manag.* **2018**, *216*, 275–284. [CrossRef]
21. Calhoun, A.; King, G.M. Regulation of root-associated methanotrophy by oxygen availability in the rhizosphere of two aquatic macrophytes. *Appl. Environ. Microbiol.* **1997**, *63*, 3051–3058.
22. Schriewer, A.; Miller, W.A.; Byrne, B.A.; Miller, M.A.; Oates, S.; Conrad, P.A.; Hardin, D.; Yang, H.; Chouicha, N.; Melli, A.; et al. Presence of Bacteroidales as a Predictor of Pathogens in Surface Waters of the Central California Coast? *Appl. Environ. Microbiol.* **2010**, *76*, 5802–5814. [CrossRef]
23. Wang, D.; Farnleitner, A.H.; Field, K.G.; Green, H.C.; Shanksd, O.C.; Boehma, A.B. *Enterococcus* and *Escherichia coli* fecal source apportionment with microbial source tracking genetic markers—Is it feasible? *Water Res.* **2013**, *47*, 6849–6861. [CrossRef]
24. Li, Y.; Wang, C.; Zhang, W.; Wang, P.; Niu, L.; Hou, J.; Wang, J.; Wang, L. Modeling the Effects of Hydrodynamic Regimes on Microbial Communities within Fluvial Biofilms: Combining Deterministic and Stochastic Processes. *Environ. Sci. Technol.* **2015**, *49*, 12869–12878. [CrossRef]
25. Brower, S.; Leff, L.; Mou, X. Effects of Short and Long-Term Hydrological Regimes on the Composition and Function of Denitrifying Bacteria in Experimental Wetlands. *Wetlands* **2017**, *37*, 573–583. [CrossRef]
26. Hua, G.; Chenga, Y.; Kongb, J.; Li, M.; Zhao, Z. High-throughput sequencing analysis of bacterial community spatiotemporal distribution in response to clogging in vertical flow constructed wetlands. *Bioresour. Technol.* **2018**, *248*, 104–112. [CrossRef]
27. Kadam, S.K.; Chandanshivea, V.V.; Raneb, N.R.; Patilb, S.M.; Gholaved, A.R.; Khandareb, R.V.; Bhosalee, A.R.; Jeonf, B.H.; Govindwar, S.P. Phytobeds with Fimbristylis dichotoma and Ammannia baccifera for treatment of real textile effluent: An in situ treatment, anatomical studies and toxicity evaluation. *Environ. Res.* **2018**, *160*, 1–11. [CrossRef]
28. Murray, M.G.; Thompson, W.F. Rapid isolation of high molecular weight plant DNA. *Nucleic Acids Res.* **1980**, *8*, 4321–4325. [CrossRef]

29. Deasy, B.M.; Rea, M.C.; Fitzgerald, G.F.; Cogan, T.M.; Beresford, T.P. A Rapid PCR Based Method to Distinguish between *Lactococcus* and *Enterococcus*. *System. Appl. Microbiol.* **2000**, *23*, 510–522. [CrossRef]
30. Bej, A.K.; Steffan, R.J.; DiCesare, J.; Haff, L.; Atlas, R.M. Detection of Coliform Bacteria in Water by Polymoerase Chain Reaction and Gene Probes. *Appl. Environ. Microbiol.* **1990**, *56*, 307–314.
31. Smith, C.J.; Danilowicz, B.S.; Clear, A.K.; Costello, F.J.; Wilson, B.; Meijer, W.G. T-Align, a web-based tool for comparison of multiple terminal restriction fragment length polymorphism profile. *FEMS Microbiol. Ecol.* **2005**, *54*, 375–380. [CrossRef]
32. The R Project for Statistical Computing. Available online: https://www.r-project.org/ (accessed on 26 February 2018).
33. Integrative Omics, Pacific Northwest National Laboratory. Venn Diagram Plotter. Available online: https://omics.pnl.gov/software/venn-diagram-plotter (accessed on 8 October 2018).
34. Tchobanoglous, G.; Burton, F.L.; Stensel, H.D. *Wastewater Engineering: Treatment and Reuse*, 4th ed.; McGraw Hill Book Co.: New York, NY, USA, 2003.
35. *Royal Decree 1620, 2007, That Stablishes the Legislative Frame for the Reuse of Recovered Wastewater*; Official Bulletin of the State: Madrid, Spain, 7 December 2007.
36. Saeed, T.; Sun, G. A comparative study on the removal of nutrients and organic matter in wetland reactors employing organic media. *Chem. Eng. J.* **2011**, *171*, 439–447. [CrossRef]
37. Torrens, A.; Molle, P.; Boutin, C.; Salgot, M. Impact of design and operation variables on the performance of vertical-flow constructed wetlands and intermittent sand filters treating pond effluent. *Water Res.* **2009**, *43*, 1851–1858. [CrossRef]
38. Brix, H. Do macrophytes play a role in constructed treatment wetlands? *Water Sci. Technol.* **1997**, *35*, 11–17. [CrossRef]
39. Teixeira, D.L.; Teixeira de Matos, A.; Pimentel de Matos, M.; Pires Vieira, D.; Dias Araújo, E.; Aparecida Ferraz, L. The influence of plant roots on the clogging process and the extractive capacity of nutrients/pollutants in horizontal subsurface flow constructed wetlands. *Ecol. Eng.* **2018**, *120*, 54–60. [CrossRef]
40. Hua, G.F.; Zhao, Z.W.; Kong, J.; Guo, R.; Zeng, Y.T.; Zhao, L.F.; Zhu, Q.D. Effects of plant roots on the hydraulic performance during the clogging process in mesocosm vertical flow constructed wetlands. *Environ. Sci. Pollut. Res.* **2014**, *21*, 13017–13026. [CrossRef]
41. Headley, T.; Nivala, J.; Kassaa, K.; Olsson, L.; Wallace, S.; Brix, H.; Afferden, M.; Müller, R. *Escherichia coli* removal and internal dynamics in subsurface flow ecotechnologies: Effects of design and plants. *Ecol. Eng.* **2013**, *61*, 564–574. [CrossRef]
42. Thies, J.E. Soil microbial community analysis using terminal restriction fragment length polymorphisms. *Soil Sci. Soc. Am. J.* **2007**, *71*, 579–591. [CrossRef]
43. Lazzaro, A.; Widmer, F.; Sperisen, C.; Frey, B. Identification of dominant bacterial phylotypes in a cadmium-treated forest soil. *FEMS Microbial. Ecol.* **2008**, *63*, 143–155. [CrossRef]
44. Bru, D.; Ramette, A.; Saby, N.P.; Dequiedt, S.; Ranjard, L.; Jolivet, C.; Arrouays, D.; Philippot, L. Determinants of the distribution of nitrogen-cycling microbial communities at the landscape scale. *ISME J.* **2011**, *5*, 532–542. [CrossRef]

Article

Suitability of Totora (*Schoenoplectus californicus* (C.A. Mey.) Soják) for Its Use in Constructed Wetlands in Areas Polluted with Heavy Metals

Juan A. Blanco

Departamento de Ciencias, Universidad Pública de Navarra, 31006 Pamplona, Spain; juan.blanco@unavarra.es; Tel.: +34-948-169859

Received: 29 October 2018; Accepted: 15 December 2018; Published: 20 December 2018

Abstract: *Schoenoplectus californicus* subsp. *tatora* (totora) is an endemic plant from wetlands in South America's Altiplano region. In the endorheic Titicaca-Desaguadero-Poopó-Salar de Coipasa system (TDPS), totora can be found along rivers, lakes, and shallow ponds. Lake Uru-Uru is a minor lake placed upstream of Lake Poopó, and it gets water inflows from the Desaguadero River, the city of Oruro and several mining and metallurgic complexes. Polluted waters from these origins, together with natural high salinity and high presence of As and Pb, make Lake Uru-Uru an ideal location to search for plant species suitable to be used in constructed and restored wetlands under pollution stress, particularly in systems with high pH and salty waters. To test if totora could meet such requirements, healthy plants were collected at two sites in Lake Uru-Uru with different exposure to polluted inflows. Chemical composition of different organs (leaves, rhizomes and roots) were compared. Results indicated totora's capacity to withstand high concentrations of a cocktail of multiple pollutants and heavy metals. Particularly, this research showed totora as a multi-hyperaccumulator (concentrations in shoots higher than 1000 mg kg^{-1}) for As, Fe and Ni. These results, combined with totora's intrinsic high rates of biomass production, slow decomposition rates and its value as raw material for local craftwork and industrial uses, support the recommendation to use totora in constructed or restored wetlands, particularly in sites polluted with heavy metals, and in waters with high salinity.

Keywords: Totora; hyperaccumulation; heavy metals; Altiplano; Lake Uru-Uru; acid mine drainage; urban wastewater

---

1. Introduction

Wetlands are one of the most productive ecosystems on Earth, usually containing complex trophic webs and high biodiversity. Wetlands also provide important ecological services such as water regulation and recycling, and particularly removal of excessive nutrients, organic matter, and pollutants. Wetlands are also an important source of income for local human populations, directly by providing water, but also indirectly by providing food (game, fisheries, grain, vegetables, etc.), fertility, energy, building materials, and so on. Unfortunately, wetlands are also an ecosystem type that faces some of the highest environmental pressures, ranging from draining to increase land for agriculture, industry, or urban development, to the reduction in water quality and quantity because of human management [1]. The situation is becoming worse as the effects of environmentally unfriendly activities, population increase, and urbanization are combined with a tradition of unsustainable management practices in wetlands [2,3].

In both developed and developing countries, the use of wetlands as disposal areas for untreated wastewater is particularly troublesome. Although wetlands have, in fact, been used as green technology to treat various wastewaters for several decades [2,3], such use must be correctly planned to not surpass the ecosystem's capacity to absorb and treat pollutants. Plants used in wetlands constructed for wastewater treatment should, therefore, have the capacity to grow in polluted sites, and even to be harvested as a way to remove pollutants stored in plant biomass. Plants ideal for such purposes should possess an intrinsic capacity to tolerate heavy metals and metalloids, concentrate them in harvestable aerial tissues, have high rates of biomass production, adsorb pollutants on the root surface, develop extensive and deep root systems, have wide distribution ranges allowing growth in many habitats, and should be easy to cultivate and harvest [4].

A particular case of polluted wetlands are those located in endorheic watersheds rich in minerals and with a high concentration of mining sites. Endorheic basins and their associated lakes can be very sensitive to variations in climate and environmentally negative human activities, such as overexploitation or pollution of water resources [5]. One of such cases is the Titicaca-Desaguadero-Poopó-Salar de Coipasa system (TDPS) in South America. The TDPS is situated in the Andes cordillera, at about 14–20° south latitude and at an average elevation of more than 3500 m above sea level (m.a.s.l.) It encloses an area of 143,900 km^2 in between Peru, Chile and Bolivia. The TDPS is a closed hydrographic basin in which the fresh water Lake Titicaca is connected to the mildly saline Lake Poopó through the Desaguadero River. The overflow of Lake Poopó reaches the hypersaline Coipasa Salt Marsh where evaporation prevents from permanent water levels.

The Poopó basin consists of two lakes: Lake Poopó and Lake Uru-Uru. Lake Uru-Uru is located upstream of Lake Poopó and receives water only from the Desaguadero River, while Lake Poopó receives water from Lake Uru-Uru, as well as from many smaller inflowing rivers. A great number of these streams are rich in dissolved solids due to mining activities as well as natural rock weathering processes [6]. Lake Uru-Uru is, therefore, an antechamber of Lake Poopó. Lake Uru-Uru is also a water body for which scarce information is available, as it has not been included in most of the research done to date in Lake Poopó.

South America's Altiplano is naturally rich in polymetallic deposits [7], being this the reason for the important mining and metallurgical activities established in the region, which play an important role in the local and national economies [8]. The TDPS region hosts important deposits of precious metals (gold, silver) and heavy metals (tin, cobalt, copper, etc.), which have been exploited since before the arrival of European settlers in the 16th century. The exploitation of such resources has traditionally been done at the expenses of water systems, as important water volumes are needed in the mines and metal processing plants. Most of these water flows continue to be returned to the watershed without proper treatment, leading to important issues of metal pollution both in the water bodies and in the surrounding areas at local and regional levels [9].

The shallow lakes and seasonal lagoons typical of the TDPS region are also the habitat of one of the most distinct plants of the Altiplano: the totora (*Schoenoplectus californicus* subsp. *tatora* (C.A. Mey.) Soják). *Schoenoplectus californicus* can be found in coastal and riparian regions from Southern North America [10] to Chile and Argentina [11]. A perennial herbaceous plant, it commonly reaches a height of 4 m, reaching 6 m occasionally. Totora is a subspecies of the giant bulrush sedge, found in South America in the Lake Titicaca basin, the middle coast of Peru, and on Easter Island in the Pacific Ocean. Totora grows in river and lake shores, at water depths up to 2.5–3 m, but occasionally also at deeper or shallower depths, particularly in areas prone to seasonal flooding. By growing in this extreme habitat, totora shows a great capacity to stand high UV levels, water and temperature fluctuations in an arid tropical tundra climate, with high levels of naturally occurring salt and arsenic [12]. In addition, totora has important economic relevance locally, as a resource for a wide range of applications, from being used as raw material in traditional boat building and basketry to be used in modern architecture and industry [13].

All these features potentially make totora an ideal species to be used either in wetland restoration or in constructed wetlands. Indeed, *S. californicus* has been reported to be used up to 28 times in constructed wetlands in North America [14]. However, such use has not been reported in the context of this species capacity to store different heavy metals. In fact, there is only one available report estimating totora's removal rates of Zn and Cu in constructed wetlands [15,16]. On the other hand, although some preliminary studies have been done to analyze pollutant levels in totora plants growing in natural wetlands [6,17], a systematic effort to define the concentration ranges of different elements in which this species can grow has not yet been conducted.

Based on my personal observations of the conditions in which totora plants can be found along the TDPS, I hypothesize that totora can stand high levels of pollutants before its growth and survival become impaired. Therefore, the particular objectives of this research were: 1) to provide the most detailed chemical composition of totora organs to date (leaves, rhizomes, and roots) that could be used as a benchmark for future work; and 2) to test if totora is suitable for its use in construction or restoration of polluted wetlands, by testing if it can withstand significant differences in chemical composition in polluted sites compared to non-polluted sites.

2. Materials and Methods

2.1. Research Sites

Lake Uru-Uru is situated in the lower reaches of the Desaguadero River, before it reaches the Lake Poopó. Lake Uru-Uru is relatively young, as it appeared in 1962 after the Desaguadero River was diverted due to works in a local mining complex. The lake has on average a length of 21 km and a width of 16 km, with 214 km^2 of surface. Its average altitude is 3686 m.a.s.l., the lake bottom is very flat (less than 1 m of topographical variation in 15 km), making the lake very shallow with only 1–2 m as average water level. This makes the lake very sensitive to seasonal and yearly water level fluctuations. The right arm of the Desaguadero River enters Lake Uru-Uru in its NW shore at Puente Español, close to the city of Oruro, the provincial capital. From Oruro, untreated polluted waters from urban and mining origin enter the lake. The continuation of the Desaguadero River drains Lake Uru-Uru through its southern tip, near the town of Machacamarca, and the overflow continues south through a network of seasonal channels and shallow creeks before reaching Lake Poopó (Figure 1). Climate is tropical xeric and cold. Mean annual precipitation is 332 mm, with precipitation from November to March accounting for 67–85% of total annual precipitation. The rest of the year is considered the dry season. Mean annual temperature is about 10.6 °C, with maximum temperatures of 23 °C during the austral summer (November and December) and the lowest (−12 °C) in June–July.

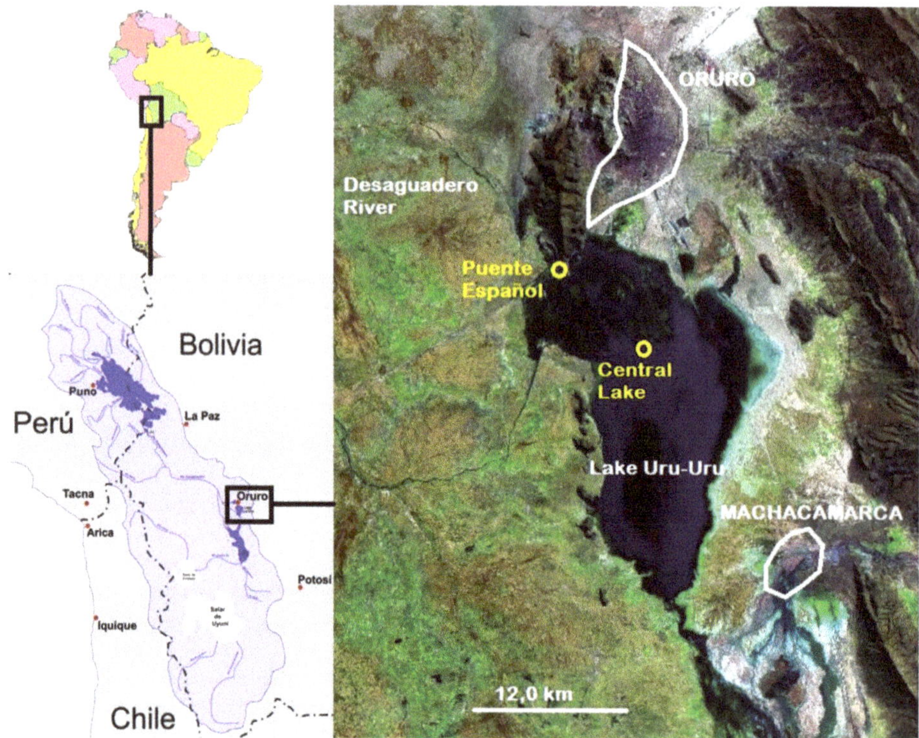

Figure 1. Situation of the endorheic Titicaca-Desaguadero-Poopó-Salar de Coipasa (TDPS) system in South America (**left**) and the sampling sites in Lake Uru-Uru (yellow circles, **right**), indicating the population centers of Oruro y Machacamarca (white lines).

Totora samples were collected from two different sites in Lake Uru-Uru. The first site was located beside the Puente Español. This site was considered as the less polluted point in Lake Uru-Uru, as it receives the water from the Desaguadero River and it is above the discharge points of the city of Oruro and the local mining facilities. The second site was located 11.5 km SE from the first, towards the central portion of the lake. The small creeks bringing effluents from the city of Oruro and acid mine drainage from the surrounding mining facilities had already reached the lake at this point, and, therefore, this was considered as a polluted site. Clear differences in water quality and sediment composition among sites have been reported before (Table 1). At each site, a 100-m long transect was defined parallel to the water shoreline. Twenty plants along each transect were randomly chosen among those that did not display signs of illness, decay or herbivore consumption. Plants were collected by digging them out with a hand shovel to extract the rhizome and most of the main roots.

Table 1. Chemical characteristics of surface waters and sediments at the two sampling sites. Values averaged from the literature [6,17–22]. A slash indicates that the element is not measured.

Variable	Water (elements in mg L^{-1})		Surface sediments (elements in ppm)	
	Puente Español	Central Lake	Puente Español	Central Lake
pH	8.47	8.96	-	-
Conductivity (µs cm^{-1})	23.18	145.00	-	-
Total suspended solids	1635.50	13530.00	-	-
Total alkali	133.85	146.50	-	-
CO_3^{2-}/CO_2	26.90/0.71	64.97/0.69	-	-
HCO_3^-	116.72	129.98	-	-
Al	0.03	0.01		63727.47
As	0.11	0.11	435.03	292.96
B	3.08	9.82	-	115.86
Be	-	0.00	-	2.06
Ca	113.20	502.34	-	44191.53
Cl	679.44	8747.17	-	-
Cd	0.04	0.04	0.78	1.86
Co	0.05	0.17	6.10	8.94
Cr	0.00	0.01	8.15	22.34
Cu	0.01	0.00	50.50	43.88
Fe	0.02	0.02	15179.40	27056.84
K	20.00	101.44	-	21266.63
Li	-	1.55	-	-
Mg	38.16	302.86	-	9815.80
Mn	0.04	0.15	422.00	546.22
Mo	-	0.01	-	2.04
Na	408.30	2991.42	-	12721.53
Ni	0.05	0.27	10.30	17.23
NO_3^-/NO_2^-	0.32/0.03	0.22/0.03	-	-
P	0.06	0.04	-	796.83
Pb	0.02	0.04	56.20	58.43
S	71.74	896.70	-	-
Sb	0.05	0.08	2.00	7.81
Se	-	0.01	-	-
Sr	-	3.32	-	0.33
Ti	-	0.02	-	2367.43
Tl	-	0.00	-	0.82
V	-	0.00	-	80.35
Zn	0.01	0.04	82.40	166.82

2.2. *Chemical Analyses*

Plants were lightly washed with distilled water to remove dirt and dust, and then oven dried at 70 °C for a week. Then, each plant was divided into different organs (leaves, roots, and rhizomes). Samples were then grounded through a 1-mm sieve (MF-10, IKA). Total N concentration was determined by the dry combustion Dumas method [23] using a TRUSPEC CN628 elemental analyzer (LECO Corporation, MI, USA). Concentration of other elements were measured using inductively-coupled plasma emission spectrometry (ICP–ICAP 6500 DUO Thermo, England), after acid digestion (HNO3–H2O2 4:1) in a microwave.

2.3. *Statistical Analysis*

Transfer factors were calculated as the ratio between the element concentration in each plant organ and the soil. Values above one indicate that plants are enriched in elements compared to their concentrations in soil, ratios around one indicate that plants are not influenced by elements, and ratios below one show that plants exclude the elements from uptake [24]. One-way analyses of equality of means were used to test for significant differences in element composition and transfer factors between plants growing in Puente Español and Central Lake. As the assumption of homogeneity of variance among groups was usually violated, the Welch's test was used [25], which is a superior option to other non-parametric tests [26]. To detect relationships among different elements, cross-correlations were carried out. Statistical significance was set at $p < 0.05$. All analysis were done with JMP 12.0 (SAS, North Carolina, USA).

3. Results

Transfer factors were significantly higher at Central Lake for As, Pb, Fe, Cu, Zn, and Cd. However, transfer factors of Mn, Ni (leaves and roots), and Cr (leaves) were significantly higher in plants collected at Puente Español. No significant differences among sites in transfer factors were found for Sb (rhizomes), Cr (roots and rhizomes), Co (roots), and Ni (rhizomes) (Table 2).

Table 2. Transfer factors (mean ± standard error) for different elements in plants collected at Puente Español and Central Lake. Asterisks indicate significantly higher value in the site for concentrations in the same organ, with Welch's tests ($p < 0.05$). A slash indicates element not measured in soil.

Element	Leaves		Roots		Rhizomes	
	Central Lake	Puente Español	Central Lake	Puente Español	Central Lake	Puente Español
Metaloids						
B	0.85 ± 0.03	-	0.82 ± 0.03	-	0.22 ± 0.01	-
As	0.09 ± 0.01 *	0.01 ± 0.01	2.95 ± 0.18 *	0.18 ± 0.01	0.11 ± 0.01 *	0.02 ± 0.01
Sb	1.14 ± 0.04	-	5.94 ± 0.56	-	1.22 ± 0.07	0.31 ± 0.01
Alkali						
Na	<0.01 ± 0.01	-	<0.01 ± 0.01	-	<0.01 ± 0.01	-
K	0.62 ± 0.02	-	0.40 ± 0.01	-	0.52 ± 0.03	-
Alkaline earth						
Be	-	-	0.54 ± 0.03	-	-	-
Mg	0.04 ± 0.01	-	0.07 ± 0.01	-	0.03 ± 0.01	-
Ca	0.04 ± 0.01	-	0.05 ± 0.01	-	0.02 ± 0.01	-
Sr	85.61 ± 2.20	-	153.43 ± 2.53	-	52.27 ± 1.65	-
Semimetals						
Al	0.01 ± 0.01	-	0.08 ± 0.01	-	0.01 ± 0.01	-
Tl	4.52 ± 0.25	-	37.04 ± 3.23	-	6.65 ± 0.41	-
Pb	0.63 ± 0.03 *	0.02 ± 0.01	6.80 ± 0.35 *	0.35 ± 0.02	0.83 ± 0.06 *	0.07 ± 0.01
Transition metals						
Ti	0.01 ± 0.01	-	0.01 ± 0.01	-	0.01 ± 0.01	-
V	0.01 ± 0.01	-	0.09 ± 0.01	-	0.01 ± 0.01	-
Cr	0.14 ± 0.01	0.77 ± 0.11 *	1.12 ± 0.06	1.48 ± 0.19	0.75 ± 0.22	1.15 ± 0.17
Mn	0.23 ± 0.01	3.39 ± 0.20 *	0.22 ± 0.01	1.68 ± 0.01 *	0.10 ± 0.01	0.89 ± 0.04 *
Fe	0.05 ± 0.01 *	0.02 ± 0.01	2.40 ± 0.15 *	0.47 ± 0.05	0.19 ± 0.02 *	0.12 ± 0.02
Co	0.06 ± 0.01	-	0.26 ± 0.03	0.19 ± 0.02	0.08 ± 0.01	-
Ni	0.18 ± 0.01	0.28 ± 0.01 *	0.45 ± 0.02	0.65 ± 0.07 *	0.71 ± 0.14	0.50 ± 0.09
Cu	0.28 ± 0.01 *	0.11 ± 0.02	3.44 ± 0.12 *	0.68 ± 0.03	0.66 ± 0.03 *	0.30 ± 0.02
Zn	10.88 ± 0.39 *	0.16 ± 0.01	52.33 ± 1.73 *	2.73 ± 0.10	17.56 ± 0.71 *	0.85 ± 0.06
Mo	0.21 ± 0.03	-	0.27 ± 0.03	-	1.02 ± 0.27	-
Cd	2.43 ± 0.10	-	73.53 ± 2.89 *	7.71 ± 0.78	13.60 ± 1.68 *	0.37 ± 0.02

Significant correlations were found among concentration of different elements, leading to the identification of six different element clusters (Figure 2). The essential macronutrients C, N, and P were grouped together as their concentrations were positively correlated. The smallest cluster was composed by K and Mn, whose concentrations were also positively correlated. The third cluster encompassed Mo, Ni, and Se. The fourth cluster was composed mostly by alkali earth elements and light transition metals (Ca, Mg, Ti, Al, V, and Sr). Their concentrations showed strong positive correlations among them, but also strong negative correlations with the main macronutrients (C, N, and P).

The biggest element cluster was composed mostly by heavier transition metals and semi-metals (As, Fe, Cu, Pb, Tl, Cd, Zn, Co, S, Cr, Bi, Sb), along with As (a metalloid) and S (a non-metal). Elements in this cluster also showed negative correlations with essential macronutrients (N, P, and K). The last cluster brought together two alkali (Li and Na), the lightest metalloid (Be), and the lightest alkali earth (Be) element. This last element also shared strong positive correlations with the transition metal/semi-metal cluster (Figure 2).

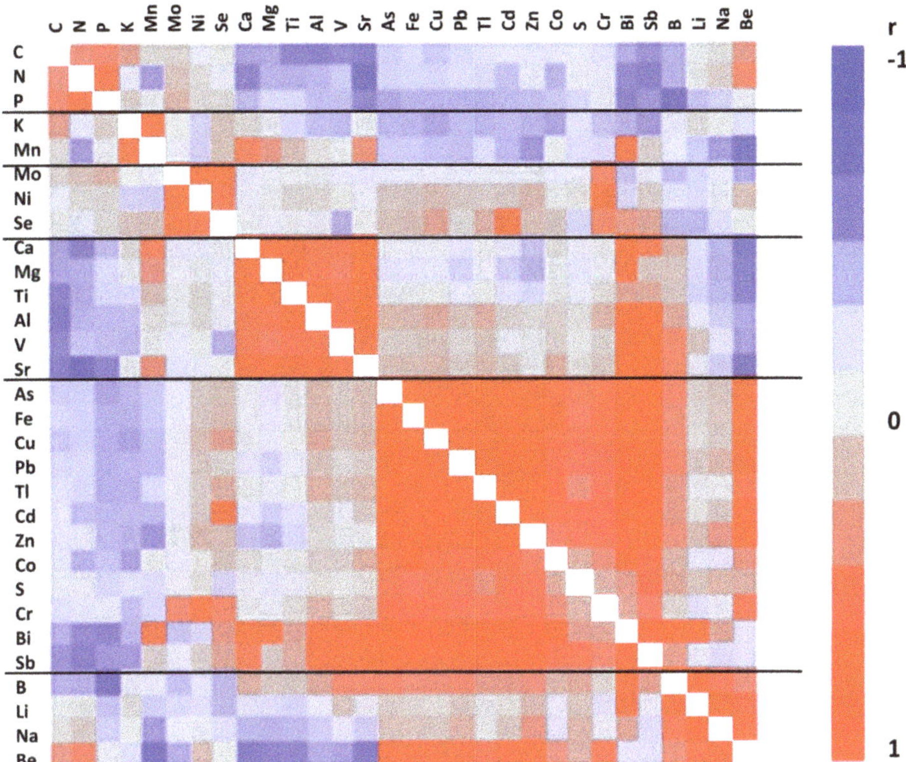

Figure 2. Heat map of Pearson's correlation coefficients among the different element concentrations analyzed.

Regarding differences among sites, for non-metal elements, no significant differences were found for the concentration of C in rhizomes and leaves, neither for P in roots and leaves. C concentration was significantly higher in roots from Central Lake plants, and P was significantly higher in rhizomes from Puente Español. For the other elements, N concentrations were higher in roots and leaves from Puente Español, but lower in rhizomes, whereas S showed much higher concentrations in all organs from Central Lake. Se was not detected in roots, but it reached significantly higher concentrations in leaves from Puente Español and in rhizomes from Central Lake (Figure A1).

For metalloids, Sb was only detected at Central Lake in all plant organs. Arsenic concentration was significantly higher at Central Lake than at Puente Español (from three times more in rhizomes to 14 times more in leaves). B also showed higher concentrations in leaves and roots from Central Lake, but no differences between sites were found for rhizomes (Figure A2).

The alkali metals Li and Na reached significantly higher concentrations in all plant organs from Central Lake. However, K concentration was significantly higher in leaves from Puente Español, and no differences were found for the other two organs.

For alkaline earth metals, Be was only detected in roots, with significantly higher concentrations at Central Lake. Mg, Ca, and Sr showed similar patterns, with significantly higher concentrations of the three elements in all organs collected at Puente Español (Figure A3).

For semi-metals, Al concentrations were higher in the organs sampled at Puente Español. However, Tl reached significantly higher concentrations in the three organs in plants from Central Lake. The same was true for Pb, which reached much higher concentrations in plants collected at Central Lake than at Puente Español (from 13 times higher in rhizomes to 30 times higher in leaves). Bi was not detected in plants from Puente Español, but it was detected in roots and rhizomes at Central Lake (Figure A4).

Transition metals showed a variety of patterns (Figures A5 and A6). Ti, V, and Mn showed significantly higher concentrations in plants from Puente Español, whereas Fe, Cu, and Zn reached much higher concentrations in plants from Central Lake. Cr and Mo showed significantly higher concentrations in leaves from Puente Español, but in roots, Cr concentration was higher in plants from Central Lake whereas Mo was higher at Puente Español. No significant differences between sites were detected for rhizomes. Co was detected in plants from Puente Español only in roots, where the concentrations were significantly lower than in Central Lake. Ni did not show significant differences among sites for leaves and roots, but its concentration in rhizomes was higher in plants from Central Lake. Cd was not detected in leaves from Puente Español, and its concentration in roots was significantly higher in Central Lake (up to 115 times).

Element distributions in totora organs followed identical patterns at both sites, with six different configurations that could be identified with different strategies to deal with toxic elements (Table 3). Most transition metals, semi-metals and metaloids followed a pattern of clear reduction in their concentrations (exclusion), as the organ was farther away from the pollution source (i.e., the highest concentrations in roots and the lowest in leaves). Mg concentrations also followed this pattern. The rest of semi-metals and alkali-earth metals, together with S, also reached their highest concentrations in roots, but concentrations in leaves were higher than in rhizomes. Two transition metals (Ni and Mo) showed higher concentrations in rhizomes than in roots, but as in the case of other elements of the same group, leaves showed the lowest concentrations. The three main macronutrients (non-metals) reached the highest concentrations in rhizomes and the lowest in roots. Three light elements (Li, Na, and B) plus a transition metal (Mn) exhibited the highest concentrations in leaves and the lowest in rhizomes. Finally, K and Se showed increasing concentrations as the distance from the pollutant source (i.e., sediments) increased (Table 3).

Table 3. Patterns in nutrient concentration along different totora tissues.

Concentration pattern	Strategy	Element group	Element
Root > Rhizome > Leaf	Excluding micronutrient	Transition metals Semi-metals Alkali-earth metals Metaloids	Cu, Zn, Cd, Ti, V, Cr, Fe, Co Al, Pb, Bi Mg As, Sb
Root > Leaf > Rhizome	Mobilizing non-labile98 macronutrient	Semi-metals Alkali-earth metals Non-metals	Tl Ca, Sr, Be [1] S
Rhizome > Root > Leaf	Storing micronutrient	Transition metals	Ni, Mo
Rhizome > Leaf > Root	Storing macronutrient	Non-metals	C, N, P
Leaf > Root > Rhizome	Mobilizing micronutrient	Transition metals Alkali metals Metaloids	Mn Li, Na B
Leaf > Rhizome > Root	Mobilizing labile macronutrient	Alkali metals Non-metals	K Se

[1] Be was detected only in roots at both sampling sites.

4. Discussion

Results presented here show totora's remarkable capacity to survive in environments with very high loads of contaminants of different types (metals, salts, alkalis) and from different origin (natural or anthropogenic). Element concentrations in totora organs at Lake Uru-Uru were higher than in other Andean lakes reported before [27,28]. In addition, Se concentrations were similar, but As concentrations were much higher than in *S. californicus* plants growing in constructed wetlands in South Carolina (USA) [29]. The Puente Español location is considered the least polluted point in Lake Uru-Uru, as it is where the Desaguadero River discharges into the lake. However, the water in the river already contains noticeable levels of salt, metals, and other pollutants, both from natural and anthropogenic origins [21]. The presence of natural high levels of toxic elements, such as As, has been widely reported in the TDPS system, implying that the natural riparian flora should have adapted to this circumstance over the ages. On the other hand, the location of Central Lake is clearly influenced by the discharge from the urban wastewaters from Oruro city, drainage from the abandoned San José's mine (which is still producing acid drainage with low pH and high content in S and other metals), and from the active mining and metallurgic operations in the vicinity, such as the Pinto metal smelter.

One way that totora plants have adapted to these potentially toxic levels of several elements is by regulating their movement from the soil to the leaves, the most physiologically important organ in plants. Plants secrete exudates that chelate metals to prevent their uptake inside the cells [30]. Metals can also be bonded in the root cell walls [31]. Such a mechanism could explain the higher transfer factors for metals found in roots, particularly at the Central Lake site, more exposed to pollutants (Tables 2 and 3).

Particularly, totora has shown its capacity to behave as an As hyperaccumulator, with average As concentrations in rhizomes reaching more than 1000 mg kg^{-1} without visible external signs of vigor loss [32]. The high presence of P in the soil can reduce As toxicity, as high phosphate concentrations in the cells outcompete As in metabolic reactions [33]. The competition between phosphate and arsenate for binding sites in transporters could also explain the lower P concentrations found in the plants growing at Central Lake. Similarly, it has been shown that when S is present in high quantities in plants growing in flooded soils, As effects are inhibited [34]. Such behavior could be related to the relationships of S with the synthesis of thiolic compounds, which in turn, are related to As accumulation and metabolism processes. This mechanism could also be working in Lake Uru-Uru, where the high S loads entering the lake with the acid mine drain could be reducing the negative effects of As from natural and anthropogenic origin. In addition, As bioavailability depends of other minerals, such as Fe and Al, as well of soil pH. In Lake Uru-Uru, the high pH and very high Fe and Al concentrations in sediments (Table 1) may be having counteracting effects, as high pH could increase the solubility of certain As forms and, therefore, increase their availability for plants, but at the same time the high levels of Fe and Al in the sediments could be sequestering an important portion of As [35]. On the other hand, high As levels are a natural feature of water sources of the TDPS system caused by its particular geology and climate patterns [6]. Therefore, it could be assumed that totora has evolved different traits to deal with high concentrations of this toxic element. Finally, As profoundly affects root growth and metabolism [36]. The lack of visible impacts in the capacity of roots to uptake P in totora could be an indicator of totora's resistance to As toxic effects, as shown by the lack of significant differences among sites for P concentrations.

Regarding other metals related to anthropogenic activities, lead is a known factor altering root morphology and N uptake [37,38]. However, Pb is naturally present in high levels in Lake Uru-Uru [6,18], and water and soils at both sites had similar levels of Pb (Table 2). Therefore, it is unlikely that this element caused the differences in N content detected in plant organs. Cd is one of the most toxic heavy metals because of its high water solubility and easiness to be uptaken by plants. Cd can disrupt Ca absorption [39] and mimic Ca activity in the cytosol, producing stomata to close independently of water status [40]. This could partially explain the much higher concentrations of Ca in plants from Puente Español, the less polluted site. Cr competes with Fe, S, and P for transporters and carriers. The high levels of Fe and S in soils and the lack of differences among sites for P concentrations could indicate some capacity to buffer Cr toxic effects. Ni may compete with Ca and Mg, and reduce P levels in plants [41,42], as well as affect water balance [43]. Ni can also displace Mg from chlorophyll, reducing photosynthesis efficiency [44,45]. Such behavior could explain the higher Mg levels in plants from Puente Español but, on the other hand, the lack of differences in P concentrations and totora's natural adaptation to high water deficits typical of the TDPS (caused either by aridity of salinity) could be protecting the plants from some of Ni's adverse effects. Cu alters the photosynthetic capacity of plants, increasing susceptibility to photoinhibition [46,47]. However, the adaptation of totora to the high radiation levels reached in the Altiplano (with an average altitude of 3686 m.a.s.l. in Lake Uru-Uru and reaching 3800 m.a.s.l. and above in the Lake Titicaca region) could also reduce its exposition to this negative effect. Fe and Mn toxicity depend on pH and water logging conditions [48,49]. In wetland plants, Fe deposits could be formed in the roots, reducing Fe absorption and constituting a mechanism that could reduce excessive Fe uptake [48]. This process could explain the high Fe levels found in roots from Central Lake. Fe could also cause the low Mn concentrations found in plants from the most polluted site [50]. Antagonism between Fe and other more toxic metals, such as Cd [51], could also have a role in preventing totora from displaying visible signs of toxicity. The lower levels of Mn found in plants growing at the most polluted site could indicate that other elements, such Pb, are impeding roots from absorbing Mn. Zn toxicity affects root development and can displace Mg from the photosynthesis apparatus in chloroplasts [52,53]. As Zn is a typical component of drainage from mines and smelters around Lake Uru-Uru, the very high Zn levels reached in the plants from Central Lake could also combine with the effects of other pollutants to explain the low Mg levels detected in plants from this site.

As for other rare elements, Tl seems to be antagonist with monovalent alkali [54,55], which are in very high concentrations in Lake Uru-Uru and could have reduced Tl absorption by roots. Bi has very low solubility and, therefore, it is not considered toxic, although very little information is available [56]. Ti has very low solubility and in moderate concentrations seems to have beneficial effects [57], but its toxicity on plants in high concentrations remains poorly studied. V has also been reported as detrimental for root growth, although its toxicity seems to be ameliorated by Ca, as they compete for transporters [58]. Mo availability is related to soil pH, but the presence of sulphate can significantly reduce Mo absorption and toxicity [59], a mechanism that may be working in Lake Uru-Uru, given the high levels of S in the sediments. Sr's toxicity in plants is related to its chemical closeness to Ca, being able to supplant this macronutrient [60]. This phenomenon could be related to the lower Ca concentrations found at the more polluted site of Central Lake compared to Puente Español. Finally, Se is easily absorbed by roots interfering with S metabolic processes [32]. Therefore, available sulfates can compete with Se and mitigate its toxicity [61]. Hence, the high S levels in waters from Central Lake (Table 2) could partially offset Se-related toxic effects. However, insoluble elemental Se is more prevalent under anaerobic conditions like the ones in Lake Uru-Uru's sediments, and, consequently, it is reasonable to assume that Se toxicity could be minor.

Mechanism of root-to-shoot transportation of heavy metals involve different compounds that chelate metals. Such compounds could facilitate translocation to vacuoles where heavy metals are then chelated and stored permanently. Chemical differences among elements prevent the use of universal transporters and, therefore, some metals are more easily transported than others are. In addition, chelate compounds vary for different elements [62]. The results shown here indicate that rhizomes play a clear role in the regulation of pollutant transfers in totora plants (Tables 2 and 3), and confirm previous results from constructed wetlands [29]. For most elements, rhizomes acted as a barrier that effectively reduced the amount of metals, semi-metals, and metaloids reaching the leaves. However, such strategy was not enough to keep the levels of all pollutants low in the aboveground biomass. On the other hand, rhizomes could be playing a storage role for the main macronutrients (C, N, and P), for which other metals such as Ni and Mo may also be benefiting. Similarly, the role of rhizomes as barriers for metals may not be adequate to control light alkalis or metalloids, such as Li, Na, and B, neither for Mn, as they have high mobility rates. Accumulation of light elements in leaves could indicate the difficulty to prevent them from entering the plant and, therefore, they are displaced to tissues away from roots to avoid physiological damage to the underground tissues [63]. Higher pollutant concentrations in roots than in other tissues suggest that metals transport to the xylem is restricted [64]. The accumulation of some elements in roots could also indicate the use of strategies to prevent the entrance of toxic pollutants into the plant [65]. Totora can also contribute to reductive conditions into the substrate [66], which could translate into more insoluble metal forms [65]. Finally, the highest K concentration found in leaves is clearly related to the important role of this macronutrient in regulating osmotic pressure, particularly in an arid and hypersaline site such as Lake Uru-Uru.

Element concentrations in totora plants also showed clear associations among them. The three most important macronutrients (C, N, P) all showed positive correlations due to their structural and physiological roles in plants. In addition, other nutrients such as K and Mn also showed positive correlations as they both are related to osmotic regulation processes. A third group of elements composed by Mo, Ni, and Se grouped micronutrients that in small concentrations are needed in different enzymatic activities. A fourth group is composed by light elements (B, Li, Na, Be) all found in salt deposits, typical of endorheic, highly evaporative systems, such as the TDPS [67]. A cluster of elements whose concentrations were positively correlated included macronutrients, such as Ca and Mg (both elements scarcely mobile with important structural functions), accompanied by elements such as V or Sr that can mimic Ca and Mg roles in plants.

The largest elements grouping was composed by the majority of metals and semi-metals. This result could indicate that, rather than being related among them by some physiological function, this group of elements usually appears together in mineral deposits. Around Lake Uru-Uru there are numerous mines and large smelters, currently active, which are an important complex source of Pb, Sn, Sb, As, Cd, and other elements [8]. These elements are deposited either in the wastewaters or in the discarded material. Mining wastes create a permanent source of pollution that eventually reaches the lake, either through liquid effluents or by wind transport [8,68–70]. However, Lake Uru-Uru also has high natural concentrations from natural weathering of minerals with high concentrations of Pb and As [6,67]. In addition, the Poopó basin in which Lake Uru-Uru is placed hosts several deposits of Sn, Bi, Zn, Pb, Cu, and Sb, combined with S, Ag, and Au, exploited by several mining and smelting complexes. These elements eventually reach the lake sediments, and explain the high correlation among them found in the organs of totora plants living in Lake Uru-Uru. Furthermore, the TDPS has several natural salt deposits mixed with clays form the lake sediments, in which high natural concentrations of Li, K, B, Mg, and Na can be found, and are exploited by different mining companies and local people [6].

Results presented here show totora capacity to accumulate very high concentrations of some pollutants. Under the strictest definition by [71], hyperaccumulator plants are those whose organs can store more than 1000 mg kg^{-1} of toxic elements in their shoots. Totora plants growing at the Central Lake crossed such threshold for As, Fe, and Zn in leaves, making totora a multi-metal hyperaccumulator. Hyperaccumulator plants are mainly endemic to metal-rich soils [72], and they have mechanisms to deal with such toxicity. In addition, totora is adapted to important change in water levels, typical from a shallow system like Lake Uru-Uru, which average depth is only 1.5 m. This feature makes the lake very prone to seasonal and year-to-year severe changes in water level, including the complete lake drying up [16,73,74]. *S. californicus* adaptability to different ecological situations is also supported by its large distribution range, from California to Southern Chile, and from sea level to the Altiplano.

Hyperaccumulator plants usually grow slowly, have limited distribution ranges, or have shallow root systems [4]. However, totora has the opposite traits and, in addition, the capacity to withstand high levels of pollutants, seasonal and periodic changes in water levels, daily swings in temperature, and high radiation levels typical from the Altiplano. Moreover, transfer factors in leaves higher than one for Sb, Sr, Tl, Zn, and Cd indicated totora's high potential for its use in decontamination and phytomining. Previous studies have shown a high capacity of *S. californicus* to remove contaminants, reaching removal rates up to 80–90% for Cu, 70% for Pb, or 47–90% for Zn [64,75]. Aboveground biomass could be harvested, and in fact, totora shoots have been traditionally used for basketry, boat building, and even artificial island building. Nowadays, totora is being suggested as a more formal building material [13]. Such uses would effectively remove the polluted elements from the wetlands and sequester them for long periods in furniture or buildings, although the fate of polluted totora products once their life span has ended should be monitored.

On the other hand, if totora biomass is left on site, part of the pollutants will return to the water as the biomass decomposes. However, totora decomposition rates are low, particularly in polluted sites [76]. In addition, previous studies have highlighted the role of *S. californicus* not only to absorb and store heavy metals in its biomass [28], but also to provide organic carbon particles to the sediments through the decomposition of dead biomass that can be used as a major carbon and energy source for microbes metabolizing the different heavy metals into sequestered, non-bioavailable and, hence, less toxic forms [15]. Consequently, an important amount of pollutants could be sequestered in totora biomass (both dead and alive). Together, all these features make totora a very attractive plant for its use in constructed wetlands, particularly for saline waters [4,75,77,78], as results presented here show its resistance both to heavy metals and to saline conditions.

5. Conclusions

Totora has been identified for the first time as a multi-hyperaccumulator of As, Fe, and Zn under the definition by [71]. As the results shown here indicate, totora can survive in environments very harsh for other aquatic plants, including high concentrations of many heavy metals and pollutants and, at the same time, high pH and salinity. Totora is a riparian, perennial tall plant, in which rhizomes are key to support its capacity to sprout and regrow after losing the aboveground biomass. Rhizomes are also important to control pollutant uptake from contaminated soils. Such resistance to high concentrations of heavy metals, together with totora's intrinsic adaptation to large variations in the environmental conditions (temperature, salinity, water levels, radiation), high growth rates, low decomposition rates, and its value as a raw material for traditional and industrial uses, make this plant species a very valuable choice for its use in constructed wetlands in polluted sites, particularly in saline waters.

Funding: The author was funded by a grant from the Spanish Agency for International Development. Funds for chemical analyses were provided by the Department of Education of the Provincial Government of Navarre, program ANABASI+D. This paper has been published with the support of the Marie Curie Alumni Association.

Acknowledgments: The author thanks Gerardo Zamora Echenique, from the Technical University of Oruro, for his arrangements and support when hosting the author's stay at UTO during the time needed for this research. The author also wants to thank the comments to the original manuscript draft by two anonymous reviewers that helped to improve this paper.

Conflicts of Interest: The author declares no conflict of interest.

Appendix A Appendix

The following figures provide data on concentrations of chemical elements in different plant organs of plants collected from the two study sites.

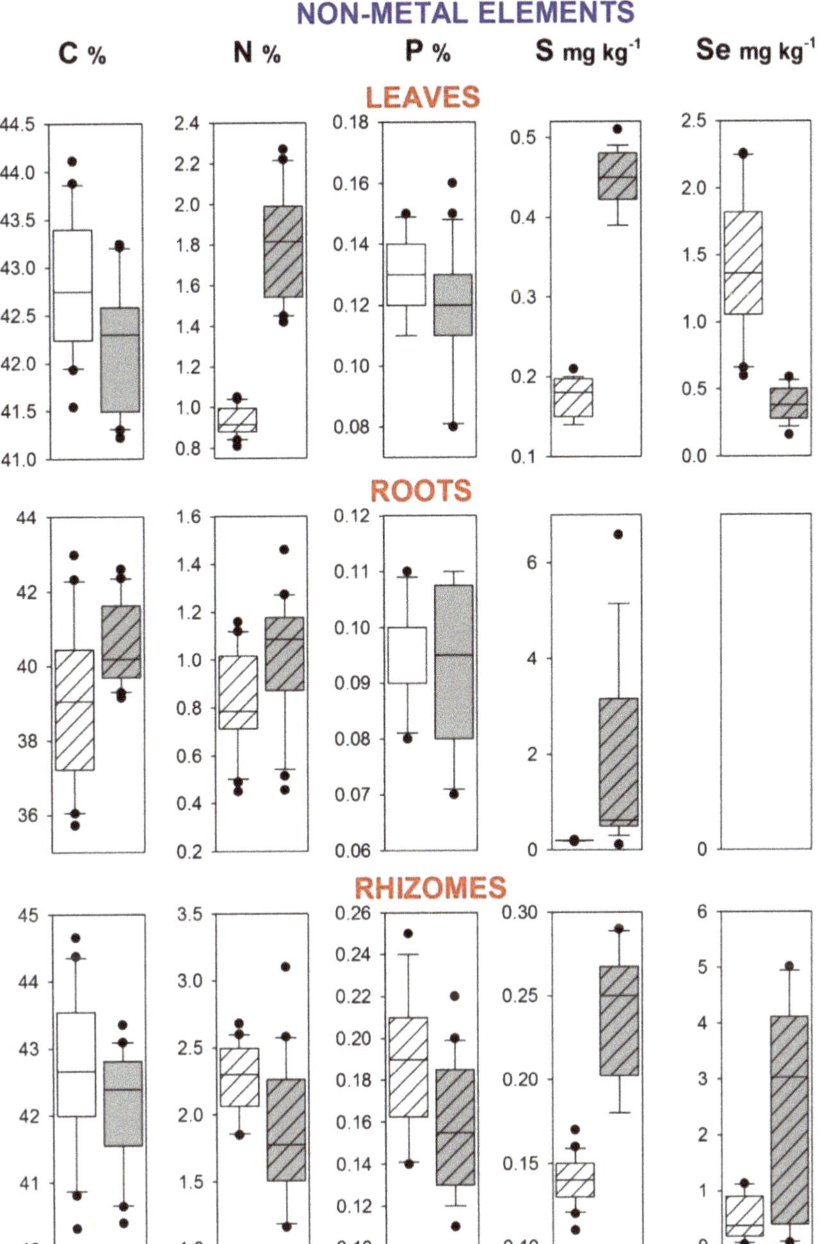

Figure A1. Box-plots of concentrations of non-metal elements (C, N, P, S, Se) in leaves, roots, and rhizomes of totora plants at two different sites: Puente Español (open box) and the lake center (grey box). Stripped boxes indicate significant differences between sites with Welch's test. No traces of Se were found in roots.

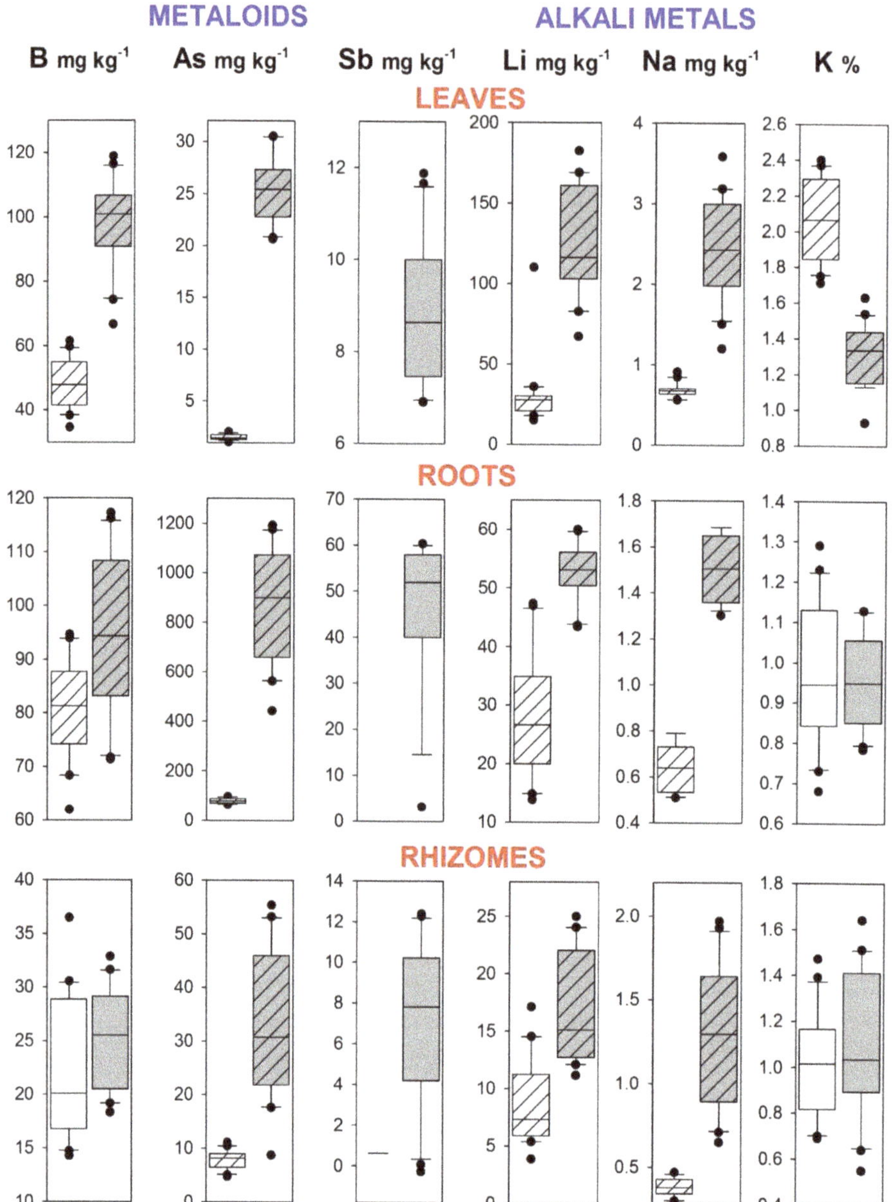

Figure A2. Box plots of concentrations of metaloids (B, As, Sb) and alkali metals (Li, Na, K) in leaves, roots, and rhizomes of totora plants at two different sites: Puente Español (open box) and the lake center (grey box). Stripped boxes indicate significant differences between sites with Welch's test.

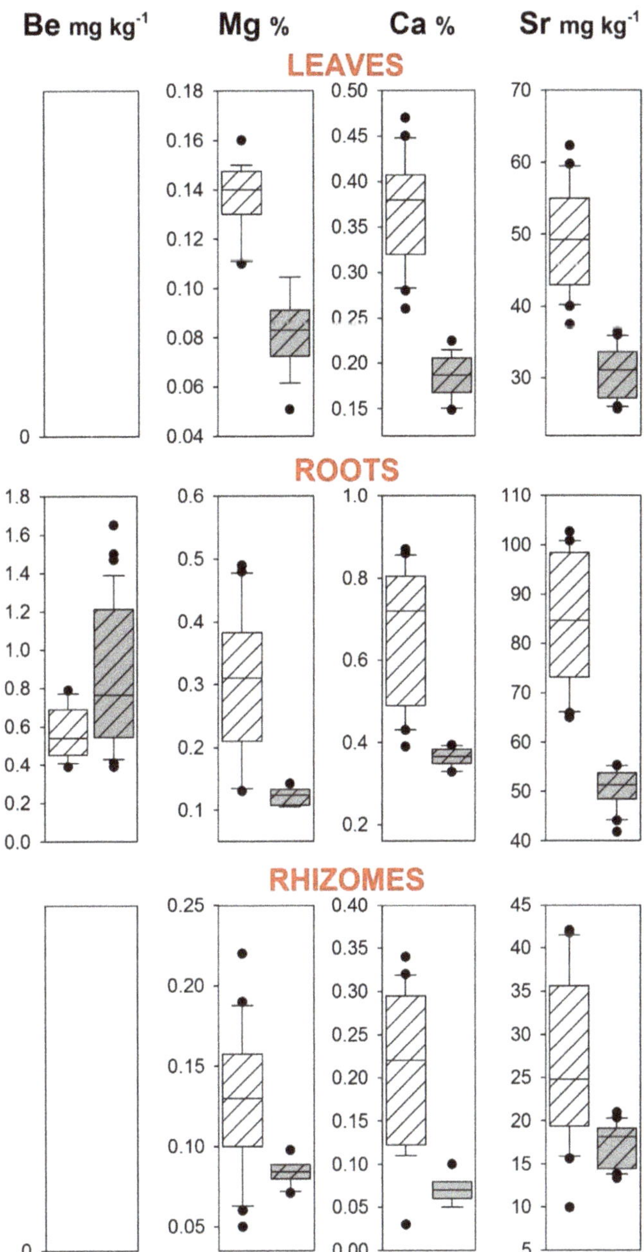

Figure A3. Box plots of concentrations of alkaline earth elements (Be, Mg, Ca, Sr) in leaves, roots, and rhizomes of totora plants at two different sites: Puente Español (open box) and the lake center (grey box). Stripped boxes indicate significant differences between sites with Welch's test. No traces of Be were found in leaves and rhizomes.

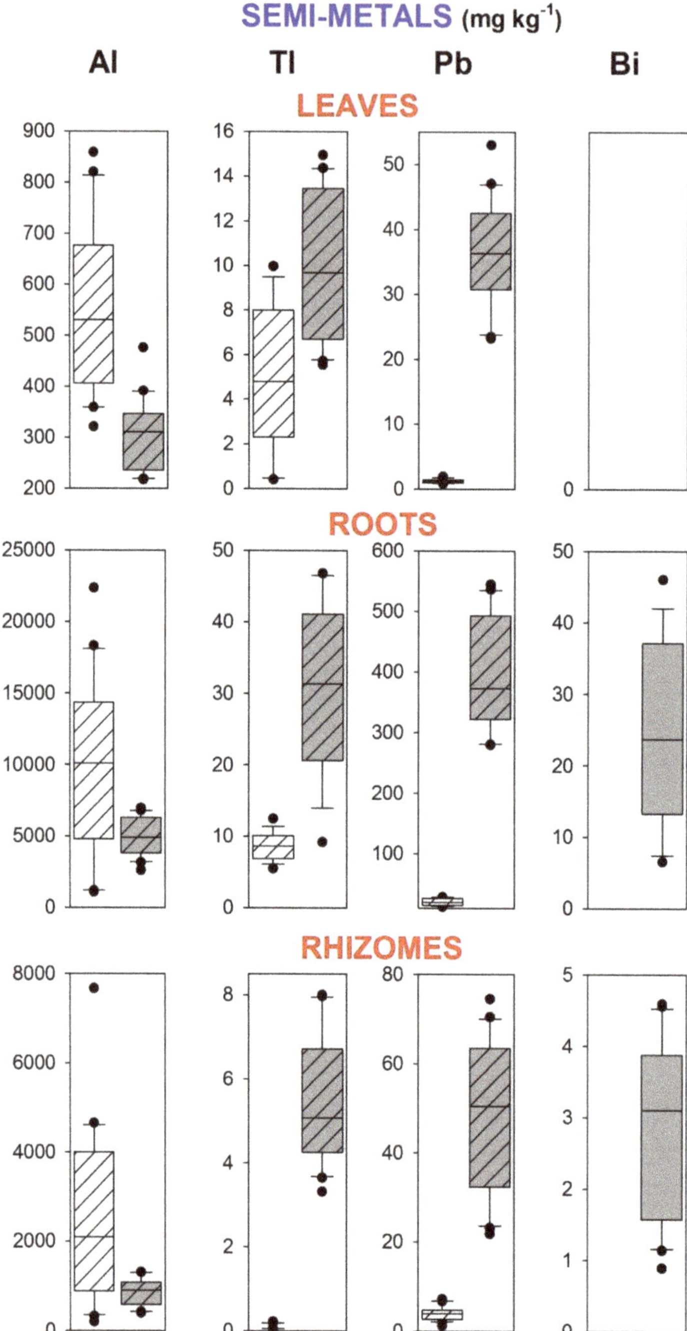

Figure A4. Box plots of concentrations of semi-metal elements (Al, Tl, Pb, and Bi) in leaves, roots, and rhizomes of totora plants at two different sites Puente Español (open box) and the lake center (grey box). Stripped boxes indicate significant differences between sites with Welch's test. No traces of Bi were found in leaves.

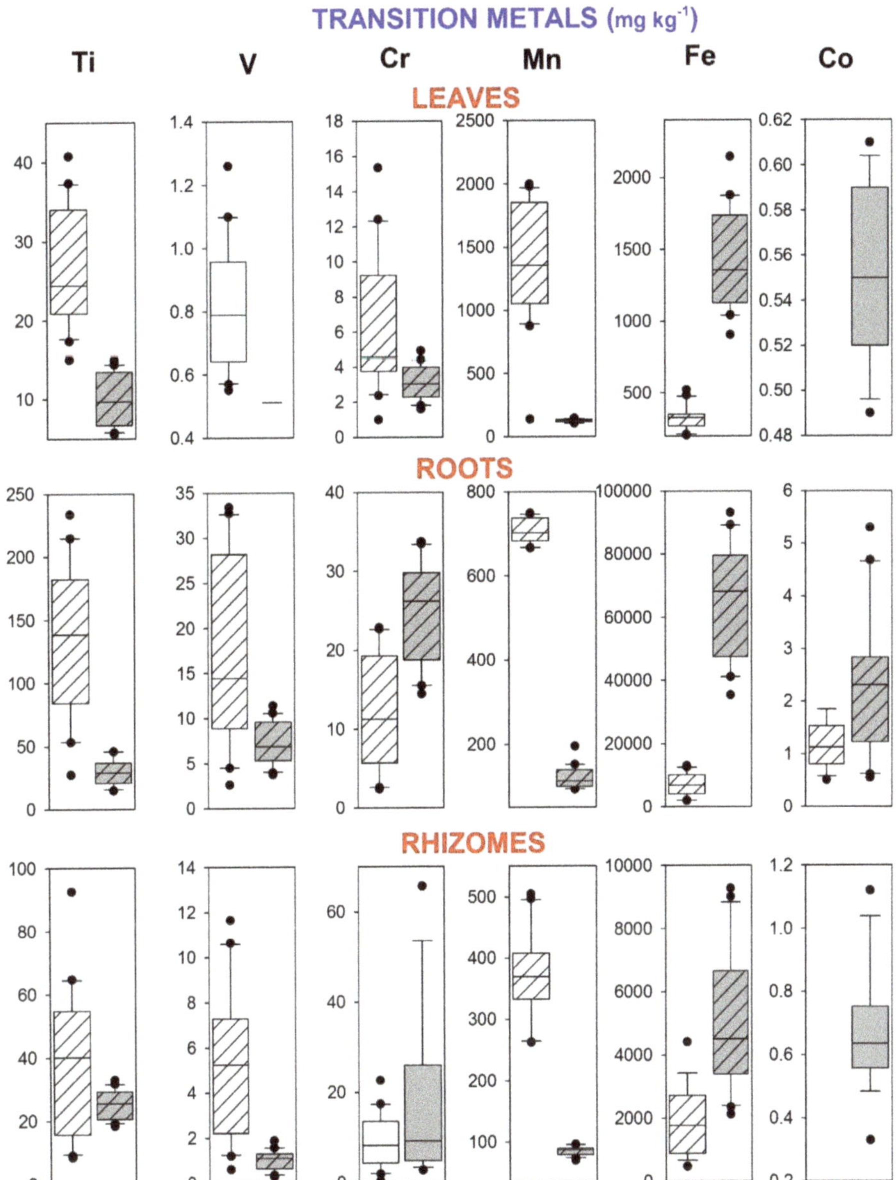

Figure A5. Box plots of concentrations of transition metals (Ti, V, Mn, Cr, Fe, Co) in leaves, roots, and rhizomes of totora plants at two different sites: Puente Español (open box) and the lake center (grey box). Stripped boxes indicate significant differences between sites with Welch's test.

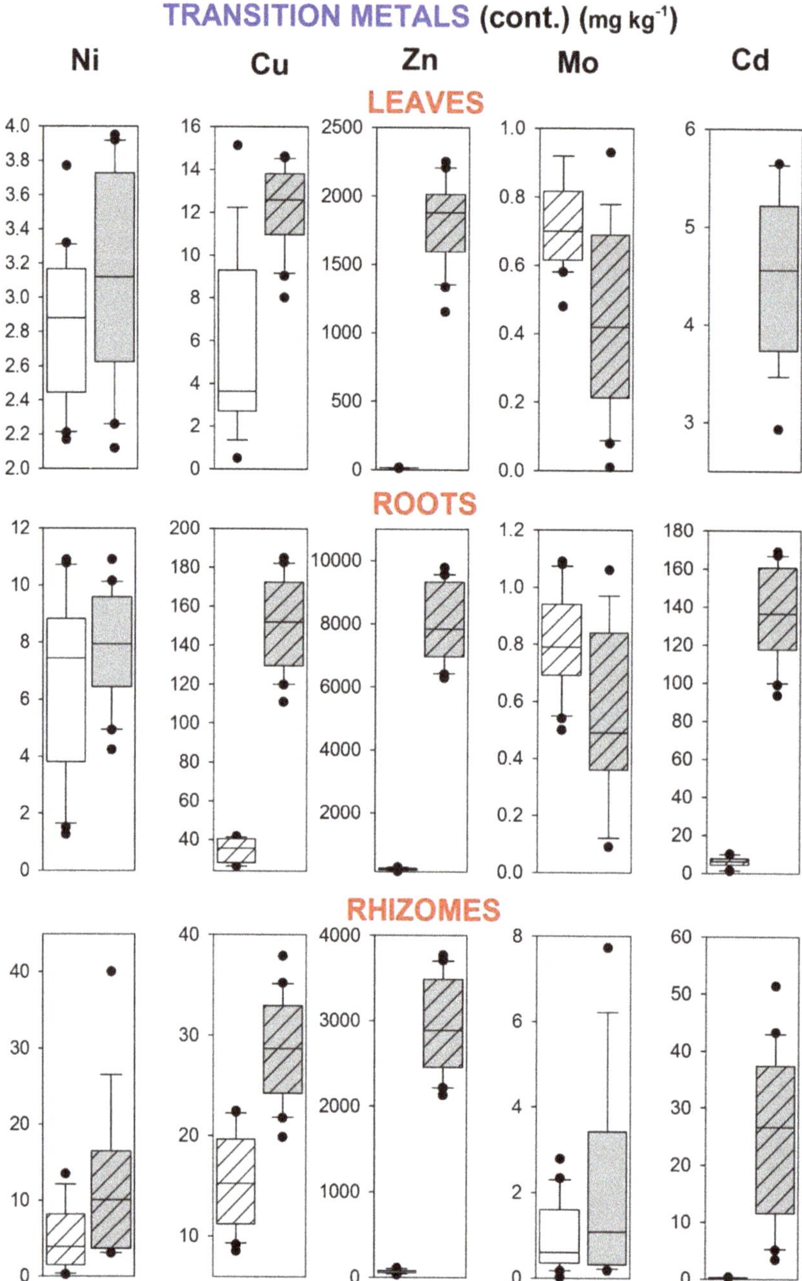

Figure A6. Box plots of concentrations of transition metals (Ni, Cu, Zn, Mb, Cd) in leaves, roots, and rhizomes of totora plants at two different sites: Puente Español (open box) and the lake center (grey box). Stripped boxes indicate significant differences between sites with Welch's test.

References

1. Whigham, D.F. Ecological issues related to wetland preservation, restoration, creation and assessment. *Sci. Total Environ.* **1999**, *240*, 1–3. [CrossRef]
2. Vyrmazal, J. Constructed wetlands for wastewater treatment: Five decades of experience. *Environ. Sci. Technol.* **2011**, *51*, 165–171. [CrossRef]
3. Wu, H.; Zhang, J.; Ngo, H.H.; Guo, W.; Hu, Z.; Liang, S.; Fan, J.; Liu, H. A review on the sustainability of constructed wetlands for wastewater treatment: Design and operation. *Biosour. Technol.* **2015**, *175*, 594–601. [CrossRef] [PubMed]
4. Nesler, A.; Furini, A. Phytoremediation: The utilization of plants to reclaim polluted sites. In *Plants and Heavy Metals*; Furini, A., Ed.; Springer: Dordrecht, The Netherlands, 2012; pp. 75–86.
5. Yapiyev, V.; Sagintayev, Z.; Inglezakis, V.J.; Samarkhanov, K.; Verhoef, A. Essesntials of endorheic basins and lakes: A review in the context of current and future water resource management and mitigation activities in Central Asia. *Water* **2017**, *9*, 798. [CrossRef]
6. García, M.E.; Bundschuh, J.; Ramos, O.; Quintanilla, J.; Persson, K.M.; Bengtsson, L.; Berndtsson, R. Heavy metals in aquatic plans and their relationship to concentrations in surface water, groundwater and sediments—A case study of Poopó basin, Bolivia. *Revista Boliviana de Química* **2005**, *22*, 11–18.
7. Clark, A.H.; Farrar, E.; Kontak, D.J.; Langridge, R.J.; Arenas, M.J.; France, L.J.; McBride, S.L.; Woodman, P.L.; Wasteneys, H.A.; Sandeman, H.A.; et al. Geologic and geochronological constrains on the metallogenic evolution of the Andes of southeastern Peru. *Econ. Geol.* **1990**, *85*, 1520–1583. [CrossRef]
8. Forturbel, F.E.; Barbieri, E.; Herbas, C.; Barbieri, F.L.; Gardon, J. Indoor metallic pollution related to mining activity in the Bolivian Altiplano. *Environ. Pollut.* **2011**, *159*, 280–2875. [CrossRef]
9. Díaz-Barriga, F.; Hamel, J.; Paz, E.; Carrizales, L.; Batres, L.; Calderón, J.; Galvao, L.; Caldas, L.Q.; McConnell, R. *Evaluación de Riesgos Para la Salud en la Población Expuesta a Metales en Bolivia*; Centro Panamericano de Ecología Humana y Salud: México DF, México, 1997.
10. Mason, H.L. *A Flora of the Marshes of California*; University of California Press: Berkeley, CA, USA, 1957.
11. Wagner, W.L.; Herbst, D.R.; Sohmer, S.H. *Manual of the Flowering Plants of Hawai'i. Vol. 2.*; Bishop Museum Special Publication 83; University of Hawai'i; Bishop Museum Press: Honolulu, HI, USA, 1990.
12. Montoya Choque, J.C.; Lovera, P.; Robert, M. *Lago Uru Uru: Evaluación de la Calidad del Agua, Sedimentos y Totora*; Centro de Ecología y Pueblos Andinos: Oruro, Bolivia, 2009.
13. Hidalgo-Cordero, J.F.; García-Navarro, J. Totora (*Schoenoplectus californicus* (C.A. Mey.) Soják) and its potential as a construction material. *Ind. Crop. Prod.* **2018**, *112*, 467–480. [CrossRef]
14. Vymazal, J. Emergent plants used in free water surface constructed wetlands: A review. *Ecol. Eng.* **2013**, *61*, 582–592. [CrossRef]
15. Murray-Gulde, C.L.; Huddleston, G.M.; Garber, K.V.; Rodgers, J.H. Contributions of *Schoenoplectus californicus* in a constructed wetland system receiving copper contaminated waste water. *Water Air Soil Pollut.* **2005**, *163*, 355–378. [CrossRef]
16. Nelson, E.A.; Specht, W.L.; Knox, A.S. Metal removal from water discharges by a constructed treatment wetland. *Eng. Life Sci.* **2006**, *6*, 26–30. [CrossRef]
17. Alanoca, L.; Amourous, D.; Momperrus, M.; Tessier, E.; Goni, M.; Guyoneaud, R.; Acha, D.; Gassie, C.; Audry, S.; Garcia, M.E.; et al. Diurnal variability and biogeochemical reactivity of mercury species in an extreme high-altitude lake ecosystem of the Bolivian Altiplano. *Environ. Sci. Pollut. Res.* **2016**, *23*, 6919–6933. [CrossRef] [PubMed]
18. García, M.E.; Bundschuh, J.; Ramos, O.; Quintanilla, J.; Persson, K.M.; Bengtsson, L.; Berndtsson, R. Heavy Metal Distribution in a Mining Region. A case Study of Lake Poopó, Bolivia. In *Proceedings of the 8th Workshop on Physical Processes in Natural Waters, Department of Water Resources Engineering, Lund, Sweden, August 2004*; Lund University: Lund, Sweden, 2004; Available online: www.academia.edu/download/45937267/maria1.doc (accessed on 17 December 2018).
19. Alanoca, L. Cycle Biogéochemique du Hg dans l'Hydrosystème Tropical d'Altitude lac Uru-Uru, Altiplano Bolivien. Ph.D. Thesis, Université Toulouse III Paul Sabatier, Toulouse, France, 2016.
20. Molina, C.I.; Ibañez, C.; Gibon, F.-M. Proceso de biomagnificación de metales pesados en un lago hiperhialino (Poopó, Oruro, Bolivia): Posible riesgo en la salud de consumidores. *Ecología en Bolivia* **2012**, *47*, 99–118.

21. Ramos Ramos, O.W.; Cáceres, L.F.; Ormaechea Muñoz, M.R.; Bhattacharya, P.; Quino, I.; Quintanilla, J.; Sracek, O.; Thunvik, R.; Bundschuh, J.; García, M.E. Sources and behavior of arsenic and trace elements in groundwater and surface water in the Poopó Lake Basin, Bolivian Altiplano. *Environ. Earth Sci.* **2012**, *66*, 793–807. [CrossRef]
22. Autoridad Binacional de Lago Titcaca. *Estudio de Hidroquímica y Contaminación. Plan Director Global Binacional de Protección—Prevención de Inundaciones y Aprovechamiento de los Recursos del Lago Titicaca, Río Desaguadero, Lago Poopó y Lago Salar de Coipasa (Sistema, T.D.P.S.)*; Autoridad Binacional de Lago Titcaca: La Paz, Bolivia, 1993.
23. Matejovic, I. Determination of carbon, hydrogen, and nitrogen in soils by automated elemental analysis (dry combustion method). *Commun. Soil Sci. Plant Anal.* **1993**, *24*, 2213–2222. [CrossRef]
24. Olowoyo, J.O.; Van Heerden, E.; Fischer, J.L.; Baker, C. Trace metals in soil and leaves of *Jacaranda mimosifolia* in Tshwane area, South Africa. *Atmos. Environ.* **2010**, *44*, 1826–1830. [CrossRef]
25. Zar, J.H. *Bioestatistical Analysis*, 4th ed.; Prentice Hall: Upper Saddle River, NJ, USA, 1999.
26. Quinn, G.P.; Keough, M.J. *Experimental Design and Data Analysis for Biologists*; Cambridge University Press: Cambridge, UK, 2002.
27. Juárez, A.; Arribére, M.A.; Arcagni, M.; Williams, N.; Rizzo, A.; Ribeiro Gueara, S. Heavy metals and trace elements in riparian vegetation and macrophytes associated with lacustrine systems in Northern Patagonia Andean Range. *Environ. Sci. Pollut. Res.* **2016**, *23*, 17995–18009. [CrossRef]
28. Rodríguez Ayala, S.; Flores, R.; Rodríguez, M.; Andocilla, M. Quantitative determination of heavy metal hyperaccumulation in a macrophyte simple of *Schoenoplectus californius* from Lago San Pablo, Imbabura-Ecuador. *Revista Ciencia* **2017**, *4*, 431–446.
29. Sundberg-Jones, S.E.; Hassan, S.M. Macrophyte sorption and bioconcentration of elements in a pilot constructed wetland for flue gas desulfurization wastewater treatment. *Water Air Soil Pollut.* **2007**, *183*, 187–200. [CrossRef]
30. Marschner, H. *Mineral. Nutrition of Higher Plants*, 2nd ed.; Academic Press: London, UK, 1995.
31. Bringezu, K.; Lichtenberger, O.; Leopold, I.; Neumann, D. Heavy metal tolerance of *Silene vulgaris*. *J. Plant Physiol.* **1999**, *154*, 536–546. [CrossRef]
32. DalCorso, G. Heavy metal toxicity in plants. In *Plants and Heavy Metals*; Furini, A., Ed.; Springer: Dordrecht, The Netherlands, 2012; pp. 1–25.
33. Christophersen, H.M.; Smith, S.E.; Pope, S.; Smith, F.A. No evidence for competition between arsenate and phosphate for uptake from soil by medic or barley. *Environ. Int.* **2009**, *35*, 485–490. [CrossRef]
34. Zhang, J.; Zhao, Q.Z.; Duan, G.L.; Huang, Y.C. Influence of sulphur on arsenic accumulation and metabolism in rice seedlings. *Environ. Exp. Bot.* **2011**, *72*, 34–40. [CrossRef]
35. Romero-Freire, A.; Sierra-Aragón, M.; Ortiz-Bernad, I.; Martín-Peinado, J. Toxicity of arsenic in relation to soil properties: Implications to regulatory purposes. *J. Soils Sediments* **2014**, *14*, 968–979. [CrossRef]
36. Duquesnoy, I.; Champeau, G.M.; Evray, G.; Ledoigt, G.; Piquet-Pissaloux, A. Enzymatic adaptations to arsenic-induced oxidative stress in *Zea mays* and genotoxic effect of arsenic in root tips of *Vicia faba* and *Zea mays*. *C. R. Biol.* **2010**, *333*, 814–824. [CrossRef]
37. Eun, S.O.; Yon, H.S.; Lee, Y. Lead disturbs microtubule organization in the root meristem of *Zea mays*. *Physiol. Plant.* **2000**, *110*, 357–365. [CrossRef]
38. Sharma, P.; Dubey, R.S. Lead toxicity in plants. *Braz. J. Plant Physiol.* **2005**, *17*, 35–52. [CrossRef]
39. DalCorso, G.; Farinati, S.; Maistri, S.; Furini, A. How plants cope with cadmium: Staking all on metabolism and gene expression. *J. Integr. Plant Biol.* **2008**, *50*, 1268–1280. [CrossRef] [PubMed]
40. Perfus-Barbeoch, L.; Leonhardt, N.; Vavaddeur, A.; Forestier, C. Heavy metal toxicity: Cadmium permeates through calcium channels and disturbs the plant water status. *Plant J.* **2002**, *32*, 539–548. [CrossRef] [PubMed]
41. Athar, R.; Ahmad, M. Heavy metal toxicity in legume-microsymbiont system. *J. Plant Nutr.* **2002**, *25*, 369–386. [CrossRef]
42. Pillay, S.V.; Rao, V.S.; Rao, K.V.N. Effect of nickel toxicity in *Hyptis suareeolens* (L.) Poit. and *Helianthus annuus* L. *Indian J. Plant Physiol.* **1996**, *1*, 153–156.
43. Chen, C.; Huang, D.; Liu, J. Functions and toxicity of nickel in plants: Recent advances and future prospects. *Clean* **2009**, *37*, 304–313. [CrossRef]
44. Ahmad, M.S.A.; Hussain, M.; Saddiq, R.; Alvi, A.K. Mungbean: A nickel indicator, accumulator or excluder? *Bull. Environ. Contam. Toxicol.* **2007**, *78*, 319–324. [CrossRef] [PubMed]

45. Alam, M.M.; Hayat, S.; Ali, B.; Ahmad, A. Effect of 28- homobrassinolide treatment on nickel toxicity in *Brassica juncea*. *Photosynthetica* **2007**, *45*, 139–142. [CrossRef]
46. Maksymiec, W.; Baszynski, T. The role of Ca^{2+} ions in modulating changes induced in bean plants by an excess of Cu^{2+} ions. Chlorophyll fluorescence measurements. *Physiol. Plant.* **1999**, *105*, 562–568. [CrossRef]
47. Pätsikkä, E.; Kairavuo, M.; Sersen, F.; Aro, E.-M.; Tyystjärvi, E. Excess copper predisposes photosystem II to photoinhibition in vivo by outcompeting iron and causing decrease in leaf chlorophyll. *Plant Physiol.* **2002**, *129*, 1359–1367. [CrossRef] [PubMed]
48. Batty, L.C.; Younger, P.L. Effects of external iron concentration upon seedling growth and uptake of Fe and phosphate by the common reed, *Phragmites australis* (Cav.) *trin* ex *steudel*. *Ann. Bot.* **2003**, *92*, 801–806. [CrossRef] [PubMed]
49. Dučic, T.; Polle, A. Transport and detoxification of manganese and copper in plants. *Braz. J. Plant Physiol.* **2005**, *17*, 103–112. [CrossRef]
50. Madejczyk, M.S.; Ballatori, N. The iron transporter ferroportin can also function as a manganese exporter. *Biochim. Biophys. Acta Biomembr.* **2012**, *1818*, 651–657. [CrossRef]
51. Zuo, Y.; Zhang, F. Soil and crop management strategies to prevent iron deficiency in crops. *Plant Soil* **2011**, *339*, 83–95. [CrossRef]
52. Broadley, M.R.; White, P.J.; Hammond, J.P.; Zelko, I.; Lux, A. Zinc in plants. *New Phytol.* **2007**, *173*, 677–702. [CrossRef]
53. Van Assche, F.; Clijsters, H. Inhibition of photosynthesis in *Phaseolus vulgaris* by treatment with toxic concentrations of zinc: Effects on electron transport and photo-phosphorylation. *Physiol. Plant.* **1986**, *66*, 717–721. [CrossRef]
54. Kaplan, D.I.; Adriano, D.C.; Sajwan, K.S. Thallium toxicity in bean. *J. Environ. Qual.* **1990**, *19*, 359–365. [CrossRef]
55. Durrant, P.J.; Durrant, B. *Introduction to Advanced Inorganic Chemistry*; Longman: London, UK, 1970.
56. Babula, P.; Adam, V.; Opatrilova, R.; Zehnalek, J.; Havel, L.; Kizek, R. Uncommon heavy metals, metalloids and their plant toxicity: A review. *Environ. Chem. Lett.* **2008**, *6*, 189–213. [CrossRef]
57. Dumon, J.C.; Erns, W.H.O. Titanium in plants. *J. Plant. Physiol.* **1988**, *133*, 203–209. [CrossRef]
58. Kaplan, D.I.; Adriano, D.C.; Carlson, C.L.; Sajwan, S. Vanadium toxicity an accumulation by beans. *Water Air Soil Pollut.* **1990**, *49*, 81–91. [CrossRef]
59. McGrath, S.P.; Micó, C.; Curdy, R.; Zhao, F.J. Predicting molybdenum toxicity to higher plants: Influence of soil properties. *Environ. Pollut.* **2010**, *158*, 3095–3102. [CrossRef]
60. Isermann, K. Uptake of Stable Strontium by Plants and Effects on Plant Growth. In *Handbook of Stable Strontium*; Skoryna, S.C., Ed.; Springer: Boston, MA, USA, 1981; pp. 65–86.
61. Terry, N.; Zayed, A.M.; De Souza, M.P.; Tarun, A.S. Selenium in higher plants. *Annu. Rev. Plant Physiol. Plant Mol. Biol.* **2000**, *51*, 401–432. [CrossRef]
62. Manara, A. Plant responses to heavy metal toxicity. In *Plants and Heavy Metals*; Furini, A., Ed.; Springer: Dordrecth, The Netherlands, 2012.
63. Jan, S.; Parray, J.A. *Approaches to Heavy Metal Tolerance in Plants*; Springer: Singapore, 2016.
64. Gallego, S.M.; Pena, L.B.; Barcia, R.A.; Azpilicueta, C.E.; Iannone, M.F.; Rosales, E.P.; Zawoznik, M.S.; Groppa, M.D.; Benavides, M.P. Unravelling cadmium toxicity and tolerance in plants: Insight into regulatory mechanisms. *Environ. Exp. Bot.* **2012**, *83*, 33–46. [CrossRef]
65. Dorman, L.; Castle, J.W.; Rodgers, J.H. Performance of a pilot-scale constructed wetland system for treating simulated ash basin water. *Chemosphere* **2009**, *75*, 939–947. [CrossRef]
66. Marchand, L.; Mench, M.; Jacob, D.L.; Otte, M.L. Metal and metalloid removal in constructed wetlands, with emphasis on the importance of plants and standardized measurements: A review. *Environ. Pollut.* **2010**, *158*, 3447–3461. [CrossRef]
67. García, M.E.; Bengtsson, L.; Persson, K.M. On the distribution of saline groundwater in the Poopó Basic, central Bolivian highland. *Vatten* **2005**, *66*, 199–203.
68. Miller, J.R.; Hudson-Edwards, K.A.; Lechler, P.J.; Preston, D.; Macklin, M.G. Heavy metal contamination of water, soil and produce within riverine communities of the Rio Pilcomayo basin, Bolivia. *Sci. Total Environ.* **2004**, *320*, 189–209. [CrossRef] [PubMed]

69. Smolders, A.J.; Lock, R.A.; Van der Velde, G.; Medina-Hoyos, R.I.; Roelofs, J.G. Effects of mining activities on heavy metal concentrations in water, sediment, and macroinvertebrates in different reaches of the Pilcomayo River, South America. *Arch. Environ. Contam. Toxicol.* **2003**, *44*, 314–323. [CrossRef] [PubMed]
70. Reif, J.S.; Ameghino, E.; Aaronson, M.J. Chronic exposure of sheep to a zinc smelter in Peru. *Environ. Res.* **1989**, *49*, 40–49. [CrossRef]
71. Fasani, E. Plants that hyperaccumulate heavy metals. In *Plants and Heavy Metals*; Furini, A., Ed.; Springer: Dordrecht, The Netherlands, 2012; pp. 55–74.
72. Baker, A.J.M.; McGrath, S.P.; Reeves, R.D.; Smith, J.A.C. Chapter 5. Metal hyperaccumulator plants: A review of the ecology and physiology of a biological resource for phytoremediation of metal-polluted soils. In *Phytoremediation of Contaminated Soil and Water*; Terry, N., Bañuelos, G., Eds.; CRC Press LLC: Boca Raton, FL, USA, 2000.
73. Pillco Zolá, R.; Bengtsson, L. Long-term and extreme water level variations of the shallow Lake Poopó, Bolivia. *Hydrol. Sci. J.* **2006**, *31*, 98–114. [CrossRef]
74. Abarca-Del-Río, R.; Crétaux, J.F.; Berge-Nguyen, M.; Maisongrande, P. Does Lake Titicaca still control the Lake Poopó system water levels? An investigation using satellite altimetry and MODIS data (2000–2009). *Remote Sens. Lett.* **2012**, *3*, 707–714. [CrossRef]
75. Villamar, C.A.; Neubauer, M.E.; Vidal, G. Distribution and availability of copper and zinc in a constructed wetland fed with treated swine slurry from an anaerobic lagoon. *Wetlands* **2014**, *34*, 583–591. [CrossRef]
76. Constantini, M.L.; Sabetta, L.; Mancinelli, G.; Rossi, L. Spatial variability of the decomposition rate of Schoenoplectus tatora in a polluted area of Lake Titicaca. *J. Trop. Ecol.* **2004**, *20*, 325–335. [CrossRef]
77. Vyrmazal, J. The use of constructed wetlands for treatment of industrial waste water: A review. *Ecol. Eng.* **2014**, *73*, 724–751. [CrossRef]
78. Liang, Y.; Zhu, H.; Bañuelos, G.; Yan, B.; Zhou, Q.; Yu, X.; Cheng, X. Constructed wetlands for saline wastewater treatment: A reivew. *Ecol. Eng.* **2017**, *98*, 275–285. [CrossRef]

Article

Nitrogen Recovery from Wastewater: Possibilities, Competition with Other Resources, and Adaptation Pathways

Jan Peter van der Hoek [1,2,*], Rogier Duijff [1] and Otto Reinstra [2]

1 Department of Water Management, Faculty of Civil Engineering and Geosciences, Delft University of Technology, Stevinweg 1, 2628 CN Delft, The Netherlands; rduijff@hotmail.com

2 Waternet, Strategic Centre, Korte Ouderkerkerdijk 7, 1096 AC Amsterdam, The Netherlands; otto.reinstra@waternet.nl

* Correspondence: j.p.vanderhoek@tudelft.nl; Tel.: +31-6-48262075

Received: 5 October 2018; Accepted: 1 December 2018; Published: 5 December 2018

Abstract: Due to increased food production, the demand for nitrogen and phosphorus as fertilizers grows. Nitrogen-based fertilizers are produced with the Haber–Bosch process through the industrial fixation of N_2 into ammonia. Through wastewater treatment, the nitrogen is finally released back to the atmosphere as N_2 gas. This nitrogen cycle is characterized by drawbacks. The energy requirement is high, and in the wastewater treatment, nitrogen is mainly converted to N_2 gas and lost to the atmosphere. In this study, technologies for nitrogen recovery from wastewater were selected based on four criteria: sustainability (energy use and N_2O emissions), the potential to recover nitrogen in an applicable form, the maturity of the technology, and the nitrogen concentration that can be handled by the technology. As in wastewater treatment, the focus is also on the recovery of other resources; the interactions of nitrogen recovery with biogas production, phosphorus recovery, and cellulose recovery were examined. The mutual interference of the several nitrogen recovery technologies was studied using adaptive policy making. The most promising mature technologies that can be incorporated into existing wastewater treatment plants include struvite precipitation, the treatment of digester reject water by air stripping, vacuum membrane filtration, hydrophobic membrane filtration, and treatment of air from thermal sludge drying, resulting respectively in 1.1%, 24%, 75%, 75%, and 2.1% nitrogen recovery for the specific case wastewater treatment plant Amsterdam-West. The effects on sustainability were limited. Higher nitrogen recovery (60%) could be realized by separate urine collection, but this requires a completely new infrastructure for wastewater collection and treatment. It was concluded that different technologies in parallel are required to reach sustainable solutions. Nitrogen recovery does not interfere with the recovery of the other resources. An adaptation pathways map is a good tool to take into account new developments, uncertainties, and different ambitions when choosing technologies for nitrogen recovery.

Keywords: nitrogen; resource recovery; wastewater treatment; energy; sustainability; adaptive policymaking

---

1. Introduction

The increase of the world population to eight to 10 billion by 2050 [1,2] will result in substantial pressure on food supply [3]. Nitrogen and phosphorus play a critical role in plant growth and supply [4]. Due to increased food production, the demand for nitrogen and phosphorus will grow. Phosphorus is a non-renewable resource. The most common form of phosphorus on Earth is locked in igneous and sedimentary deposits, with the mining of these rocks being the most viable method of extraction. With the current rate of extraction and consumption, these "readily exploitable" sources of phosphorus will be depleted within the next 45–100 years [5]. The reserve of this resource is getting smaller, and now phosphate is on the European Union (EU) list of critical raw material [6]. Driven by future shortages, a paradigm shift is currently underway from an attitude that considers wastewater as a waste to be treated, to a proactive interest in recovering materials and energy from these streams [7]. Much research is being carried out into phosphorus removal from wastewater [8–10], and technologies are now applied at full-scale [11].

Nitrogen is abundantly present in the atmosphere (almost 80%) in a highly stable and non-reactive form, N_2 gas. Nitrogen in its reactive forms (ammonium, nitrite, and nitrate) is essential for plant growth, and its content is limited in soils. Most naturally occurring reactive nitrogen comes from lightning (2%) and biological fixation (98%) [4]. Since the Haber–Bosch process was invented in 1909, which provides an industrial fixation of N_2 into ammonia, the production of N-based fertilizers supported the largest historical increase in food production capacity [12]. The Haber–Bosch process more than quadrupled the productivity of agricultural crops [13].

However, the introduction of the Haber–Bosch process affected the nitrogen cycle. The increased food production by use of N-based fertilizers produced by the Haber–Bosch process is excreted mainly as urea and NH_4^+ by human metabolism, and discharged to the sewer. To avoid the eutrophication of water, in the current wastewater treatment technology based on the conventional activated sludge process, the reduced reactive nitrogen is biologically converted to its nonreactive N_2 gas form through the nitrification/denitrification or deammonification (anammox) process [14], and then released back into the atmosphere.

Although the nitrogen cycle is closed through the combination of industrial fixation of N_2 into ammonia by the Haber–Bosch process and the enhanced microbiological conversion of reduced reactive nitrogen to N_2 gas, it is also characterized by serious environmental drawbacks. Firstly, nitrogen entering waste streams is mainly converted to N_2 gas and lost to the atmosphere rather than reused. Secondly, the processes of N-fixation for fertilizers' production and N-dissipation for wastewater treatment require much energy. Thirdly, the biological removal of nitrogen from the wastewater results in nitrous oxide (N_2O) gas emissions representing an intermediate of increasing concern in terms of greenhouse gas emissions from wastewater treatment plants: the emission is relatively small (3% of the estimated total anthropogenic N_2O emission), but is a significant factor (26%) in the greenhouse gas footprint of the total water chain [15].

For these reasons, it is relevant to examine more sustainable pathways for nitrogen, which consist of interventions in the present (anthropogenic) nitrogen cycle, such as the direct recovery of nitrogen from wastewater and reuse. Up until now, there has been only limited experience with nitrogen recovery from wastewater combined with nitrogen reuse at full scale. Ammonia precipitation as struvite is applied in practice, but the main focus of this process is phosphorus recovery [11]. In a household-scale wastewater treatment system operated with domestic sewage, gardening/irrigation water was recovered from raw sewage or secondary effluent by low-pressure ultrafiltration [16]. In the European MEMORY project ("membranes for energy and water recovery"), the technical and economic feasibility of a submerged anaerobic membrane bioreactor, treating urban wastewater, is demonstrated at an industrial scale. Instead of consuming electricity to destroy organic matter and nitrogen, methane is generated directly from the raw wastewater, and the membranes produce disinfected reusable water that is rich in fertilizers [17].

At the same time, there are many other initiatives than nitrogen recovery and nitrogen reuse to make the wastewater treatment more sustainable. Many of these focus on resource recovery. A transition in wastewater treatment plants toward the reuse of wastewater-derived resources is recognized as a promising solution to shift wastewater treatment from standard treatment to the current emphasis on sustainability [18]. In addition to water, energy and nutrient recovery (phosphorus and nitrogen) emerging options are e.g., the recovery of cellulose fibers [19], biopolymers [20], bioplastics [21], and protein [22]. In the Netherlands, there is a special program, the Energy and Raw Materials Factory, focusing on the recovery of materials and energy from wastewater to contribute to the circular economy. The program involves resources such as cellulose, bioplastics, phosphate, alginate-like exopolymers from aerobic granular sludge, and biomass [23]. Due to its many possibilities, the challenge is how to develop a coherent policy and strategy, and how to make the right choices [24].

Within the possibilities for nitrogen recovery and nitrogen reuse, competing synergistic or neutral interventions and technologies may also exist, resulting in lock-ins (measures that are mutually exclusive), no-regret measures (measures that do not limit the number of options after a decision), and win–win measures (measures that are significant for more than one strategy).

This study has three specific objectives. Firstly, it explores alternatives to recover and reuse nitrogen from wastewater in a more sustainable way (Section 3.2). Secondly, the selected alternatives are placed beside other alternatives for resource recovery from wastewater to judge the exclusion or synergy with these other resource recovery alternatives (Section 3.3). Thirdly, the alternatives for nitrogen recovery and reuse are compared with each other to identify lock-ins, win–win, and no-regret measures (Section 3.3).

2. Materials and Methods

2.1. Wastewater Treatment Plant Amsterdam-West

The wastewater treatment plant (WWTP) Amsterdam-West was used as a specific case in this study. This plant is operated by the water utility Waternet, which is the public water service of the City of Amsterdam and the Regional Water Authority Amstel, Gooi, and Vecht. Figure 1 schematically shows the process configuration of this plant. After primary treatment, the wastewater is transferred to a series of biological treatment tanks. Together, these form the modified University of Cape Town (mUTC) process with biological phosphorus and nitrogen removal. Finally, the wastewater passes the secondary settling tank. Primary sludge and waste sludge are digested. Digested sludge is dewatered, after which the dewatered sludge is transported to a struvite installation to produce struvite.

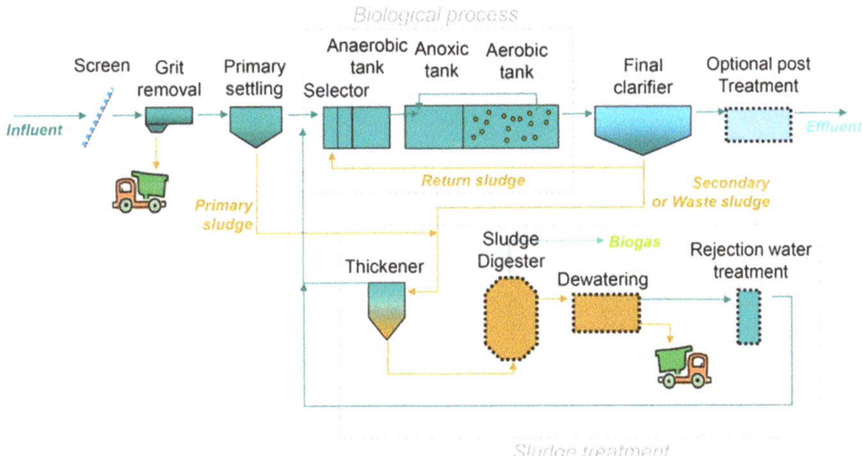

Figure 1. Wastewater treatment plant Amsterdam-West.

This WWTP was chosen for analysis because it has a large capacity of 1,014,000 people equivalents (PEs). The N-load to this plant through Amsterdam's wastewater is 3932 ton N/year [25], which is 4.4% of the total N-load in sewer water in the Netherlands. In addition, sludge from the other WWTPs operated by Waternet is transported to this plant for digestion, by which the total N-load to this plant equals 4705 ton N/year, which is 5.3% of the total N-load in wastewater treatment in the Netherlands. During the digestion, nitrogen is released in the form of NH_3/NH_4^+, which can be recovered by several technologies. These characteristics make WWTP Amsterdam-West potentially attractive for nitrogen recovery.

2.2. Nitrogen Balance and Water Balance

A nitrogen mass balance was made for the whole treatment process of WWTP Amsterdam-West. Also, a water balance was made for the whole treatment process. The nitrogen balance shows where nitrogen is present and in which quantities in the treatment process. Combination with the water balance shows the nitrogen concentrations in the treatment process. Concentration is an important parameter, as many recovery techniques work more efficiently at higher concentrations. Locations with high nitrogen mass and a high nitrogen concentration are attractive for nitrogen recovery.

2.3. Selection of Alternatives

Based on a literature review, alternatives were identified. By the use of four specific criteria, alternatives were selected for further evaluation. The criteria were:

1. The alternative has to be more sustainable with respect to energy use and N_2O emissions;
2. The alternative has to focus on the recovery of nitrogen in an applicable form;
3. The alternative must be applicable in practice;
4. The alternative has to be able to cope with nitrogen in the concentration range that is present in the wastewater treatment process (60–8800 mg/L, see Section 3: Results and Discussion).

For criterion 1, the combination Haber–Bosch – deammonification was considered as a benchmark. This implies that the alternative requires lower energy consumption as compared with the combination Haber–Bosch – deammonification, and should result in N_2O emissions during the wastewater treatment that are far below the conventional nitrification–denitrification process and below the deammonification process. To quantify this, the nitrogen cycle as shown in Figure 2 has to be considered.

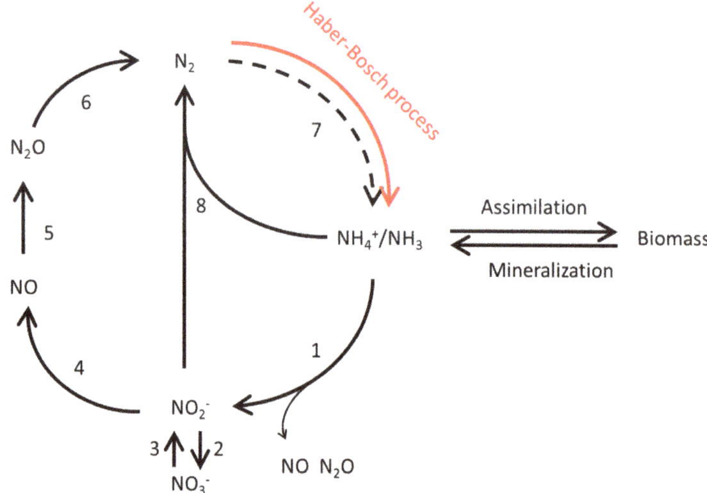

Figure 2. The nitrogen cycles. (1) Aerobic ammonium oxidation, (2) aerobic nitrite oxidation, (3) nitrate reduction to nitrite, (4) nitrite reduction to nitric oxide, (5) nitric oxide reduction to nitrous oxide, (6) nitrous oxide reduction to dinitrogen gas, (7) nitrogen fixation (not relevant in most wastewater treatment plants), (8) ammonium oxidation with nitrite (anammox). Complete nitrification comprises steps 1 and 2, complete denitrification comprises steps 3–6 (adapted from Kampschreur et al. [15]).

The primary energy requirement of N-fixation in the Haber–Bosch process is in the range of 37–45 MJ/kg-N, while the nitrification–denitrification wastewater treatment process (steps 1–2 and steps 3–6 in Figure 2) requires about 42.2–45 MJ/kg-N [26,27]. So, the total primary energy requirement for N-fixation and N-removal reaches 90 MJ/kg-N. N-removal by the deammonification process (a two-step process where ammonia-oxidizing bacteria aerobically convert half of the ammonia to nitrite, and anammox bacteria anaerobically oxidize the residual ammonia using nitrite to produce nitrogen gas without the organic carbon substrate required for conventional heterotrophic denitrification, as shown in step 1 and step 8 in Figure 2) requires 3.1 MJ/kg-N [27] to 16 MJ/kg-N [26], and reduces the total energy use of N-fixation and N-removal to less than 61 MJ/kg-N, which is the benchmark value. With respect to N_2O emissions, in the conventional nitrification–denitrification process, N_2O is produced in step 1 (aerobic ammonia oxidation), while in the denitrification (steps 3–6), incomplete denitrification can lead to N_2O emissions [15]. N-removal by the deammonification process results in less N_2O emission, as can be seen in Figure 2. The aerobic ammonium oxidation results in N_2O (step 1), but the anaerobic oxidation of ammonia to nitrogen gas (step 8) does not emit N_2O. The global warming potential of the deammonification process is only 40%, as compared with the conventional nitrification–denitrification process [28], which is considered as the benchmark value.

By means of these criteria, the alternatives were scored qualitatively, as shown in Table 1.

Table 1. Preselection of alternatives on four criteria.

	Sustainability		Recovery of Nitrogen in an Applicable Form		Maturity of the Alternative		Concentration Range
		++	Specific product	++	Mature technology		
+	Lower energy use and lower N_2O emissions	+	Concentrated stream separated from the wastewater	+	Available on the market	+	Within the range of 60–8800 mg/L and capable of treating large quantities
+−	Lower energy use or lower N_2O emissions	+−	Concentrated wastewater stream	+−	Successful pilot plant		
−	No lower energy use, no lower N_2O emissions	−	Transfer to N_2 gas in combination with energy production	−	Successful proof of concept	−	Outside the range of 60–8800 mg/L and/or not capable of treating large quantities
		−−	Transfer to N_2 gas	−−	In conceptual phase		

Note: ++ very positive score; + positive score; +− neutral score; − negative score; −− very negative score

2.4. Relation to Other Alternatives for Resource Recovery from Wastewater

In the Dutch program "The Energy and Raw Materials Factory", the focus is on the recovery of energy and the materials phosphorus, cellulose, bio-ALE (alginate-like exopolymers from aerobic granular sludge), and bioplastics from wastewater [23]. In this study, the relation of nitrogen recovery with biogas production, phosphorus recovery, and cellulose recovery was analyzed. Bio-ALE was excluded, because the recovery of bio-ALE requires the application of the Nereda aerobic granular sludge technology as wastewater treatment [29], and this technology is not applied at the WWTP Amsterdam-West. Bioplastic was excluded, because the production costs of this material are currently still rather high; it is twice as much as the regular market prices. In addition, there is no available stable industrial production process yet [23].

2.5. Interdependencies between Nitrogen Recovery Alternatives

There is a wide variety of available alternatives for nitrogen recovery and reuse. External factors, which may change over time due to technological, environmental, economic, and market developments, influence the choice for an optimal alternative. Adaptive policy making is an approach to make decisions at this moment, taking into account future developments. It considers uncertainties and complex dynamics, and adaptation pathways show which interventions can be done in which sequence and at which time [30]. This approach was applied to see interdependencies between the nitrogen recovery alternatives, as represented in adaptation pathway maps.

3. Results and Discussion

3.1. Nitrogen Flow through the Wastewater Treatment Process

The water balance of the WWTP Amsterdam-West is shown in Figure 3. The first step was a black box approach to close the water balance over the system. There was a slight unbalance of 1.8% over the whole system, which was probably due to evaporation. Therefore, 1.8% was added to the effluent flow. The incoming flow (1,044,548 inhabitants) consists of the flushing water of toilets (31.7 L/person/day), grey water (99.6 L/person/day), urine (0.94 L/person/day), feces (1.4 L/person/day), and rainwater. For rainwater, it was assumed that it contributed 20% to the total incoming flow [31–33].

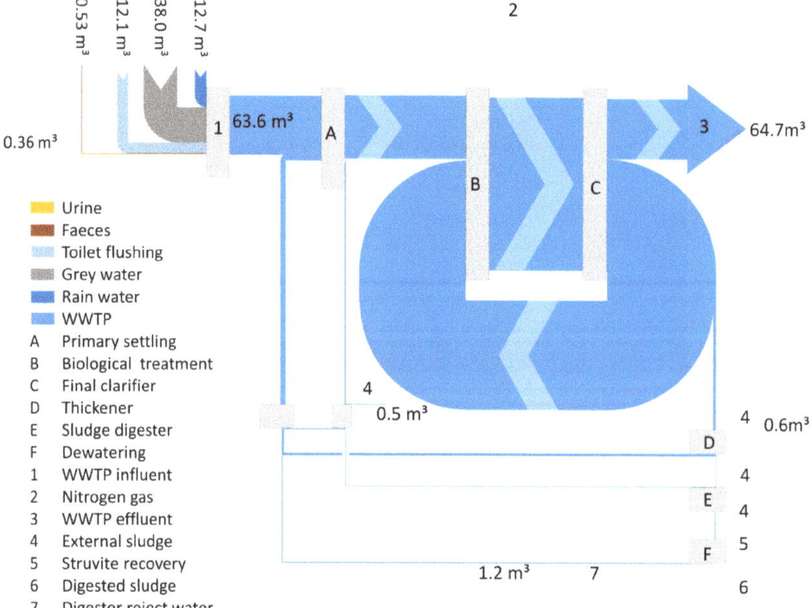

Figure 3. The water balance of the wastewater treatment plant (WWTP) Amsterdam-West (volume flows in 10^6 m^3/year).

The nitrogen balance of the WWTP Amsterdam-West is shown in Figure 4. Also, for this balance, the first step was a black box approach, based on the measured nitrogen concentration in the influent and effluent. Nitrogen in surplus sludge was determined at the plant (75 g N/kg ds). For primary sludge, digested sludge, and external sludge, the same value was assumed. Due to the low-volume flows, the impact of this assumption is very limited. The nitrogen content in the digester reject water was determined at 1030 mg/L, but showed large variations (750–1700 mg N/L). The balance was closed by the assumption that all other outflow concerned nitrogen gas. The total incoming nitrogen mass (exclusive of the incoming external sludge) was divided over urine, feces, flushing water of toilets, greywater, and rainwater, with the following assumptions: urine contributes 80% to the total incoming mass [34,35], the contribution of feces is based on 1.4 g N/person/day [32], while rainwater and the flushing water of the toilets do not contribute.

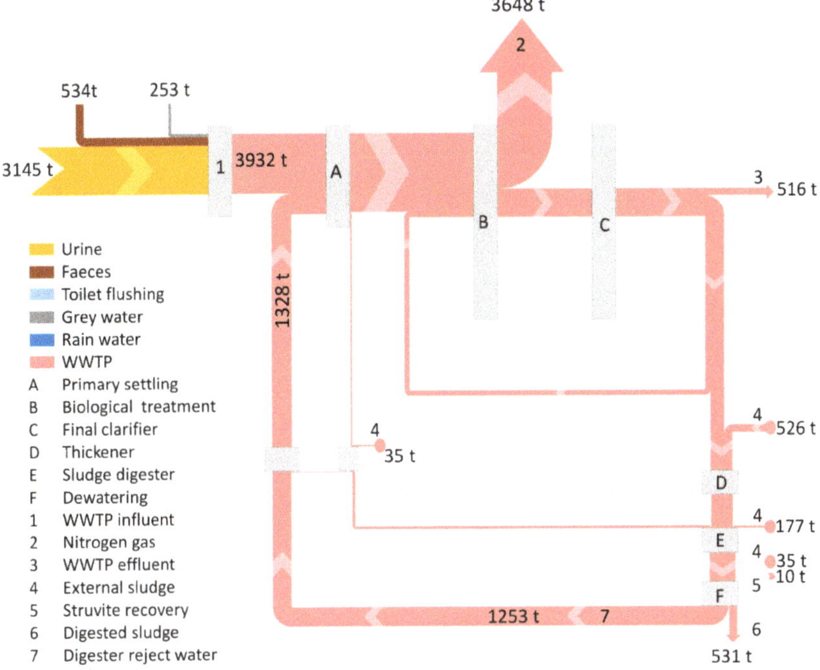

Figure 4. The nitrogen balance of the WWTP Amsterdam-West (mass flow in tons/year).

Based on these balances, the concentrations in specific flows can be calculated and related to the total nitrogen inflow through the system (3932 ton N in the influent, 773 ton N from external sludge, comprising in total 4705 ton N). Table 2 shows the results. Urine has the highest contribution and the highest concentration. Based on the urine volume and the assumed mass contribution to the influent (80%), the concentration is 8800 mg N/L, which is close to the concentration of 8830 mg N/L mentioned in Stowa [32]. The second flow with a high concentration is the digester rejects water. At a concentration of 1030 mg N/L, this flow contributes 27% to the total nitrogen inflow.

Table 2. Nitrogen concentration and relative nitrogen mass in four specific flows.

Flow	Concentration (mg N/L)	Relative Contribution to Total N Inflow (%)
Urine	8800	67
Digester reject water	1030	27
WWTP influent	61	84
WWTP effluent	8.1	11

Both the high concentrations and the relatively high contributions may be attractive to take these flows into account when considering nitrogen recovery and reuse. In addition, nitrogen recovery from these flows will lower the nitrogen load of the WWTP, and thus result in a lower energy use and a lower N_2O emission. Table 2 also shows the nitrogen concentrations in the influent and effluent of the treatment plant, and the relative contribution to the total nitrogen flow. The influent has a large contribution at a relatively low concentration.

3.2. Nitrogen Recovery and Reuse: Technologies and Strategies

At present, many technologies are available to recover nitrogen from wastewater [4,36–38]. In principle, these technologies can be divided in four strategies to recover and reuse nitrogen:

- technologies with the aim of recovering nitrogen directly from wastewater or digester reject water;
- technologies with the aim of concentrating nitrogen in wastewater or digester reject water to enhance recovery technologies;
- technologies to treat urine or sludge;
- technologies with the aim of incorporating nitrogen in biomass.

Figure 5 shows an overview of strategies with related technologies.

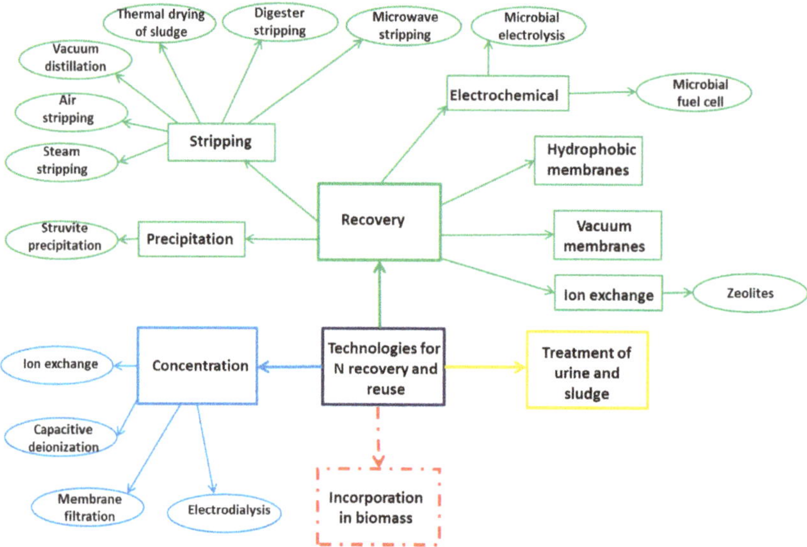

Figure 5. Overview of strategies with related technologies for nitrogen recovery and reuse.

The technologies for further evaluation were selected based on the four criteria. The fourth strategy, incorporation of nitrogen in biomass, was not considered, as this strategy focuses on recovery and/or the production of biomass from wastewater in general, and not on the recovery of nitrogen in specific. The results of the selection are shown in Table 3. A description of the technologies and the detailed scores on the criteria are presented in Supplemental Material 1 and Supplemental Material 2.

The selection shows that it is not possible for the technologies to reach a high score on the criterion "sustainability", because in all of the cases, N_2O emissions still take place on a level above the N_2O emissions of the benchmark process (Haber–Bosch and deammonification). Most of the technologies recover nitrogen from the digester reject. This reduces the N-load of the wastewater treatment system (maximum 27%, based on Table 2), but without a radical change of the wastewater treatment system, emissions will remain too high. A 27% reduction in N-load while maintaining the conventional nitrification–denitrification process will not result in a 60% decrease of global warming potential as can be achieved by introduction of the deammonification process. Only urine treatment (maximum 67% reduction in N-load) is close to the benchmark with respect to N_2O emissions. For that reason, it was decided to select the technologies for further evaluation based on a positive score on the other three criteria: recovery of nitrogen in an applicable form, maturity, and concentration range. Based on that, the technologies for further evaluation are struvite precipitation, air stripping, thermal drying of sludge with subsequent air treatment, hydrophobic membranes, vacuum membranes, urine treatment, and sludge reuse. Table 4 shows a first estimate of the nitrogen that can be recovered at the WWTP Amsterdam-West. The struvite recovery is based on the full-scale design of the WWTP Amsterdam-West and the operational experiences with this plant [11]. For air stripping, an efficiency of 90% was assumed [38]. The nitrogen recovery by the thermal drying of sludge is based on the nitrogen content in the sludge of WWTP Amsterdam-West and the maximum efficiency, as described in Horttanainen et al. [39]. As hydrophobic membranes for the treatment of digester reject water, polytetrafluoroethylene (PTFE) membranes (flat-sheet, hollow fiber, and spiral wound) and its expanded form (ePTFE) are preferred for NH_3 extraction due to their hydrophobic characters, excellent organic resistance, and chemical stability with acidic and alkaline solutions [40]. Efficiency depends strongly on the process conditions; an efficiency of 75% was assumed. Conventional flat-sheet porous PTFE membranes have been applied for vacuum membrane distillation for ammonia removal, with efficiencies varying between 70–90% [41]. The treatment of human urine for nitrogen recovery can be achieved with evaporation, electrodialysis, and reverse osmosis with at least 90% recovery [42]. With respect to sludge reuse, it was assumed that 100% of the digested sewage sludge is applied.

Table 3. Selection of technologies for nitrogen recovery and reuse.

Technology	Sustainability	Recovery of Nitrogen in an Applicable Form	Maturity	Concentration Range	Selected for Further Evaluation
Membrane filtration	+−	+−	++	+	No
Capacitive deionization	−	+−	−	+	No
Struvite precipitation	−	++	++	+	Yes
Steam stripping	−	++	+-	−	No
Air stripping	−	++	+	+	Yes
Vacuum distillation	−	++	−	+	No
Thermal sludge drying with subsequent air treatment	−	++	++	+	Yes
Digester stripping	−	++	−	+	No
Microwave stripping	−	++	+	+	No
Electrodialysis	−	+	+−	+	No
Microbial electrolysis	−	++	−	+	No
Microbial fuel cell	−	++	−	+	No
Hydrophobic membranes	−	++	+	+	Yes
Vacuum membranes	−	++	+	+	Yes
Ion exchange	−	++	+	−	No
Urine treatment	+−	++	+	+	Yes
Sludge reuse	−	++	+−	+	Yes

Note: ++ very positive score; + positive score; +− neutral score; − negative score; −− very negative score

Based on these estimates, it can be seen that especially air stripping, hydrophobic membranes, vacuum membranes, and urine treatment result in an increase of sustainability when the present wastewater treatment process of the WWTP Amsterdam-West is considered as a benchmark: The N-load of the wastewater treatment system reduces by 20–60%, resulting in a lower N_2O emission. Whether also the energy use will be reduced strongly depends on the energy use of the nitrogen recovery technology and the system boundaries. For example, air stripping requires

90 MJ/kg-N [26], which is much more than the nitrification–denitrification process (42.2–45 MJ/kg-N), but it is comparable to the total primary energy requirement of N-fixation and N-removal by nitrification–denitrification process (90 MJ/kg-N). Table 4 also shows that only technologies in parallel will result in a substantial nitrogen recovery. The use of technologies in parallel will be addressed in Section 3.4.

Table 4. Recovery of nitrogen at WWTP Amsterdam-West with the selected technologies.

Technology	Application at Stream	Process Conditions	N Recovery	
			Mass (tons)	% of Total N Flow
Struvite recovery	Digested sludge	Production of 900 ton struvite with 5.7% N	51.3	1.1
Air stripping	Digester reject water	90% efficiency	1128	24
Thermal drying of sludge	Digested sludge	531 ton N in sludge, 19% as ammonia, efficiency 99%	99.9	2.1
Hydrophobic membranes	Digester reject water	75% efficiency	940	20
Vacuum membranes	Digester reject water	75% efficiency	940	20
Urine treatment	Incoming urine	90% recovery	2831	60
Sludge reuse	Digested sewage sludge	100% application	531	11

3.3. Competition with Biogas Production and Recovery of Phosphorus and Cellulose

Biogas production, the recovery of phosphorus, and the recovery of cellulose are part of the Dutch program "The Energy and Raw Materials Factory" [23]. Nitrogen recovery is not a part of this program, so it is important to determine how the selected options for nitrogen recovery interact with biogas production, phosphorus recovery, and cellulose recovery. For biogas production, it is assumed that anaerobic sludge digestion is applied [43]; for phosphorus recovery, it is assumed that struvite precipitation in the digested sludge is applied [8–10], and for cellulose recovery, it is assumed that fine-mesh sieves are applied as pretreatment for biological municipal wastewater treatment [19]. Table 5 shows the interactions. In fact, all of the nitrogen recovery technologies are no-regret measures, except for the reuse of sludge. The reuse of sludge has an effect on biogas production. In case it is acceptable to reuse sludge with a lower organic carbon content, there is no interaction between nitrogen recovery through sludge reuse and the Dutch program "The Energy and Raw Materials Factory" at all. As nitrogen recovery on the one hand, and biogas production, phosphorus recovery, and cellulose recovery, on the other, do not exclude each other, biogas production, phosphorus recovery, and cellulose recovery were not taken into account for the adaptation pathways of nitrogen recovery alternatives.

In addition to the effects of nitrogen recovery on biogas production, phosphorus recovery, and cellulose recovery, it is also important to determine the effects vice-versa. Table 6 shows the results. It can be concluded that biogas production has an effect. With respect to the nitrogen recovery technologies—struvite precipitation, air stripping, and thermal drying of sludge—it is a win–win measure, as it enhances nitrogen recovery. With respect to sludge reuse, it is a lock-in measure: it reduces the total amount of sludge and the nitrogen content of the sludge. Also, phosphorus recovery has an effect: it reduces the N-content and P-content of the sludge. However, as in the Netherlands, there is a surplus of manure with especially a surplus of phosphorus, the removal of phosphorus from the wastewater treatment sludge may be beneficial to market this material in agriculture [44].

Table 5. Effect of selected nitrogen recovery technologies on biogas production, phosphorus recovery, and cellulose recovery from the Dutch program "The Energy and Raw Materials Factory".

N-Recovery Technology	Effect on		
	Biogas Production	**Phosphorus Recovery**	**Cellulose Recovery**
Struvite precipitation	Nitrogen is recovered as struvite from the sludge after digestion, and does not affect the digestion of sludge and biogas production	Nitrogen and phosphorus are simultaneously removed as struvite, no interference	Nitrogen is recovered as struvite from the digested sludge and does not affect cellulose recovery as pretreatment
Air stripping	Air stripping is applied on the digester reject water, and does not affect the digestion of sludge and biogas production	Air stripping is applied on the digester reject water and does not affect the recovery of phosphorus as struvite from the digested sludge	Air stripping is applied on the digester reject water and does not affect cellulose recovery as pretreatment
Thermal drying of sludge	Thermal drying of sludge is applied after sludge digestion, and does not affect biogas production	Thermal drying of sludge is applied after struvite recovery and does not affect phosphorus recovery	Thermal drying of sludge takes place at the end of the treatment process and does not affect cellulose recovery as pretreatment
Hydrophobic membranes and vacuum membranes	Hydrophobic and vacuum membranes are applied on the digester reject water, and do not affect the digestion of sludge and biogas production	Hydrophobic and vacuum membranes are applied on the digester reject water and do not affect the recovery of phosphorus as struvite from the digested sludge	Hydrophobic membranes and vacuum membranes are applied on the digester reject water and do not affect cellulose recovery as pretreatment
Urine treatment	Urine hardly contains any organic material; separate urine collection and treatment does not affect biogas production	The total nitrogen load to the wastewater treatment system is that high (urine contributes for 80% to nitrogen mass in the influent, still 20% in other incoming flows) that the separate collection and treatment of urine does not affect phosphorus recovery through struvite precipitation	Urine contains no cellulose, so the separate collection and treatment of urine does not affect cellulose recovery
Sludge reuse	In case the aim is to use sludge with a high organic carbon content, sludge digestion is not preferred, so it does affect biogas production	Sludge is used as a residual product, so it does not affect preceding phosphorus recovery	Sludge is used as a residual product so it does not affect cellulose recovery as pretreatment

Table 6. Effects of biogas production, phosphorus recovery, and cellulose recovery from the Dutch program "The Energy and Raw Materials Factory" (TERMF) on selected nitrogen recovery technologies.

TERMF Recovery	Effect on N-Recovery Technology					
	Struvite Precipitation	**Air Stripping**	**Thermal Drying of Sludge**	**Hydrophobic and Vacuum Membranes**	**Urine Treatment**	**Sludge Reuse**
Biogas production	Through the digestion of sludge, P and N are released in high concentrations, which is advantageous for struvite precipitation	Through the digestion of sludge, N is released in high concentrations as ammonium/ammonia, which is advantageous for air stripping	Through the digestion of sludge, N is released in high concentrations as ammonium/ammonia, which is advantageous for recovery during the drying of sludge	Through the digestion of sludge, N is released in high concentrations as ammonium/ammonia, which is advantageous for recovery during membrane filtration	Biogas is produced during sludge digestion and does not affect the separate collection and treatment of urine as the first step in the wastewater treatment system	Sludge digestion for biogas production reduces the amount of sludge and transfers nitrogen to the digester reject water, resulting in a lower N-content of the sludge
Phosphorus recovery	Nitrogen recovery and phosphorus are simultaneously removed as struvite, no interference	Phosphorus recovery as struvite precipitation is applied after sludge digestion, and thus does not affect N-recovery through the air stripping of digester reject water	Phosphorus recovery through struvite precipitation lowers both N and P-concentrations in the sludge, so the N-recovery through sludge drying after struvite precipitation is lower	Phosphorus recovery as struvite precipitation is applied after sludge digestion, and thus does not affect N-recovery from digester reject water through membrane filtration	Phosphorus is recovered from the digested sludge, and does not affect the separate collection and treatment of urine as the first step in the wastewater treatment system	Phosphorus recovery through struvite precipitation lowers the N- and P-content of the sludge, but a low P-content may be attractive to market the product in agriculture
Cellulose recovery	N and P are not recovered through cellulose recovery, so there is no effect on N recovery through struvite precipitation	N is not recovered through cellulose recovery, so there is no effect on N-recovery through the air stripping of digester reject water	The total amount of organic material that is introduced in the wastewater treatment system is reduced, so the amount of sludge is reduced. However, the N-mass in the sludge is not reduced	N is not recovered through cellulose recovery, so no effect on N-recovery through the membrane filtration of digester reject water	Urine is collected and treated prior to cellulose recovery, so no effect	The total amount of organic material that is introduced in the subsequent wastewater treatment system after cellulose recovery is reduced, so the amount of sludge is reduced. However, the N-mass in the sludge is not reduced

3.4. Adaptation Pathway Maps for Nitrogen Recovery Alternatives

To construct the adaptation pathways, the alternatives were grouped into three specific actions: (1) the recovery of nitrogen; (2) the treatment of specific waste streams; and (3) other alternatives that may affect nitrogen recovery.

The first group contains struvite precipitation, air stripping, the thermal drying of sludge, and hydrophilic and vacuum membranes to recover nitrogen. These technologies can be applied in the wastewater treatment system, but can also applied on pure urine that is separately collected. The treatment of specific streams (group 2) concerns urine treatment to reuse this stream directly (e.g., hydrolysis of urea or stabilization of urine) and sludge reuse. Other alternatives that may affect nitrogen recovery (group 3) are an increase of the nitrogen content in the digester reject water e.g., through thermal hydrolysis pretreatment of sludge [45,46], the addition of urine to the existing wastewater treatment plant, and the separate collection of urine.

The adaptation pathways map, as shown in Figure 6, presents an overview of the relevant pathways to reach the desired shared goal: nitrogen reuse from wastewater. All of the alternatives are represented by a colored horizontal line, and can be considered as 'different ways leading to Rome'. A vertical line with the same color indicates that after the choice of a specific alternative (with that color), switches are possible to other alternatives via transfer stations. A terminal station represents the moment of an adaptation tipping point: the alternative is effective until this moment. Transfer stations show the available alternatives after this point. Transparent pathways and transfer stations represent unnecessarily complicated ways to achieve a measure.

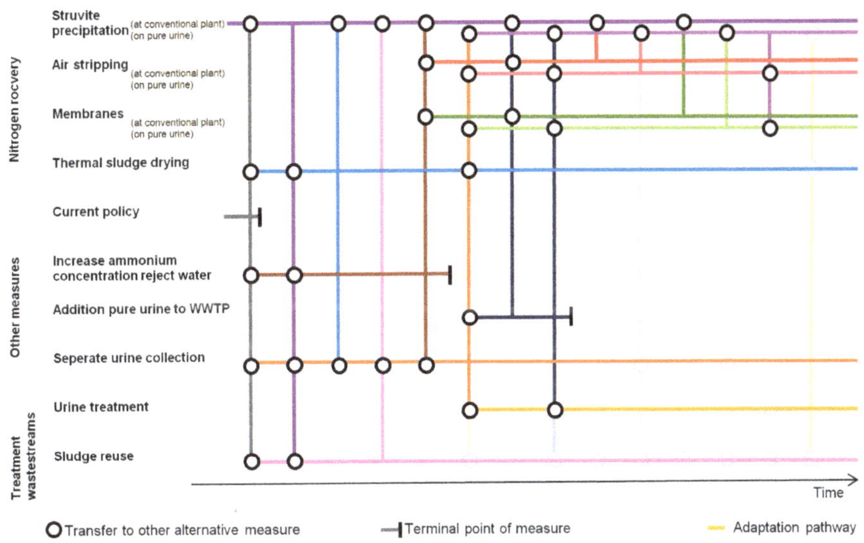

Figure 6. Adaptation pathways map of alternatives for nitrogen recovery from wastewater.

As an example: if the choice is made for struvite precipitation, the purple line is followed. From this line, a vertical purple line originates. This means that after the choice for struvite precipitation, a switch can be made to the thermal drying of sludge through the transfer station. On the other hand, no switch can be made from thermal sludge drying to struvite precipitation at the same moment in time, in case the initial choice was the thermal drying of sludge: the vertical line has another color. Later in time, the switch is possible (crossing blue lines).

The map shows an indication of time on the X-axis, which is not absolute. It indicates that some alternatives are not directly applicable, and some other measures are needed first. For example, the treatment of urine and/or addition of urine to the existing treatment require new sanitation concepts. The introduction of new sanitation is only possible in new housing estates, and requires time. However, urine can already been collected separately on an ad hoc basis, e.g., at festivals, and this urine can be used in many alternatives. The application of hydrophobic and vacuum membranes require high N concentrations in the digester reject water, so the first step is to develop methods to increase this concentration, and after this development, membranes are applicable.

Although the adaptation pathways map is complex, it is a very helpful tool to determine which pathways have to be followed to realize a specific scenario with a specific goal. Figures 7–10 show four specific scenarios that decision makers could follow.

Figure 7 shows the pathways that can be followed when the goal is to recover a limited amount of nitrogen with alternatives that have little impact on the existing wastewater treatment systems, and with a high level of feasibility. Recovery through the thermal drying of sludge and struvite precipitation seems attractive.

Figure 8 shows the pathways that can be followed when the ambition is to recover more nitrogen, and more risks can be accepted. In that case, technologies to increase the concentration of nitrogen in the digester reject water with subsequent air stripping of the digester reject water can be chosen.

In case a high impact is allowed, new sanitation can be chosen. The corresponding pathways are shown in Figure 9.

Finally, the goal can be to recover the maximum amount of nitrogen from wastewater. This scenario with corresponding pathways is presented in Figure 10. Many alternatives have to be introduced in parallel: nitrogen is recovered from pure urine, as well as from the sludge and digester reject water at the wastewater treatment plant.

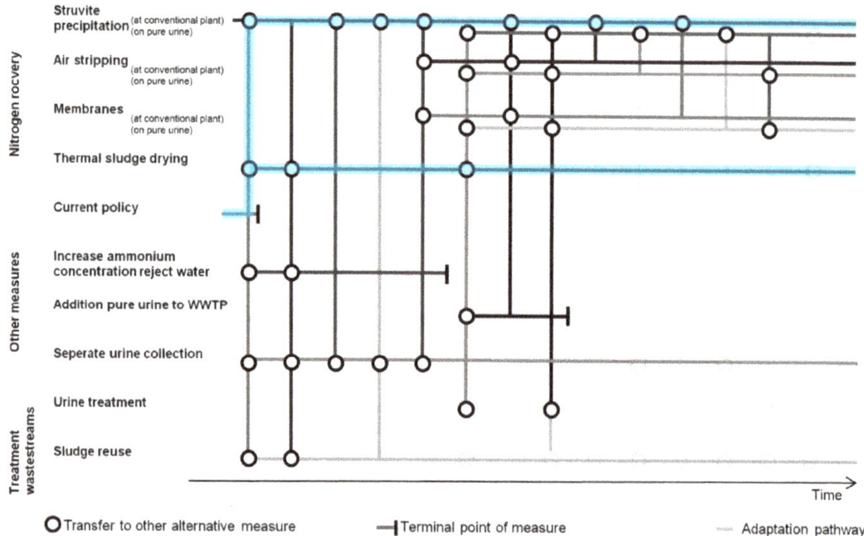

Figure 7. Adaptation pathways for nitrogen recovery from wastewater for scenario A: limited N-recovery with limited impact.

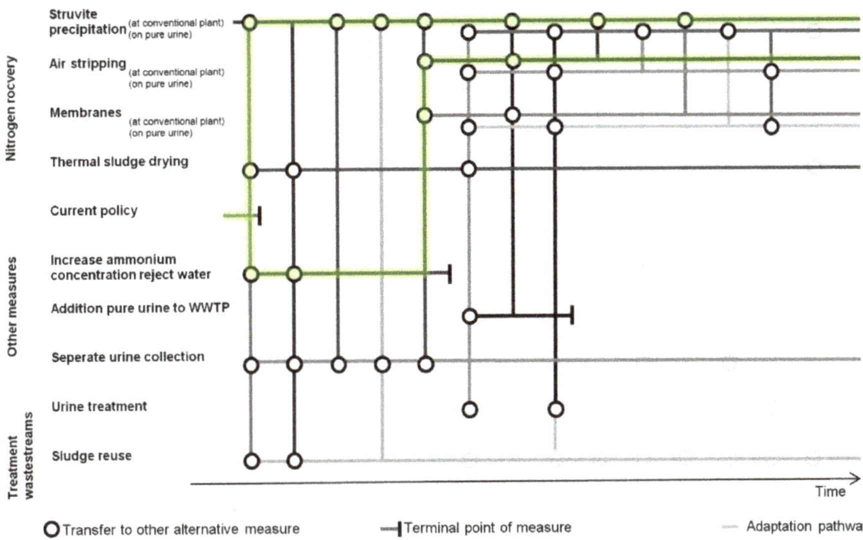

Figure 8. Adaptation pathways for nitrogen recovery from wastewater for scenario B: moderate N-recovery with acceptable risks.

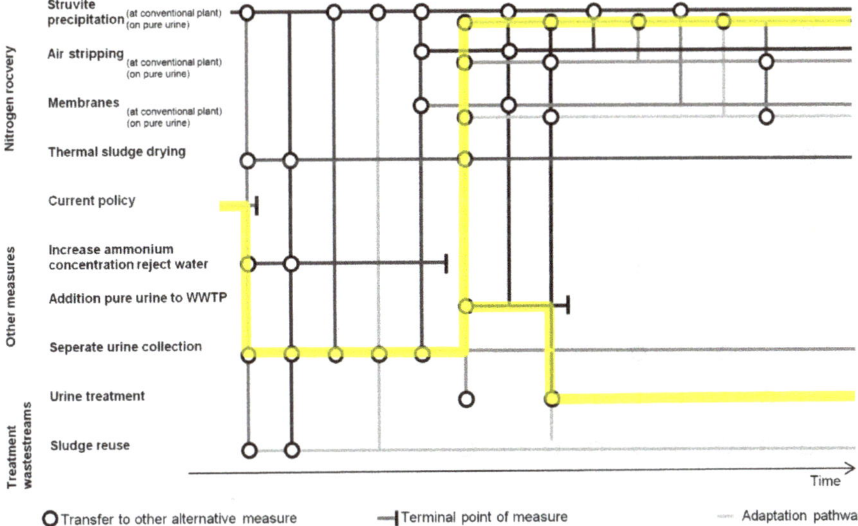

Figure 9. Adaptation pathways for nitrogen recovery from wastewater for scenario C: new sanitation with high impact.

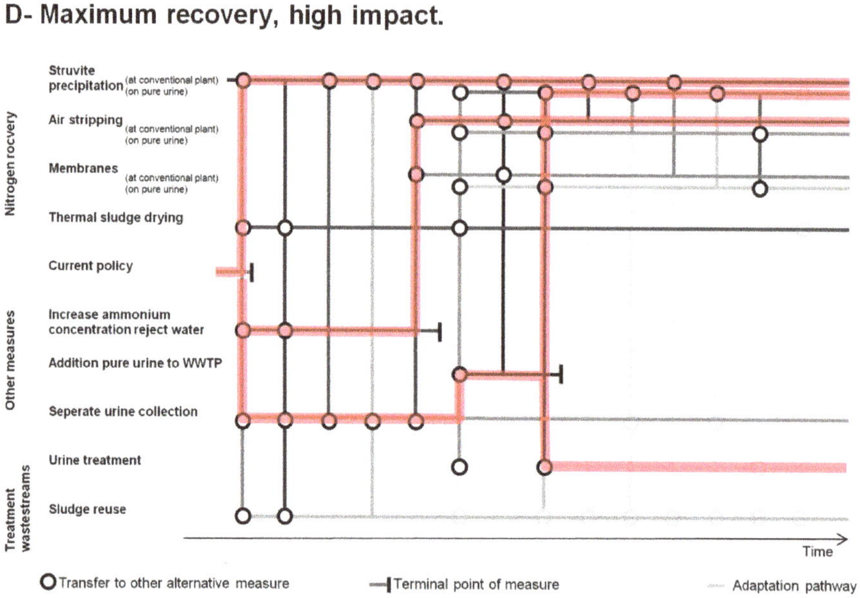

D- Maximum recovery, high impact.

Figure 10. Adaptation pathways for nitrogen recovery from wastewater for scenario D: maximum N-recovery with high impact.

4. Conclusions

- Nitrogen recovery from wastewater with the existing wastewater treatment system as a starting point results in only limited improvement in sustainability.
- Radical changes in wastewater treatment, and the application of several nitrogen recovery technologies in parallel, are required to improve sustainability substantially. The separate collection and treatment of urine is an attractive option, but requires a completely new infrastructure for wastewater collection and wastewater treatment.
- Nitrogen recovery from wastewater does not negatively affect biogas production from wastewater, phosphorus recovery from wastewater, and cellulose recovery from wastewater.
- The use of adaptation pathways maps is an attractive method to compare and judge several combinations of nitrogen recovery technologies, especially when different strategies have to be analyzed, and technological and market developments are uncertain.

Supplementary Materials: The following are available online at http://www.mdpi.com/2071-1050/10/12/4605/s1, Supplemental Material 1: Description of technologies; Supplemental Material 2: Evaluation of technologies.

Author Contributions: The work is conceived and supervised by J.P.v.d.H. and O.R. R.D. worked on the literature review, data collection and wrote the original draft as a report. All three authors contributed towards the preparation and review of the manuscript.

Funding: This research received no external funding.

Conflicts of Interest: The authors declare no conflict of interest.

References

1. Ezeh, A.C.; Bongaarts, J.; Mberu, B. Global population trends and policy options. *Lancet* **2012**, *380*, 142–148. [CrossRef]
2. United Nations. *World Urbanization Prospects: The 2015 Revison*; United Nations: New York, NY, USA, 2015.

3. Alexandratos, N.; Bruinsma, J. *World Agriculture towards 2030/2050: The 2012 Revision*; ESA Working Paper No. 12-03; FAO: Rome, Italy, 2012. Available online: http://www.fao.org/docrep/016/ap106e/ap106e.pdf (accessed on 14 August 2018).
4. Sengupta, S.; Nawaz, T.; Beaudry, J. Nitrogen and Phosphorus Recovery from Wastewater. *Curr. Pollut. Rep.* **2015**, *1*, 155–166. [CrossRef]
5. Duley, B. *Recycling Phosphorus by Recovery from Sewage*; Centre Europeen dÉtudes des Polyphosphates: Brussels, Belgium, 1998; pp. 1–17.
6. European Commission. *The European Critical Raw Materials Review*; MEMO/14/37726/05/2014; European Commission: Brussels, Belgium, 2014. Available online: http://europa.eu/rapid/press-release_MEMO-14-377_en.htm (accessed on 14 August 2018).
7. Puchongkawarin, C.; Gomez-Mont, C.; Stuckey, D.C.; Chachuat, B. Optimization-based methodology for the development of wastewater facilities for energy and nutrient recovery. *Chemosphere* **2015**, *140*, 150–158. [CrossRef] [PubMed]
8. Doyle, J.D.; Parsons, S.A. Struvite formation, control and recovery. *Water Res.* **2002**, *36*, 3925–3940. [CrossRef]
9. Garcia-Belinchón, C.; Rieck, T.; Bouchy, L.; Galí, A.; Rougé, P.; Fàbregas, C. Struvite recovery: Pilot-scale results and economic assessment of different scenarios. *Water Pract. Technol.* **2013**, *8*, 119–130. [CrossRef]
10. Bergmans, B.J.C.; Veltman, A.M.; van Loosdrecht, M.C.M.; van Lier, J.B.; Rietveld, L.C. Struvite formation for enhanced dewaterability of digested wastewater sludge. *Environ. Technol.* **2014**, *35*, 549–555. [CrossRef]
11. van der Hoek, J.P.; Struker, A.; de Danschutter, J.E.M. Amsterdam as a sustainable European metropolis: Integration of water, energy and material flows. *Urban Water J.* **2017**, *14*, 61–68. [CrossRef]
12. Erisman, J.W.; Sutton, M.A.; Galloway, J.; Klimont, Z.; Winiwarter, W. How a century of ammonia synthesis changed the world. *Nat. Geosci.* **2008**, *1*, 636–639. [CrossRef]
13. Stein, L.Y.; Klotz, M.G. The nitrogen cycle. *Curr. Biol.* **2016**, *26*, R94–R98. [CrossRef]
14. Paredes, D.; Kuschk, P.; Mbwette, T.S.A.; Stange, F.; Müller, R.; Köser, H. New aspects of microbial nitrogen transformations in the context of wastewater treatment—A review. *Eng. Life Sci.* **2007**, *7*, 13–25. [CrossRef]
15. Kampschreur, M.J.; Temmink, H.; Kleerebezen, R.; Jetten, M.S.M.; van Loosdrecht, M.C.M. Nitrous oxide emission during wastewater treatment. *Water Res.* **2009**, *43*, 4093–4103. [CrossRef] [PubMed]
16. Diamantis, V.I.; Anagnostopoulos, K.; Melidis, P.; Ntougias, S.; Aivasidis, A. Intermittent operation of low pressure UF membranes for sewage reuse at household level. *Water Sci. Technol.* **2013**, *68*, 799–806. [CrossRef] [PubMed]
17. Membranes for Energy and Water Recovery. The MEMORY Project: Co-Funded by the European Community under the Life+ Financial Instrument with the Grant Agreement n.LIFE13ENV/ES/001353. Available online: www.life-memory.eu (accessed on 14 November 2018).
18. Wang, X.; McCarty, P.L.; Liu, J.; Ren, N.-Q.; Lee, D.-J.; Yu, H.-Q.; Qian, J. Probalistic evaluation of integrating resource recovery into wastewater treatment to improve environmental sustainability. *Proc. Natl. Acad. Sci. USA* **2015**, *112*, 1630–1635. [CrossRef] [PubMed]
19. Ruiken, C.J.; Breurer, G.; Klaversma, E.; Santiago, T.; van Loosdrecht, M.C.M. Sieving wastewater cellulose recovery, economic and energetic evaluation. *Water Res.* **2013**, *47*, 43–48. [CrossRef] [PubMed]
20. Tamis, J.; Marang, L.; Jiang, Y.; van Loosdrecht, M.C.M.; Kleerebezem, R. Modeling PHA-producing microbial enrichment cultures—Towards a generalized model with predictive power. *New Biotechnol.* **2014**, *31*, 324–334. [CrossRef] [PubMed]
21. Kleerebezem, R.; van Loosdrecht, M.C.M. Mixed culture biotechnology for bioengineering production. *Curr. Opin. Biotechnol.* **2007**, *18*, 207–212. [CrossRef] [PubMed]
22. Matassa, S.; Verstraete, W.; Pikaar, I.; Boon, N. Autotrophic nitrogen assimilation and carbon capture for microbial protein production by a novel enrichment of hydrogen-oxidizing bacteria. *Water Res.* **2016**, *101*, 137–146. [CrossRef]
23. van Leeuwen, K.; de Vries, E.; Koop, S.; Roest, K. The Energy & Raw Materials Factory: Role and Potential Contribution to the Circular Economy of the Netherlands. *Environ. Manag.* **2018**, *61*, 786–795.
24. van der Hoek, J.P.; de Fooij, H.; Struker, A. Wastewater as a resource: Strategies to recover resources from Amsterdam's wastewater. *Resour. Conserv. Recycl.* **2016**, *113*, 53–56. [CrossRef]
25. Waternet. *Technical Year Report Wastewater Treatment 2017*; Internal Report; Waternet: Amsterdam, The Netherlands, 2018. (In Dutch)

26. Maurer, M.; Schwegler, P.; Larsen, T.A. Nutrients in urine: Energetic aspects of the removal and recovery. *Water Sci. Technol.* **2003**, *48*, 37–46. [CrossRef]
27. Mulder, A. The quest for sustainable nitrogen removal technologies. *Water Sci. Technol.* **2003**, *48*, 67–75. [CrossRef] [PubMed]
28. Lin, Y.; Guo, M.; Shah, N.; Stuckey, D.C. Economic and environmental evaluation of nitrogen removal and recovery methods from wastewater. *Bioresour. Technol.* **2016**, *215*, 227–238. [CrossRef] [PubMed]
29. van der Roest, H.; van Loosdrecht, M.C.M.; Langkamp, E.J.; Uijterlinde, C. Recovery and reuse of Bio-ALE from granular Nereda sludge. *Water* **2015**, *21*, 48.
30. Haasnoot, M.; Kwakkel, J.H.; Walker, W.E.; ter Maat, J. Dynamic adaptive policy pathways: A method for crafting robust decisions for a deeply uncertain world. *Glob. Environ. Chang.* **2013**, *23*, 485–498. [CrossRef]
31. de Fooij, H. Wastewater as a Resource—Strategies to Recover Resources from Amsterdam's Wastewater. Master's Thesis, University of Twente, Enschede, The Netherlands, 25 January 2015.
32. Stowa. *Treatment of Urine*; Report Stowa 2010-W02; Foundation for Applied Water Research: Amersfoort, The Netherlands, 2010. (In Dutch)
33. Rose, C.; Parker, A.; Jefferson, B.; Cartmell, E. The Characterization of Feces and Urine: A Review of the Literature to Inform Advanced Treatment Technology. *Crit. Rev. Environ. Sci. Technol.* **2015**, *45*, 1827–1879. [CrossRef] [PubMed]
34. Stowa. *Separate Urine Collection and Treatment. Options for Sustainable Wastewater Treatment Systems and Mineral Recovery*; Report Stowa 2001-39; Foundation for Applied Water Research: Amersfoort, The Netherlands, 2001.
35. Stowa. *Options for Separate Treatment of Urine*; Report Stowa 2005-11; Foundation for Applied Water Research: Amersfoort, The Netherlands, 2005.
36. Maurer, M.; Muncke, J.; Larsen, T.A. Technologies for nitrogen recovery and reuse. In *Water Recycling and Resource Recovery in Industry*; Lens, P., Pol, L.H., Wilderer, P.A., Asano, T., Eds.; IWA Publishing: London, UK, 2002; pp. 491–510.
37. Stowa. *Explorative Research on Innovative Nitrogen Recovery*; Report 2012-51; Foundation for Applied Water Research: Amersfoort, The Netherlands, 2012.
38. Capodaglio, A.G.; Hlavínek, P.; Raboni, M. Physico-chemical technologies for nitrogen removal from wastewater: A review. *Rev. Ambient. Agua* **2015**, *10*, 481–489.
39. Horttanainen, M.; Deviatkin, I.; Havukainen, J. Nitrogen release from mechanically dewatered sewage sludge during thermal drying and potential for recovery. *J. Clean. Prod.* **2017**, *142*, 1819–1826. [CrossRef]
40. Kunz, A.; Mukhtar, S. Hydrophobic membrane technology for ammonia extraction from wastewaters. *J. Braz. Assoc. Agric. Eng.* **2016**, *36*, 377–386. [CrossRef]
41. EL-Bourawi, M.S.; Khayet, M.; Ra, M.; Ding, Z.; Li, Z.; Zhang, X. Application of vacuum membrane distillation for ammonia removal. *J. Membr. Sci.* **2007**, *301*, 200–209. [CrossRef]
42. Maurer, M.; Pronk, W.; Larsen, T.A. Treatment processes for source-separated urine. *Water Res.* **2006**, *40*, 3151–3166. [CrossRef]
43. McCarty, P.L.; Bae, J.; Kim, J. Domestic wastewater treatment as a net energy producer—Can this be achieved? *Environ. Sci. Technol.* **2011**, *45*, 7100–7106. [CrossRef] [PubMed]
44. Regelink, I.C.; Ehlert, P.A.I.; Römkens, P.F.A.M. *Perspectives of the Supply of (Phosphorus Reduced) Wastewater Treatment Sludge to the Agricultural Sector*; Report 2819; Wageningen Environmental Research: Wageningen, The Netherlands, 2017. (In Dutch)
45. van Dijk, L.; de Man, A. Continuous thermal sludge hydrolysis: Lower costs and more energy at the wastewater treatment plant. *J. H_2O* **2010**, *43*, 8–9. (In Dutch)
46. Phothilangka, P.; Schoen, M.A.; Huber, M.; Luchetta, P.; Winkler, T.; Wett, B. Prediction of thermal hydrolysis pretreatment on anaerobic digestion of waste activated sludge. *Water Sci. Technol.* **2008**, *58*, 1467–1473. [CrossRef] [PubMed]

Article

Effect of Nutrient Removal and Resource Recovery on Life Cycle Cost and Environmental Impacts of a Small Scale Water Resource Recovery Facility

Ben Morelli [1], Sarah Cashman [1], Xin (Cissy) Ma [2,*], Jay Garland [3], Jason Turgeon [4], Lauren Fillmore [5], Diana Bless [2] and Michael Nye [3]

1 Eastern Research Group, 110 Hartwell Ave., Lexington, MA 02421, USA; ben.morelli@erg.com (B.M.); sarah.cashman@erg.com (S.C.)

2 United States Environmental Protection Agency, National Risk Management Research Laboratory, 26 West Martin Luther King Drive, Cincinnati, OH 45268, USA; bless.diana@epa.gov

3 United States Environmental Protection Agency, National Exposure Research Laboratory, 26 West Martin Luther King Drive, Cincinnati, OH 45268, USA; garland.jay@epa.gov (J.G.); nye.michael@epa.gov (M.N.)

4 United States Environmental Protection Agency, Region 1, 5 Post Office Square, Suite 100, OEP 5-2, Boston, MA 02109, USA; turgeon.jason@epa.gov

5 Water Research Foundation, 1199 N Fairfax Street, Suite 900, Alexandria, VA 22314, USA; laurenfillmore@wildblue.net

* Correspondence: ma.cissy@epa.gov; Tel.: +1-513-569-7828

Received: 6 September 2018; Accepted: 28 September 2018; Published: 3 October 2018

Abstract: To limit effluent impacts on eutrophication in receiving waterbodies, a small community water resource recovery facility (WRRF) upgraded its conventional activated sludge treatment process for biological nutrient removal, and considered enhanced primary settling and anaerobic digestion (AD) with co-digestion of high strength organic waste (HSOW). The community initiated the resource recovery hub concept with the intention of converting an energy-consuming wastewater treatment plant into a facility that generates energy and nutrients and reuses water. We applied life cycle assessment and life cycle cost assessment to evaluate the net impact of the potential conversion. The upgraded WRRF reduced eutrophication impacts by 40% compared to the legacy system. Other environmental impacts such as global climate change potential (GCCP) and cumulative energy demand (CED) were strongly affected by AD and composting assumptions. The scenario analysis showed that HSOW co-digestion with energy recovery can lead to reductions in GCCP and CED of 7% and 108%, respectively, for the upgraded WRRF (high feedstock-base AD performance scenarios) relative to the legacy system. The cost analysis showed that using the full digester capacity and achieving high digester performance can reduce the life cycle cost of WRRF upgrades by 15% over a 30-year period.

Keywords: LCA; LCCA; wastewater treatment; anaerobic digestion; biogas; resource recovery; nutrient removal; water-energy-nutrient nexus

1. Introduction

Urban water systems have been evolving as the industrial market economy grows. During the last century, cities in the U.S. and around the world have implemented municipality-run water management approaches to resolve sanitary and freshwater supply issues [1]. Wastewater has been treated as waste to be eliminated with the investment of large amounts of energy and materials, regardless of the potential values of wastewater constituents. In recent decades, many municipalities are facing deteriorating water quality in water bodies due to eutrophication and pollution from point-sources such as effluents from wastewater treatment facilities. In response, the U.S. Environmental Protection Agency (U.S. EPA) has implemented more stringent effluent quality standards [2]. In addition, much of the wastewater treatment infrastructure is in dire need of improvement due to age, wear, and tear. In 2013, the American Society of Civil Engineers' Infrastructure Report Card assigned both drinking water and wastewater infrastructures a grade of D^+, indicating a considerable backlog of overdue maintenance and a pressing need for modernization [3]. With a growing population facing increased regulatory requirements, resource constraints, and financial challenges, communities are seeking more comprehensive and sustainable solutions to address multiple environmental challenges and maximize the recovery of water, energy, nutrients, and materials [1,4,5]. Municipal wastewater and other high strength organic wastes (HSOW) generated in cities are now regarded as a resource for water, energy, and nutrients [6–10].

However, the environmental sustainability of wastewater systems goes beyond the treatment plants. It has been argued that many impacts occur at a larger watershed level or along upstream supply chains during energy, chemical, and material production [11,12]. These complex water issues are inherently intertwined and cannot be solved by traditional siloed water management approaches [1]. It is necessary to apply system-based tools or metrics and integrated assessment frameworks such as life cycle assessment (LCA) and life cycle cost assessment (LCCA) to measure trade-offs and develop optimized solutions [5,13].

With increased environmental and financial resources at stake in these more complex processes, the U.S. EPA research team intends to explore how best to meet a set of goals that often seem contradictory. The effluent quality improvement associated with more stringent standards for nutrient removal may often be achieved at the expense of increases in energy use, chemical inputs, and system costs. When communities are required to improve nutrient removal to reduce eutrophication in receiving waterbodies, system analyses can help identify environmentally efficient and cost-effective options and incentivize utilities to strive for energy self-sufficiency.

Driven by the changes in regulatory nutrient permit limits instituted by the Chesapeake Bay Watershed Cleanup Initiative, the Village of Bath, New York and Bath Electric, Gas & Water Systems (BEGWS) proposed the resource recovery hub concept to move its traditional wastewater industry in a new cost-effective direction. BEGWS' goal was conversion of an energy-consuming wastewater treatment plant into a water resource recovery facility (WRRF) [14] that generates energy and reuses water and nutrients to improve the resiliency, reliability and revenue of the local electrical and water supply while reducing the environmental footprint of the facility and the watershed.

BEGWS proposed a system upgrade ("upgraded system") from a "legacy system" for enhanced nutrient removal to reach a summer time permit limit of 3.6 mg/L ammonia nitrogen. The potential retrofitted WRRF in Bath, NY provided a unique opportunity as a testbed to evaluate transformative technologies and holistic solutions to water, energy, nutrient, and solid waste resource challenges. In this study, LCA and LCCA tools were used: (1) to evaluate the comparative environmental benefits and burdens associated with the legacy and upgraded systems; and (2) to determine the energy recovery potential of the upgraded systems including co-digestion of HSOW and the cost benefits of offsetting energy use.

The focus of this study was the WRRF in Bath, NY (approximately 5600 people). The facility has a permitted flow capacity of 1 million gallon per day (MGD) (3800 m^3/day) [15] and discharges effluent into the Cohocton River, which empties into the Chesapeake Bay potentially impacting sensitive habitats [16]. Figure 1 shows the LCA system boundaries for the legacy and upgraded WRRFs. Foreground unit processes refer to WRRF and end-of-life (EOL) unit processes for which life cycle inventory (LCI) data were developed as a part of this project. Background unit processes refer to upstream material and energy production processes for which LCI data were drawn from existing databases, as discussed in Section 2.2.

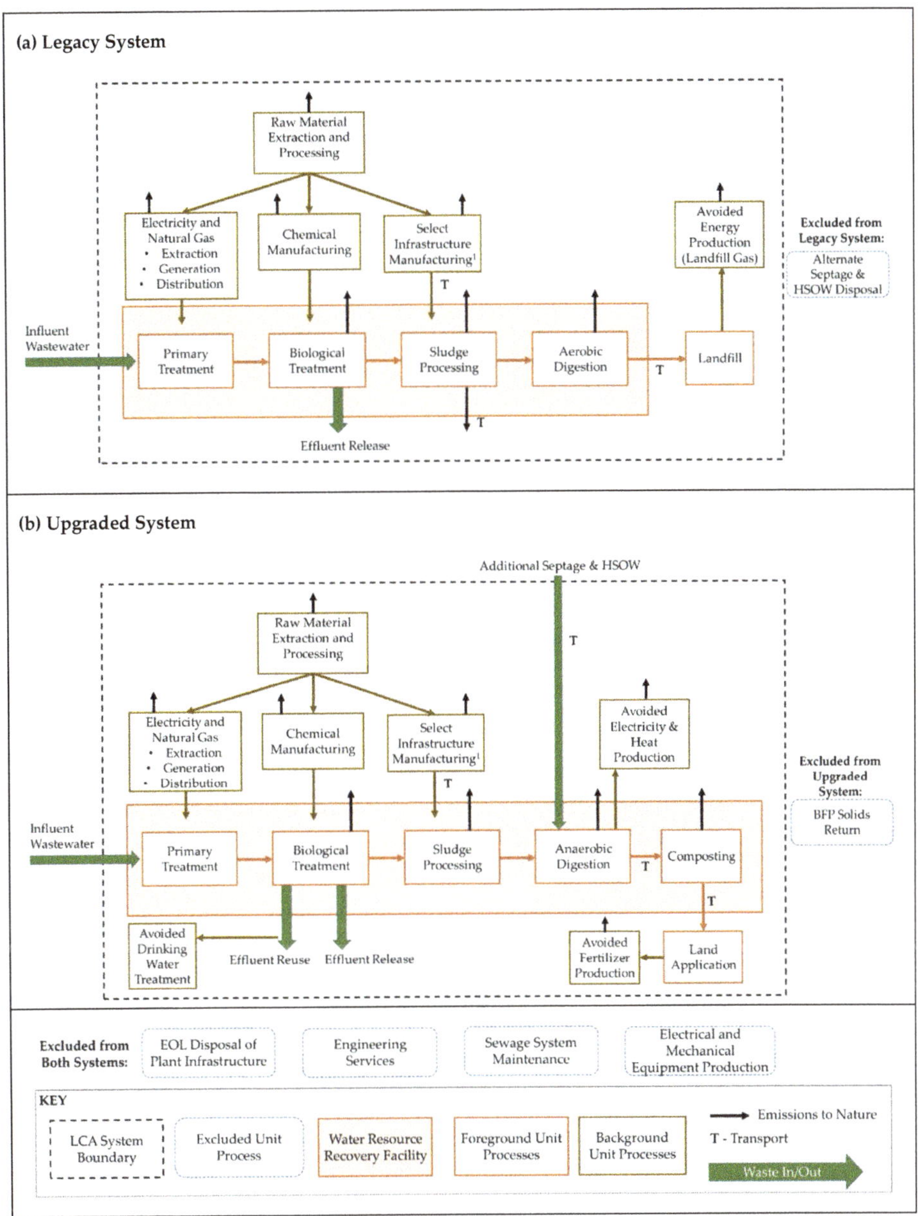

Figure 1. System diagrams for the (**a**) Legacy and (**b**) Upgraded water resource recovery facilities. The infrastructure constitutes the concrete, structural steel, and aggregate required for unit process and building construction. Abbreviations: BFP, belt filter press; EOL, end-of-life; HSOW, high strength organic waste; LCA, life cycle assessment.

The legacy WRRF configuration represents the treatment process that had been in place since 1993, and it is a typical example of conventional activated sludge (CAS) treatment in use throughout the U.S. (Figure S1). The plant headworks consisted of a mechanical bar screen, comminutors, a grit well, and a Parshall flume. Solids were removed via primary settling and secondary clarification, aided by a polyaluminum chloride (PAC) flocculent, added prior to aeration. The effluent was discharged without disinfection. Primary solids and waste activated sludge (WAS) was subjected to gravity thickening, which was in turn aerobically digested for solids stabilization and dewatering via a belt filter press (BFP) before trucking and disposal in a local landfill [17]. The Bath regional landfill is equipped with a methane capture system designed to achieve a 95% capture rate for use in a waste-to-energy (WTE) facility [18].

The upgraded WRRF is a renovation of the legacy facility. BEGWS is exploring the option of chemically enhanced primary settling in combination with a Modified-Ludzack Ettinger (MLE) biological treatment system for nitrogen removal. The facility is also considering expanding its receipt of residential septage and HSOW for processing in the proposed anaerobic digester (AD). Locally available forms of HSOW include cheese processing and slaughterhouse waste from regional food processing facilities. Septage waste is trucked to the facility by companies collecting septic tank and portable toilet waste in the Bath region. These proposed options in combination with the currently installed MLE system compose the modeled upgraded WRRF. Ferric chloride is used to enhance solids captured in the primary clarification system and is paired with a screen compaction press for grit removal and gravity belt thickening (GBT) for dewatering of primary solids and WAS (Figure S2). The legacy aerobic digester was repurposed to provide a pre-anoxic and swing tank for nitrogen removal prior to aeration, creating an MLE biological treatment process. The MLE treatment process is a common method of biological nutrient removal that relies on facultative bacteria in the anoxic zone to remove nitrogen by using organic matter as an electron donor for denitrification [19]. The swing tank can be run in aerobic or anoxic mode depending upon operational requirements.

The combined municipal sludge stream is blended with incoming septage and HSOW before introduction to the two-stage mesophilic AD facility. The first AD tank is completely mixed and heated to 35 °C. An internal combustion engine based combined heat and power (CHP) system is used to generate electricity and thermal energy for subsequent sale to the local grid and for internal facility use [17]. Digested solids are thickened via BFP and trucked a short distance (0.8 km) to an on-site windrow composting facility where solids are combined with locally available organic materials to achieve the appropriate C:N ratio and moisture content for successful composting. Finished Class A compost is screened and cured prior to trucking to local agricultural fields where it is used as an agricultural amendment and fertilizer replacement.

While LCAs have been previously conducted on small scale wastewater treatment facilities [20,21] and sewage sludge co-digestion with HSOW [22–25], our study is the first effort that comprehensively explores how a small traditional wastewater facility in the U.S. can be retrofitted into a more sustainable resource recovery hub. According to U.S. EPA's estimate, approximately 72% of the operational public wastewater treatment facilities (over 11,000) in the U.S. are considered "small systems" [26]. This study applied a range of input and output parameters likely to correspond to those observed in other communities. The results of the scenarios considered here may have implications for these utilities. Our intent is to provide a comprehensive and flexible suite of environmental and cost indicators to encourage small scale wastewater utility decision makers to start thinking more systematically when dealing with issues unique to their cities or communities.

2. Methods

This study used the LCA methodology, following guidelines outlined in ISO 14044 [27], to generate comparative LCA results for two WRRF configurations under several operational scenarios. The basis of the system comparison, known as the functional unit, was one cubic meter of wastewater treated to the effluent qualities defined in Table 1. The comparative scenarios represent before and after conditions for a 1 MGD (3800 m^3/day) CAS WRRF that underwent the proposed upgrades to its secondary biological treatment unit to reach enhanced effluent quality, while simultaneously employing resource recovery strategies such as AD and composting. Upgrades were required because the legacy WRRF was not able to meet updated nitrogen permit limits. Differences in effluent quality are reflected in the environmental analysis. For the upgraded WRRF, feedstock scenarios were developed to understand the environmental and cost impact of co-digesting varying quantities of septage and HSOW. For all scenarios, the LCI was developed by dividing annual flows of input materials (e.g., chemicals and energy) by the influent flowrate of 1 MGD (3800 m^3/day), standardizing LCI values per cubic meter of treated wastewater. The change in facility flowrate associated with HSOW feedstock scenarios was excluded from the standardization calculation as HSOW contributes minorly (<2.5%) to overall volumetric flow. Infrastructure material inputs were also originally calculated on an annual basis and then standardized per cubic meter of treated wastewater by allocating equally over their useful lifespan. We did not evaluate the environmental impacts of alternative treatment options for HSOW within the legacy scenario, due to a lack of information about current treatment methods. This exclusion could affect relative impact between the legacy and upgraded WRRFs. The functional unit formulation facilitates comparison of facility level treatment impacts encompassing the changes in system layout, waste acceptance, and operational performance. LCCA was used to evaluate the net present value (NPV) of the upgraded WRRF across all feedstock and AD performance scenarios.

Table 1. Influent and effluent wastewater characterization.

Characteristic	Influent [28]	Effluent, Legacy [28]	Effluent, Upgraded [17]	Effluent, Permitted [15]	Units
Suspended Solids	437	7.9	5	30	mg/L
$CBOD_5$ [1]	279	7.4	2	25	mg/L
Total Kjeldahl Nitrogen	56	16	4.4	n.a. [2]	mg/L N
Ammonia	32	6.7	3.6	3.6 [3]	mg/L NH_3
Total Phosphorus	8	0.7	0.6	0.6	mg/L P
Nitrite	<1	2.8	0.8	n.a. [2]	mg/L N
Nitrate	<1	13	14	n.a. [2]	mg/L N
Organic Nitrogen	29	9	0.8	n.a. [2]	mg/L N
Total Nitrogen	57	31	20	20	mg/L N

[1] Carbonaceous biochemical oxygen demand; [2] n.a., not applicable, not a permitted value; [3] monthly average summertime permit limit. Winter value is 8.4 mg/L ammonia.

2.1. Scenario Analysis

The study employed scenario and sensitivity analyses to test LCA and LCCA results over a range of input and output parameters likely to correspond to those observed in similar communities. We defined low (base), medium, and high scenarios for the acceptance of AD co-digestion feedstock (Table 2). Low, base, and high scenarios were defined for AD operational performance and EOL emission rates, with associated parameter values listed in Tables 3 and 4. The EOL emission scenarios only pertain to greenhouse gases (GHG) generated in the landfill (legacy WRRF) or during composting (upgraded WRRF). In total, results were generated for 6 and 54 scenario combinations for the legacy and upgraded treatment systems, respectively (Tables S1 and S2).

Table 2. Feedstock scenario parameter values.

Feedstock Type [1]	Legacy [2]	Low (Base) *	Medium	High	Units
	Feedstock Quantity				
Primary Sludge	67	71	71	71	m^3/day
Waste Activated Sludge	290	300	300	300	m^3/day
Septage	30	61	61	61	m^3/day
High Strength Organic Waste	-	-	15	30	m^3/day

[1] Feedstock quantity values are presented prior to dewatering. [2] The quantity of solids processed by the legacy system remains constant. Legacy values are provided as a comparison to the upgraded WRRF AD feedstock scenarios. * Designates the parameter values associated with the base, upgraded results scenario.

The AD system designed for the treatment plant upgrade is oversized for the solids processing needs for municipal wastewater alone. The feedstock scenarios test the effect of using the excess AD capacity to process additional septage and HSOW. The legacy treatment system did not include AD. Table 2 shows the quantity of waste feedstock prior to dewatering from each source category destined for digestion. HSOW does not require dewatering and consists of industrial cheese waste, slaughterhouse waste, and winery waste available in the region. HSOW characteristics are documented in Table S3.

Low, base, and high estimates of AD performance were examined for their effects on biogas production and associated environmental and cost benefits and burdens. We selected parameter values to represent a reasonable range of AD operational performance as documented in the literature and summarized in Table 2. Biogas yield values were calculated as a weighted average of feedstock-specific biogas yield estimates reported in Table S4.

The carbon and nitrogen content of AD biosolids impact GHG emissions produced during landfilling or composting and land application of these materials. Low, base, and high estimates of potential composting and landfill emission rates were evaluated. The analysis evaluates both windrow and aerated static pile (ASP) composting systems. We calculated impacts assuming methane capture rates characteristic of the Bath regional landfill, 95% [18], and a national average landfill methane capture rate of 57% [29]. Landfill methane emissions were calculated over a 100-year period using a first-order decay equation and parameter values presented in Table 3 [30]. A landfill carbon storage credit was applied to the non-degradable carbon and the fraction of degradable carbon that remains after 100 years. Degradation rates approximate conditions that range from cold and dry (low) to warm and moist (high), depending on the emission scenario [30].

Table 3. Upgraded WRRF anaerobic digester performance scenarios.

Anaerobic Digestion Parameter	Low	Base *	High	Units
	AD Performance			
Loading Rate [1]	220	270	350	kg VS/m^3/day
Biogas Yield [2]	0.75	0.94	2.2	m^3/kg VS destroyed
Volatile Solids Reduction [31]	45	60	65	%
Methane Content of Biogas [32]	60	65	70	% v/v
Biogas Heat Content [32]	0.55	0.59	0.61	MJ/ft^3
CHP Electrical Efficiency	30 [32]	36 [17]	42 [32]	%
CHP Thermal Efficiency	41 [32]	51 [17]	43 [32]	%

[1] Low, base, and high AD loading rates correspond to the low (base), medium, and high feedstock scenarios. [2] Biogas yield values vary depending on the feedstock scenario. Values presented in Table 3 correspond to the low (base) feedstock scenario. * Designates the parameter values associated with the base, upgraded results scenario. Abbreviations: CHP, combined heat and power; VS, volatile solids; v, volume.

Base results for the legacy treatment system are defined by the base EOL emission assumptions and a methane capture rate characteristic of the Bath regional landfill. Base results for the upgraded treatment system are defined by base parameter values for the feedstock, AD performance, and EOL emission scenarios assuming the use of windrow composting. Base scenarios are intended to serve as a benchmark for comparison against the sensitivity scenarios. The base scenario for the upgraded treatment plant does not include co-digestion of HSOW and assumes typical, achievable values for AD performance and biogas generation. Base EOL emission estimates are in the middle of the reported range.

Table 4. EOL greenhouse gas emissions scenarios for the legacy and upgraded WRRFs.

EOL Treatment Option	Low	Base *	High	
Landfill—with Methane Capture [1]		**GHG Emissions**		**Units**
Carbon Content of Dry Solids	38 [33]	48	57 [33]	%
Degradable Organic Carbon [30]	5.0	5.0	5.0	% wet mass [2]
Degradable Carbon Decomposed [33]	50	65	80	%
Degraded Carbon to CH_4 [30,33]	50	50	50	%
Methane to CO_2 in landfill cover [33]	25	10 [30]	3	%
k, degradation rate	0.10	0.18	0.23	unitless
EOL Treatment Option	**Low**	**Base ***	**High**	
Composting [3]		**GHG Emissions**		**Units**
CH_4 Emissions [4]	0.11 [34]	0.82	2.5 [35]	% incoming C
N_2O Emissions [5]	0.34 [34]	2.7	4.7 [36]	% incoming N

[1] The landfill EOL treatment option is only associated with the legacy WRRF. [2] The BFP produces dewatered biosolids with a solids content of 20% for trucking to the landfill. [3] The compost EOL treatment option is only associated with the upgraded WRRF. [4] Methane emissions are assumed to be zero for covered aerated static pile (ASP) system with biofilter [33]. [5] Nitrous oxide emissions are the same for both ASP and windrow composting. * Designates the parameter values associated with the base results scenarios. Abbreviations: C, carbon; EOL, end-of-life; N, nitrogen.

2.2. Life Cycle Inventory Model

Construction and operation LCI data for the legacy WRRF were provided by facility staff. Historic data for an average annual flowrate of 0.67 MGD (2500 m^3/day) were used. Electricity and chemical use were scaled to account for the 1 MGD (3800 m^3/day) basis of this analysis. We used the NY regional electrical grid mix to estimate the environmental impact of electricity consumption and the environmental benefit of avoided electricity production (Table S5). Electricity use for plant equipment was calculated based on mechanical equipment horsepower or voltage and current draw for each piece of equipment as demonstrated, in (Supporting Information (SI) Section S1.5. Estimates of material use for major infrastructure elements were calculated using dimensions from facility blueprints. Quantities of earthwork, concrete, steel, polyvinyl chloride (PVC), and gravel were estimated for the following units: (1) Parshall flume; (2) primary settling tank; (3) wet well; (4) aeration basins; (5) aerobic digester; (6) sludge thickener; (7) inter-unit piping; (8) control buildings; and (9) collection system piping. Materials associated with mechanical equipment such as pumps, mixers, and boilers were excluded from the analysis.

Equipment and infrastructure information carried over from the legacy WRRF was supplemented with additional equipment and energy requirements available in engineering design documents to represent the upgraded WRRF [17]. We estimated infrastructure for the chemically enhanced primary clarification unit and of the receiving station using basic unit dimensions and assuming similar construction methods required by other units. Infrastructure estimates for the primary and secondary digesters were calculated using CAPDETWorks™ engineering design and costing software [37].

We scaled equipment electricity use for the upgraded WRRF for sludge processing units affected by the feedstock scenarios based on the change in volumetric flow, mass flow rate, or loading rate as appropriate. For example, the increase in BFP electricity use required for the high feedstock scenario was based on the additional equipment operation time required to process the increased volume of digested sludge. Table S6 lists the scaling factors, applied to the base electricity consumption values, used to estimate electricity consumption for each feedstock-AD scenario. The legacy and upgraded treatment systems both consume PAC and polymer as flocculation and dewatering aids, respectively. The upgraded treatment system also uses ferric chloride for chemically enhanced primary clarification. Polymer consumption was calculated from dosage rates applied per the quantity of solids processed, which varies for the upgraded WRRF according to feedstock scenario (Tables S7 and S8). SI Section S1.6 details the calculations used to estimate LCI chemical quantities. We calculated the increased energy requirement for secondary aeration based on the additional BOD and N loading to secondary treatment associated with AD and BFP supernatant. An oxygen demand factor of 4.25 kg O_2 per kg of NH_3 nitrified was used to calculate the energy requirement of N loading [38].

We determined GHG emissions from legacy and upgraded secondary treatment processes and aerobic digestion from influent TKN and BOD concentrations, which vary across feedstock scenarios. We assumed 0.035% of nitrogen influent to each unit is released as nitrous oxide [39]. Methane emissions were calculated using a theoretical maximum methane generation rate of 0.6 kg CH_4/kg influent BOD, which reflects methane emissions under anaerobic conditions [40]. We adjusted the theoretical maximum downwards using a methane correction factor of 0.005, which reflects the potential for methane production in small pockets of anaerobic conditions. The CAS methane correction factor is on the same order of magnitude as that calculated based on data from Czepiel et al. (1993) [41], and is on the low end of the IPCC's reported uncertainty range of 0 to 0.1 for a well-managed aerobic treatment plant [40]. For the MLE system with zones for both nitrification and denitrification, the same approach was applied using a methane correction factor of 0.05 [42]. For the MLE unit, we assumed that 0.16% of influent nitrogen is lost as nitrous oxide [43]. Nitrous oxide emissions from receiving streams for both systems were calculated based on the IPCC guideline that 0.005 kg of N_2O-N are emitted per kg of nitrogen discharged to the aquatic environment [40]. The methodology used to calculate landfill GHG emissions is presented in Section 2.1, and further details on each GHG calculation method can be found in SI Section S1.7.

The legacy WRRF was credited with avoided electricity use for the share of landfill methane delivered to the WTE facility assuming an electrical conversion efficiency of 29%. Table S9 lists additional statistics concerning methane capture or release at the Bath regional and national average landfills. Avoided electricity and natural gas use associated with the upgraded WRRF is a product of biogas recovery (Table S10) and varies depending upon the feedstock-AD scenario under consideration (Tables S11–S13). Table S14 lists the additional quantity of natural gas that is required to provide facility and digester heat beyond that provided by biogas combustion. AD fugitive methane emissions were estimated assuming a 1% methane loss both from the digesters and CHP engine (Tables S15 and S16). An avoided fertilizer production credit was applied to the upgraded WRRF based on N and P content of the composted biosolids assuming a fertilizer replacement value of 73% [44]. Approximately 4% of annual treated effluent was assumed to be reused for irrigation at a local golf course during the summer months.

We constructed the LCI model using unit process and elementary flow data drawn from the U.S. EPA's harmonized version of the U.S. LCI [45] and Ecoinvent 2.2 LCI databases [46]. Elementary flows are flows of material (emissions) and energy both to and from nature. They include process and combustion emissions associated with upstream material extraction, industrial processing, manufacturing, transportation, and EOL disposal in addition to the specific elements of the foreground LCI model described here. Summarized system-level LCI quantities for the legacy and upgraded treatment systems are presented in Table 5.

Table 5. Inventory of principal material and energy flows by treatment stage for legacy and upgraded base scenarios.

Treatment Stage	Material and Energy Inputs [1]				Process Emissions		Avoided Products			
	Electricity	Natural Gas	Transport	Chemicals	Methane	Nitrous Oxide	Electricity	Natural Gas	Fertilizer	Effluent Reuse
	kWh/m^3	MJ/m^3	tkm/m^3	kg/m^3	kg/m^3	kg/m^3	kWh/m^3	MJ/m^3	kg/m^3	m^3/m^3
Legacy WRRF (Base Scenario)										
Primary Treatment	0.15	-	-	0.55	-	-	-	-	-	-
Secondary Treatment	0.36	-	-	-	6.3×10^{-4}	2.5×10^{-5}	-	-	-	-
Sludge Processing	0.39	-	-	5.4×10^{-4}	2.4×10^{-3}	5.7×10^{-5}	-	-	-	-
End-of-Life	-	-	0.05	-	0.02	6.1×10^{-4}	0.09	-	-	-
Effluent Release	9.3×10^{-5}	-	-	-	-	2.5×10^{-4}	-	-	-	-
Facilities	-	0.71	-	-	-	-	-	-	-	-
Total	0.90	0.71	0.05	0.55	0.02	9.4×10^{-4}	0.09	-	-	-
Upgraded WRRF (Base Scenario)										
Primary Treatment	0.32	-	-	0.03	-	-	-	-	-	-
Secondary Treatment	0.61	-	-	4.8×10^{-3}	5.3×10^{-3}	1.1×10^{-4}	-	-	-	0.04
Sludge Processing	0.38	0.56	0.40	5.0×10^{-3}	1.8×10^{-3}	-	0.45	2.2	-	-
End-of-Life	5.5×10^{-4}	-	0.05	-	0.01	1.7×10^{-3}	-	-	0.03	-
Effluent Release	0.03	-	-	-	-	1.5×10^{-4}	-	-	-	-
Facilities	-	1.1	-	-	-	-	-	-	-	-
Total	1.3	1.7	0.45	0.04	0.02	2.0×10^{-3}	0.45	2.2	0.03	0.04

[1] This table excludes infrastructure materials and aggregates chemical and fertilizer use on a mass basis.

2.3. *Life Cycle Impact Assessment*

Life cycle impact assessment (LCIA) is the stage in an LCA study in which elementary flows in the LCI associated with the entire supply chain are characterized to estimate environmental impacts. This paper presents impact results for eutrophication potential (EP), 100-year global climate change potential (GCCP), water use (WU), and cumulative energy demand (CED). WU and CED are inventory indicators and do not require characterization, but still include both direct and indirect resource use. GCCP and EP are calculated per the TRACI 2.1 method for the U.S. geographic context [47,48]. EP characterizes both nitrogen and phosphorus releases and is therefore generalized to be relevant for both freshwater and marine contexts. Water use was calculated using ReCiPe 2008 [49] and excluded turbine water use and evaporative loss from hydroelectricity. CED is calculated using the methodology developed for implementation in Ecoinvent 2.2 [50]. Because biogas enters the facility as a waste product, its energy content is not included in the CED calculation as CED is intended to estimate energy withdrawn from nature.

All process carbon dioxide emissions during wastewater or EOL treatment were modeled as biogenic in origin and do not contribute to GCCP. We did, however, estimate the fraction of carbon that is converted to methane, which contributes to GCCP. Carbon credits were attributed to both the legacy and upgraded WRRFs for the fraction of carbon either in landfill or agricultural fields that remains sequestered for over 100 years. Avoided products associated with energy recovery, compost use, and treated effluent reuse generate environmental credits that reduce net environmental impact and were tracked as negative values. The term gross environmental impact is used to refer to the total impact results in the absence of environmental credits.

LCIA results were also generated for particulate matter formation potential, smog formation potential, acidification potential, and fossil depletion potential. LCIA methods used for these additional impact categories are listed in Table S17. We provide results for the additional impact categories in a supplementary Bath WRRF Results File.

2.4. *Life Cycle Cost Assessment*

The LCCA tabulates financial expenditures and revenue over a 30-year period using a NPV calculation to determine total project costs in present dollars. Life cycle costs for the upgraded WRRF and payback period of the AD system were calculated using three sets of LCCA parameters yielding low, base, and high estimates of system NPV (Table S18). No LCCA is performed for the legacy system since it is no longer a viable design due to changes in permit requirements. We applied the three cost scenarios to each of the nine feedstock-AD performance scenarios to estimate a reasonable NPV range for each LCA scenario. A payback period was only calculated for the AD system as other unit processes generate no direct revenue. In the case of Bath NY, installation of MLE advanced secondary treatment is required to meet effluent quality standards and is not based on economic rationale.

Main data sources for the analysis include historical budget data for the legacy WRRF. A previous LCCA study was carried out by GHD Inc. engineering for upgrades to primary clarification and secondary treatment, and relevant cost data were incorporated into this assessment [51]. We used CAPDETWorks^{TM} engineering design and costing software to estimate AD construction and staffing costs [37]. Supplementary costs were referenced from the RSMeans building construction cost database [52], or via personal communication with WRRF staff.

All costs were classified as either annual or capital costs. Total capital costs are the sum of purchased equipment costs and the associated direct and indirect costs. Direct costs include mobilization, site preparation and electrical, piping, instrumentation, and building requirements. Indirect costs include professional services, miscellaneous and contingency costs, and profit. Direct costs were estimated as a percentage of the purchased equipment price. Indirect costs were estimated as a percentage of the sum of purchased equipment and direct costs. Total annual costs are the sum of operation costs, material costs, chemical costs, and energy costs. Escalation factors were applied to current labor, material, operation, and energy prices to provide an estimate for future years. Escalation factors vary between

0% and 4% annually depending upon the cost category and scenario. The discount rate utilized for the NPV calculation varies between 3% and 6% in the high and low cost scenarios, respectively. Assumed revenues from generated electricity, septage disposal, HSOW disposal, and compost sales also vary between cost scenarios. Direct and indirect cost factors, escalation rates, and formulas used to calculate cost escalation, net present value, and payback period are detailed in SI Section S1.11.

3. Results

3.1. LCA Results

Figure 2 shows LCIA results by treatment stage for the base legacy WRRF, base upgraded WRRF, and an optimized upgraded WRRF scenario for GCCP, CED, EP, and WU. The optimized scenario corresponds to the high feedstock, high AD performance, and the low EOL emission scenarios. Results are also presented by process contribution for both GCCP and CED. Treatment stage figures aggregate impact according to groupings of unit processes and associated equipment (Table S21) within the WRRF, while process contribution figures aggregate impact according to underlying drivers such as energy use, chemical use, or process emissions. Whiskers bounding the net impact results for the base upgraded WRRF represent the full range of impact results generated by the feedstock-AD-EOL emission scenarios, with scenario parameter values as defined in Table 2 through Table 4. All upgraded results in the figure are representative of the windrow composting system. GCCP results for the ASP system are included in Figure 3. All scenario LCA results are provided by unit processes in Table S22 and S23 for the legacy WRRF and the upgraded WRRF, respectively.

Study findings in Figure 2c demonstrate that effluent nutrient discharges dominate EP, particularly for the legacy WRRF. Upgrading the WRRF to include MLE biological nutrient removal yields a 37% reduction in EP results when comparing base scenario results for both systems. Both net CED and GCCP impact results increase in response to system upgrades in the base scenario, by 5% and 25%, respectively. WU decreases dramatically in response to the small amount of effluent reuse (4% of influent) that accompanies the WRRF upgrade, producing a net environmental benefit in this impact category. In addition to avoided water extraction associated with reuse, avoided fertilizer production also yields a net reduction in water use by avoiding water consumption for chemical fertilizer production.

The extent to which environmental benefits are achieved by the optimized scenario is dependent on several management practices. First, it is necessary to fully utilize the available capacity of the AD unit and to supplement municipal sludge with HSOW. Additionally, the high biogas yields that lead to large avoided energy credits in the high AD scenario are predicated on the use of chemically enhanced primary clarification to maximize the quantity and digestibility of volatile solids available in the digester.

The whisker range demonstrates that GCCP and CED impact results of the upgraded WRRF and their relative relationship to those of the legacy system are strongly dependent upon the feedstock-AD-EOL scenario under consideration. For these two impact categories, the maximum reductions in impact results, relative to the legacy base scenario, are 180% for GCCP and 210% for CED. In all cases, the minimum impact result is associated with the high feedstock-high AD-low EOL emissions scenario, which is included as the optimized upgraded scenario in Figure 2. Eutrophication potential is less sensitive to the feedstock-AD scenario, demonstrating only a 23% difference between the maximum and minimum impact results. Water use results respond negligibly to the sensitivity scenarios as effluent reuse dominates results and remains constant across scenarios.

Figure 2e,f shows the reductions in net GCCP and CED, attributable to avoided electricity and natural gas from biogas recovery, avoided drinking water treatment from wastewater reuse, and avoided fertilizer production from compost land application. GCCP of the upgraded WRRF is reduced via the carbon sequestration credit, attributable to compost land application. Collectively, the carbon credit and avoided product benefits reduce gross GCCP and CED of the upgraded WRRF by 46%

and 42%, respectively, in the upgraded base scenario. In the absence of compost land application and avoided product benefits, the environmental burdens associated with WRRF upgrades exceed those of the legacy system by 130% and 70% for GCCP and CED.

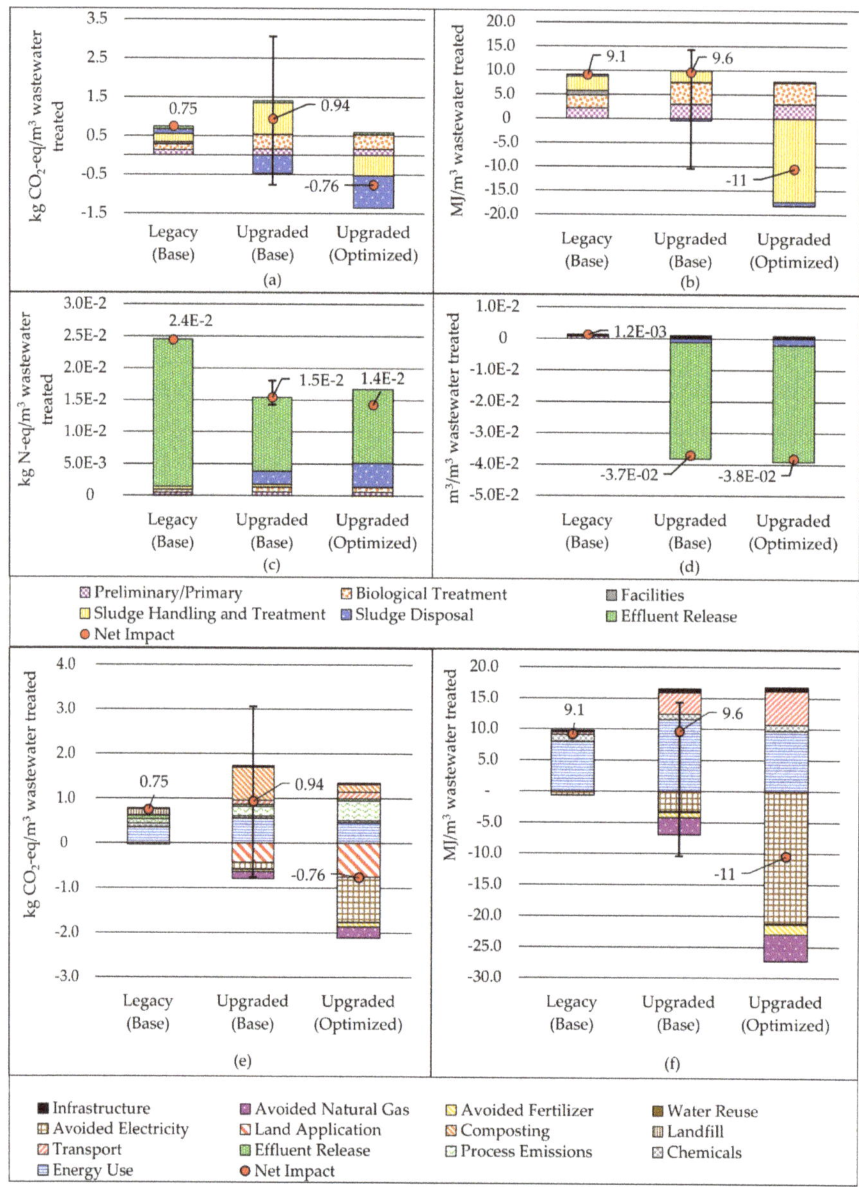

Figure 2. Results by treatment stage for: (**a**) Global climate change potential; (**b**) Cumulative energy demand; (**c**) Eutrophication potential; and (**d**) Water use; and by process contribution for: (**e**) Global climate change potential; and (**f**) Cumulative energy demand. Facilities refers to infrastructure and energy use of administration buildings. Optimized: high feedstock (Table 2)-high AD performance (Table 3)-low EOL emission scenario (Table 4).

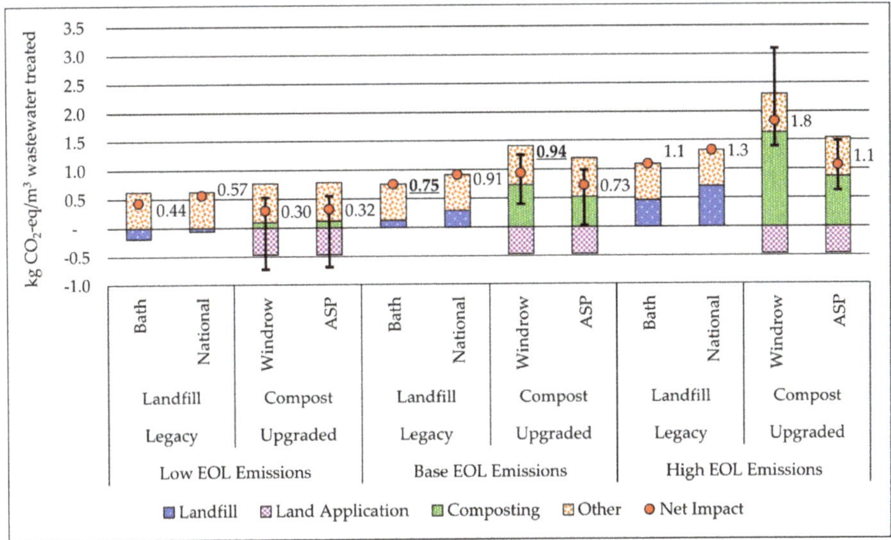

Figure 3. Impact of end-of-life strategy on global climate change potential of legacy and upgraded WRRFs. Whisker bars show the range of results for all feedstock-AD scenarios where applicable. Underlined impact results identify the base scenario results (Tables 2–4). Abbreviations: ASP, aerated static pile; EOL, end-of-life.

Figure 3 shows the effect of the low, base, and high EOL emission scenarios and compost system selection on GCCP impact for the legacy and upgraded WRRFs. The results depicted are associated with the base feedstock-base AD scenario for the upgraded WRRF. We calculated legacy system results assuming methane capture and WTE performance of both the Bath regional and national average landfill. The "other" category in Figure 3 includes all additional GCCP impact associated with treatment processes not associated with EOL processes.

Compost emissions in both the base and high EOL emission scenarios are a dominant contributor to life cycle GHG emissions. The upgraded WRRF reliant upon windrow composting is particularly sensitive to the methane emission factor, as demonstrated by both the range and maximum height of the whisker bars within the high EOL emission scenario. The ASP system demonstrates a more limited sensitivity due to the biofilter, which eliminates methane emissions by oxidizing the methane back to carbon dioxide [53]. Carbon sequestration associated with all systems reduces net GCCP impact and remains relatively consistent across scenarios. The impact of waste disposal for the additional septage processed by the upgraded WRRF is not included for the legacy system and could affect relative impact results between the legacy and upgraded WRRFs. The range of compost emission factors presented in Table 4 is sufficient to cause a shift in composting GCCP impact that pivots between provision of a net benefit to be the single largest contributor to GCCP impact.

A range of EOL emissions is also possible for the Bath regional and national average landfill, which can lead to a similar swing between the provision of net GCCP benefits or considerable relative impact. Benefits attributable to landfilling include avoided energy from landfill gas WTE systems and sequestration of landfilled carbon. All four EOL treatment options yield a net reduction in GCCP impact under the low EOL emission scenario. ASP composting provides the best GCCP performance within the base and high EOL emission scenarios followed by the Bath regional landfill option.

3.2. LCCA Results

Life cycle costing reflects the NPV of operating the upgraded WRRF over a 30-year period. Figure 4 presents NPV, in million dollars, for all feedstock-AD scenarios broken down into five cost categories: construction, operations, materials, chemicals, and energy. NPV of the upgraded system ranges from 31 to 38 million dollars depending upon utilization and performance of the AD system under base LCCA assumptions. Whisker bars on each column reflect the range of NPV estimates generated by the low and high LCCA scenarios. System NPV is more sensitive to LCCA parameters, particularly revenue fees, as more HSOW is accepted and capacity utilization of the AD units increases. Across all feedstock-AD scenarios, the low cost LCCA scenario yields a percent difference that is 13–44% below the base cost LCCA NPV.

The NPV of construction costs is approximately 17 million dollars regardless of scenario and is the largest contributor to NPV followed by operations. Operations includes plant staffing costs, ancillary material and facility costs such as waste disposal and professional services. The net contribution of energy costs to system NPV is reduced due to revenue generated via electricity sales and avoided natural gas purchases. The table at the top of Figure 4 lists the AD payback period for all scenarios. The columns in the table correspond to feedstock-AD scenarios in the column chart. Only the low LCCA scenario can generate a payback period for the AD and CHP investment that is less than the 30-year system lifespan. Apart from the discount rate, which is a primary determinant of NPV resulting from the analysis, the payback period is strongly determined by revenue per gallon of septage and HSOW. Assumed revenue ranges from 7.9 to 39 dollars per m^3 for HSOW [31] and from 1.9 to 2.6 dollars per m^3 for septage [51], and will be determined by local market conditions.

For the base scenario, the life cycle cost of installing and operating the AD, CHP, and composting systems is approximately 11 million dollars, and constitutes 31% of the systems total NPV. For the high feedstock-high AD-base cost scenario, the life cycle cost of installing and operating these unit processes drops to approximately 5 million dollars due to revenue associated with waste tipping fees and energy cost savings. Installation of these unit processes is not required to meet the new effluent permit guidelines. Other life cycle costs are necessary for continued operation of the WRRF and for upgrades associated with installing biological nutrient removal.

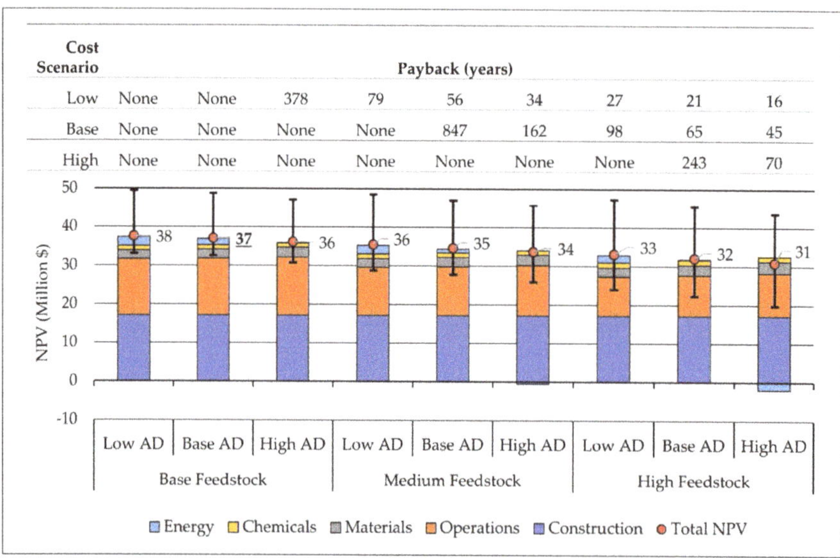

Cost Scenario	Payback (years)								
Low	None	None	378	79	56	34	27	21	16
Base	None	None	None	None	847	162	98	65	45
High	None	None	None	None	None	None	None	243	70

Figure 4. NPV of upgraded WRRF by cost category and AD payback period. Bars show low to high NPV estimate for each feedstock-AD scenario. The underlined impact result identifies the base scenario result (Tables 2–4). Abbreviations: AD, anaerobic digester; NPV, net present value.

4. Discussion

Results demonstrate that capacity utilization and performance of the AD system are determining factors in whether a relative decrease or increase in environmental impact is associated with the WRRF upgrade. Gross LCIA results (Figure 2e,f) reflect the increases in CED and GCCP that accompany increased material and energy requirements for advanced secondary treatment, AD, and composting. The environmental credits attributable to avoided energy products from biogas recovery, avoided fertilizer production, and treated effluent reuse, that characterize the concept of a resource recovery hub, serve to limit the relative increase in net CED and GCCP associated with achieving improved effluent quality within the upgraded base scenario. Results of the scenario analysis demonstrate the potential to realize equivalent GCCP and environmental benefits in CED and WU categories if the full capacity of the AD system (high feedstock scenario) is utilized assuming base AD performance.

If high AD performance is achieved, GCCP of the upgraded treatment plant is less than that of the legacy system, and environmental benefits in the CED and WU impact categories increase. The high AD performance scenario assumes a marked increase in biogas production associated with high biogas yield estimates for primary sludge from chemically enhanced primary clarification. The pairing of this unit with AD is intentional in that a higher fraction of volatile solids is removed from the wastewater prior to secondary treatment, which serves the dual purpose of limiting loading to the MLE and preserving those volatile solids for degradation in the AD where they produce environmental benefits and revenue via energy recovery. While the high estimate of biogas yield associated with chemically enhanced primary clarifier sludge is based on a pilot study, the theoretical performance of this system is promising from an environmental perspective at the 1 MGD scale.

GCCP is the most volatile of the four impact categories and demonstrates multiple dependencies on energy use, EOL emissions, and process emissions from WRRF unit operation and effluent release. Results are particularly sensitive to EOL emission scenarios indicating the importance of carefully considering selection of these treatment options and the subsequent management of composting systems. The ASP composting system demonstrates the lowest sensitivity to EOL emission scenarios and is likely to be the most reliable means of achieving low GCCP impact for WRRFs looking to rely on composting. Results reflect positively on efforts to increase methane capture and WTE capture at landfills around the country.

The LCCA illustrates the difficulty of achieving desirable economic performance of AD at small scales. An AD payback period of less than the 30-year system lifespan was only demonstrated for the three high feedstock-low cost scenarios, due to the higher tipping fees and energy value assumed in the low cost scenario. Despite the challenge of realizing system payback, the trend discovered highlights the benefit, in terms of reduced system NPV, that can result from supplementing municipal solids with HSOW to boost biogas production. Indeed, at larger plant scales with existing digester capacity, economic viability can be achieved more readily [24,54]. At this scale of implementation, however, the environmental benefits associated with AD come at the expense of increased life cycle costs beyond what would be required only to meet the updated permit nitrogen standards.

Results from comparable LCAs generally support the findings of this study, mainly that the environmental and economic benefits of AD are best realized when capacity is increased through co-digestion of HSOW. In a study of two larger Australian municipalities, Edwards et al. [23] found that compared to the separate management of sewage sludge and food waste, anaerobic co-digestion could result in a smaller global warming, acidification and eutrophication potential. At a smaller scale, Remy and Jekel [55] showed that the recovery of energy from the co-digestion of source-separated blackwater and household biowaste could lower the cumulative energy demand of waste service provision by 13–26% compared to conventional activated sludge treatment and organics incineration. Evaluating the entire municipal water cycle, Xue et al. [5] compared traditional centralized water and wastewater services to several "fit-for-purpose" scenarios. Their results showed that the electricity generated from blackwater and food waste co-digestion could offset at least 40% of life cycle energy consumption and result in lower eutrophication impacts. All studies identified significant sensitivities to AD capacity and organics loading. This research provides guidance for other small communities as they consider WRRF upgrades and in their long-term sustainability efforts. The analysis demonstrates the potential environmental benefits of strategies aimed at resource recovery. Findings highlight the importance of conscientious process selection and management of WRRF operations if relative reductions in environmental impact are to be reliably achieved.

5. Conclusions

- Installation of AD and operation as a resource recovery hub yielded reductions in GCCP and CED that can offset increased energy and material requirements of enhanced nutrient removal.
- Sensitivity results demonstrated the environmental benefit of utilizing the full capacity of AD facilities by accepting HSOW and pursuing best management practices to achieve high AD operation performance while minimizing potential for GHG generation at composting facilities.
- The pairing of chemically enhance primary clarification with AD demonstrated the potential to increase biogas production, reducing overall plant environmental burdens.
- At the 1 MGD (3800 m^3/day) scale, realization of environmental benefits from WRRF upgrades and a focus on resource recovery strategies are more reliably attainable than a monetary return on investment.

Supplementary Materials: The following are available online at http://www.mdpi.com/2071-1050/10/10/3546/s1, Word File: Bath WRRF Supplementary Tables and Figures; Excel File: Bath WRRF Results File.

Author Contributions: This paper is a result of the collaborative efforts of the listed authors. The following contributions were made: Conceptualization, J.G., X.C.M. and J.T.; Methodology, S.C., B.M., X.C.M. and J.T.; Validation, S.C., X.C.M., J.T., M.N. and L.F.; Formal Analysis, B.M. and S.C.; Investigation, B.M. and S.C.; Writing—Original Draft Preparation, B.M., S.C. and X.C.M.; Writing—Review and Editing, all listed authors; Visualization, B.M. and S.C.; Supervision, J.G. and X.C.M.; and Project Administration, X.C.M., D.B. and J.G.

Funding: This research was funded by the EPA Office of Research and Development's Safe and Sustainable Water Resources Program. The research was supported by EPA contracts EP-C-12-021 and EP-C-16-015.

Disclaimer: The views expressed herein are strictly the opinions of the authors and in no manner represent or reflect current or planned policy by the federal agencies. Mention of trade names or commercial products does not constitute endorsement or recommendation for use. The information and data presented in this product were obtained from sources that are believed to be reliable. The authors declare no competing financial interest.

Acknowledgments: Kim Miller and Guy Hallgren provided insight and primary data on the Bath, NY wastewater treatment plant operations and infrastructure for both the legacy and upgraded systems investigated. Engineering design of treatment plant upgrades was performed by personnel from Conestoga-Rovers & Associates, now a division of GHD Group. Lori Stone of Water Environment & Reuse Foundation as well as Pradeep Jangbari of New York State Department of Environmental Conservation provided technical review comments. Janet Mosley and Jessica Gray of Eastern Research Group provided technical input and review of the life cycle inventory and cost analysis. Sam Arden of Eastern Research group provided a literature review and summary of past LCA work looking at wastewater treatment, anaerobic digestion, and co-digestion.

Conflicts of Interest: The authors declare no conflict of interest.

References

1. Ma, C.; Xue, X.; Gonzalez-Mejia, A.; Garland, J.; Cashdollar, J. Sustainable Water Systems for the City of Tomorrow—A Conceptual Framework. *Sustainability* **2015**, *7*, 12071–12105. [CrossRef]
2. U.S. Environmental Protection Agency National Pollutant Discharge Elimination System (NPDES): Nutrient Permitting. Available online: https://www.epa.gov/npdes/nutrient-permitting (accessed on 6 September 2018).
3. ASCE (American Society of Civil Engineers) 2017 Infrastructure Report Card: America's Grades. Available online: https://www.infrastructurereportcard.org/americas-grades/ (accessed on 6 September 2018).
4. Schoen, M.; Hawkins, T.; Xue, X.; Ma, C.; Garland, J.; Ashbolt, N.J. Technologic resilience assessment of coastal community water and wastewater service options. *Sustain. Water Qual. Ecol.* **2015**, *6*, 75–87. [CrossRef]

5. Xue, X.; Schoen, M.E.; Ma, X.; Hawkins, T.R.; Ashbolt, N.J.; Cashdollar, J.; Garland, J. Critical insights for a sustainability framework to address integrated community water services: Technical metrics and approaches. *Water Res.* **2015**, *77*, 155–169. [CrossRef] [PubMed]
6. Augustin, K.; Skambraks, A.K.; Li, Z.; Giese, T.; Rakelmann, U.; Meinzinger, F.; Schonlau, H.; Günner, C. Towards sustainable sanitation—The Hamburg Water Cycle in the settlement Jenfelder Au. *Water Sci. Technol.* **2014**, *14*, 13–21. [CrossRef]
7. Grant, S.B.; Saphores, J.D.; Feldman, D.L.; Hamilton, A.J.; Fletcher, T.D.; Cook, P.L.M. Taking the "Waste" Out of "Wastewater" for human water security and ecosystem sustainability. *Science* **2012**, *337*, 681–686. [CrossRef] [PubMed]
8. Guest, J.S.; Skerlos, S.J.; Barnard, J.L.; Beck, M.B.; Daigger, G.T.; Hilger, H.; Jackson, S.J.; Karvazy, K.; Kelly, L.; Macpherson, L.; et al. A new planning and design paradigm to achieve sustainable resource recovery from wastewater. *Environ. Sci. Technol.* **2009**, *43*, 6126–6130. [CrossRef] [PubMed]
9. McCarty, P.L.; Bae, J.; Kim, J. Domestic wastewater treatment as a net energy producer—Can this be achieved? *Environ. Sci. Technol.* **2011**, *45*, 7100–7106. [CrossRef] [PubMed]
10. U.S. Environmental Protection Agency. *Case Study Primer for Participant Discussion: Biodigesters and Biogas*; U.S. Environmental Protection Agency: Washington, DC, USA, 2012.
11. Berger, M.; Finkbeiner, M. Water footprinting: How to address water use in life cycle assessment? *Sustainability* **2010**, *2*, 919–944. [CrossRef]
12. Muñoz, I.; Rodríquez, A.; Rosal, R.; Fernández-Alba, A.R. Life cycle assessment of urban wastewater reuse with ozonation as tertiary treatment: A focus on toxicity-related impacts. *Sci. Total Environ.* **2009**, *407*, 1245–1256. [CrossRef] [PubMed]
13. Hester, E.T.; Little, J.C. Measuring environmental sustainability of water in watersheds. *Environ. Sci. Technol.* **2013**, *47*, 8083–8090. [CrossRef] [PubMed]
14. Fulcher, J. Changing the Terms. Available online: https://news.wef.org/changing-the-terms/ (accessed on 3 August 2018).
15. New York State Department of Environmental Conservation. *State Pollutant Discharge Elimination System Permit for Village of Bath Wastewater Treatment Plant*; New York State Department of Environmental Conservation: New York, NY, USA, 2014.
16. Chesapeake Bay Foundation about the Bay: The Issues. Available online: http://www.cbf.org/about-the-bay/issues (accessed on 24 May 2017).
17. Conestoga-Rovers and Associates. *Village of Bath WWTP Upgrades Final Engineering Report*; Bath Electric, Gas & Water Systems: Bath, NY, USA, 2015.
18. Comela, R. *Navalis Construction Announces Completion of Renewable Landfill Gas-to-Energy Plant in Upstate New York*; Navalis Construction: Rochester, NY, USA, 2010.
19. Tchobanoglous, G.; Burton, F.L.; Stensel, H.D.; Tsuchihashi, R.; Abu-Orf, M.; Bowden, G.; Pfrang, W. *Wastewater Engineering: Treatment and Resource Recovery*, 5th ed.; McGraw-Hill Education: New York, NY, USA, 2014; ISBN 978-0-07-340118-8.
20. Cornejo, P.K.; Zhang, Q.; Mihelcic, J.R. How does scale of implementation impact the environmental sustainability of wastewater treatment integrated with resource recovery? *Environ. Sci. Technol.* **2016**, *50*, 6680–6689. [CrossRef] [PubMed]
21. Machado, A.P.; Urbano, L.; Brito, A.G.; Janknecht, P.; Salas, J.J.; Nogueira, R. Life cycle assessment of wastewater treatment options for small and decentralized communities. *Water Sci. Technol.* **2007**, *56*, 15–22. [CrossRef] [PubMed]
22. Chiu, S.L.H.; Lo, I.M.C.; Woon, K.S.; Yan, D.Y.S. Life cycle assessment of waste treatment strategy for sewage sludge and food waste in Macau: Perspectives on environmental and energy production performance. *Int. J. Life Cycle Assess.* **2016**, *21*, 176–189. [CrossRef]
23. Edwards, J.; Othman, M.; Crossin, E.; Burn, S. Anaerobic co-digestion of municipal food waste and sewage sludge: A comparative life cycle assessment in the context of a waste service provision. *Bioresour. Technol.* **2017**, *223*, 237–249. [CrossRef] [PubMed]
24. Krupp, M.; Schubert, J.; Widmann, R. Feasibility study for co-digestion of sewage sludge with OFMSW on two wastewater treatment plants in Germany. *Waste Manag.* **2005**, *25*, 393–399. [CrossRef] [PubMed]

25. Righi, S.; Oliviero, L.; Pedrini, M.; Buscaroli, A.; Della Casa, C. Life Cycle Assessment of management systems for sewage sludge and food waste: Centralized and decentralized approaches. *J. Clean. Prod.* **2013**, *44*, 8–17. [CrossRef]
26. US EPA, O. Small Wastewater Systems Research. Available online: https://www.epa.gov/water-research/small-wastewater-systems-research-0 (accessed on 25 September 2018).
27. ISO. *ISO 14044: 2006 Environmental Management—Life Cycle Assessment—Requirements and Guidelines;* ISO 14044; The International Organization for Standardization: Geneva, Switzerland, 2006; p. 54.
28. BEGWS (Bath Electric Gas & Water Systems). *Bath Wastewater Treatment Plant Influent and Effluent Water Quality Data: October 2011 to November 2015;* BEGWS: Bath, NY, USA, 2011–2015.
29. U.S. Environmental Protection Agency. *Inventory of U.S. Greenhouse Gas Emissions and Sinks: 1990–2013;* U.S. Environmental Protection Agency: Washington, DC, USA, 2015.
30. Research Triangle Institute. *Greenhouse Gas Emissions Estimation Methodologies for Biogenic Emissions from Selected Source Categories: Solid Waste Disposal, Wastewater Treatment, Ethanol Fermentation;* U.S. Environmental Protection Agency: Washington, DC, USA, 2010.
31. Appleton, A.R.; Rauch-Williams, T. *Co-Digestion of Organic Waste Addressing Operational Side-Effects;* Water Environment Research Foundation: Alexandria, VA, USA, 2017.
32. Wiser, J.R.; Schettler, J.W.; Willis, J.L. *Evaluation of Combined Heat and Power Technologies for Wastewater Treatment Facilities;* U.S. Environmental Protection Agency: Washington, DC, USA, 2010.
33. SYLVIS Environmental. *The Biosolids Emissions Assessment Model (BEAM);* Canadian Council of Ministers of the Environment: Winnipeg, MB, Canada, 2011.
34. Hellmann, B.; Zelles, L.; Palojarvi, A.; Bai, Q. Emission of climate-Relevant trace gases and succession of microbial communities during open-Windrow composting. *Appl. Environ. Microbiol.* **1997**, *63*, 1011–1018. [PubMed]
35. Hellebrand, H.J. Emission of nitrous oxide and other trace gases during composting of grass and green waste. *J. Agric. Eng. Res.* **1998**, *69*, 365–375. [CrossRef]
36. Fukumoto, Y.; Osada, T.; Hanajima, D.; Haga, K. Patterns and quantities of NH_3, N_2O and CH_4 emissions during swine manure composting without forced aeration—-Effect of compost pile scale. *Bioresour. Technol.* **2003**, *89*, 109–114. [CrossRef]
37. Hydromantis. *CAPDETWorks Version 3.0 Software: Rapid Design and Costing Solution for Wastewater Treatment Plants;* Hydromantis Environmental Software Solution: Hamilton, ON, Canada, 2014.
38. CDM Smith. *Technical Memorandum Report Greater Lawrence Sanitary District Organics to Energy Feasibility Study;* CDM Smith: Boston, MA, USA, 2013; p. 95.
39. Czepiel, P.M.; Crill, P.M.; Harriss, R.C. Nitrous oxide emissions from municipal wastewater treatment. *Environ. Sci. Technol.* **1995**, *29*, 2352–2356. [CrossRef] [PubMed]
40. Intergovernmental Panel on Climate Change. *IPCC Guidelines for National Greenhouse Gas Inventories, Prepared by the IPCC, National Greenhouse Gas Inventories Programme;* IGES: Kanagawa, Japan, 2006; ISBN 4-88788-032-4.
41. Czepiel, P.M.; Crill, P.M.; Harriss, R.C. Methane emissions from municipal wastewater treatment processes. *Environ. Sci. Technol.* **1993**, *27*, 2472–2477. [CrossRef]
42. Daelman, M.R.; van Voorthuizen, E.M.; van Dongen, L.G.; Volcke, E.; van Loosdrecht, M.C. Methane and nitrous oxide emissions from municipal wastewater treatment–Results from a long-Term study. *Water Sci. Technol.* **2013**, *67*, 2350–2355. [CrossRef] [PubMed]
43. Chandran, K. *Greenhouse Nitrogen Emissions from Wastewater Treatment Operation: Phase I;* Water Environment Research Foundation: Alexandria, VA, USA, 2012; ISBN 978-1-78040-481-3.
44. Smith, K.A.; Jeffrey, W.A.; Metcalfe, J.P.; Sinclair, A.H.; Williams, J.R. *Nutrient Value of Digestate from Farm-Based Biogas Plants;* Scottish Executive Environmental and Rural Affairs Department: Edinburgh, UK, 2007.
45. NREL U.S. Life Cycle Inventory Database 2012. Available online: https://www.lcacommons.gov/lca-collaboration/search/page=1&group=National_Renewable_Energy_Laboratory (accessed on 1 October 2018).
46. Frischknecht, R.; Jungbluth, N.; Althaus, H.-J.; Doka, G.; Dones, R.; Heck, T.; Hellweg, S.; Hischier, R.; Nemecek, T.; Rebitzer, G.; et al. The ecoinvent database: Overview and methodological framework. *Int. J. Life Cycle Assess.* **2005**, *10*, 3–9. [CrossRef]
47. Bare, J. TRACI 2.0: The Tool for the reduction and assessment of chemical and other environmental impacts. *Clean Technol. Environ. Policy* **2011**, *13*, 687–696. [CrossRef]

48. Bare, J.; Norris, G.A.; Pennington, D.W.; McKone, T. TRACI: The Tool for the reduction and assessment of chemical and other environmental Impacts. *J. Ind. Ecol.* **2002**, *6*, 49–78. [CrossRef]
49. Goedkoop, M.; Heijungs, R.; Huijbregts, M.; De Schryver, A.; Struijs, J.; van Zelm, R. *ReCiPe 2008: A Life Cycle Impact Assessment Method Which Comprises Harmonised Category Indicators at the Midpoint and the Endpoint Level: Report 1 Characterization*; ReCiPe: Bilthoven, The Netherlands, 2009.
50. Althaus, H.-J.; Bauer, C.; Doka, G.; Dones, R.; Frischknecht, R.; Hellweg, S.; Humbert, S.; Jungbluth, N.; Kollner, T.; Loerincik, Y.; et al. *Implementation of Life Cycle Impact Assessment Methods: Data v2.2 (2010)*; Ecoinvent Centre: St. Gallen, Switzerland, 2010.
51. GHD Engineering. *Life Cycle Cost Analysis Evaluation: Preliminary and Primary Treatment*; Bath Electric, Gas & Water Systems: Bath, NY, USA, 2016.
52. RSMeans; Building Construction Cost Data 2016. Available online: https://www.rsmeans.com/ (accessed on 1 October 2018).
53. Canadian Council of Ministers of the Environment. *Biosolids Emissions Assessment Model (BEAM): User Guide*; Canadian Council of Ministers of the Environment: Winnipeg, MB, Canada, 2009.
54. Bolzonella, D.; Battistoni, P.; Susini, C.; Cecchi, F. Anaerobic codigestion of waste activated sludge and OFMSW: The experiences of Viareggio and Treviso plants (Italy). *Water Sci. Technol.* **2006**, *53*, 203–211. [CrossRef] [PubMed]
55. Remy, C.; Jekel, M. Energy analysis of conventional and source-separation systems for urban wastewater management using life cycle assessment. *Water Sci. Technol.* **2012**, *65*, 22–29. [CrossRef] [PubMed]

Communication

Sustainable Exploitation of Coffee Silverskin in Water Remediation

Angela Malara [1], Emilia Paone [1], Patrizia Frontera [1,2,*], Lucio Bonaccorsi [1], Giuseppe Panzera [1] and Francesco Mauriello [1,*]

1 Dipartimento di Ingegneria Civile, dell'Energia, dell'Ambiente e dei Materiali, Università "Mediterranea", 89122 Reggio Calabria, Italy; angela.malara@unirc.it (A.M.); emilia.paone@unirc.it (E.P.); lucio.bonaccorsi@unirc.it (L.B.); giuseppe.panzera@unirc.it (G.P.)

2 Consorzio Interuniversitario per la Scienza e la Tecnologia dei Materiali (INSTM), 50121 Firenze, Italy

* Correspondence: patrizia.frontera@unirc.it (P.F.); francesco.mauriello@unirc.it (F.M.)

Received: 12 September 2018; Accepted: 1 October 2018; Published: 3 October 2018

Abstract: Coffee silverskin (CS), the main solid waste produced from the coffee industry, has efficiently been used as adsorbent material to remove potential toxic metals (PTMs). In order to assess its suitability in water remediation, kinetic adsorption experiments of Cu^{2+}, Zn^{2+}, and Ni^{2+} ions from wastewater were carried out and the adsorption performance of the waste material was compared with that of another well-known waste from coffee industry, spent coffee grounds (SCG). By using CS as sorbent material, ion removal follows the order $Cu^{2+} > Zn^{2+} > Ni^{2+}$ with the adsorption equilibrium occurring after about 20 min. The adsorption efficiency of Ni^{2+} ions is the same for both investigated materials, while Cu^{2+} and Zn^{2+} ions are removed to a lesser extent by using CS. Equilibrium-adsorption data were analyzed using two different isotherm models (Langmuir and Freundlich), demonstrating that monolayer-type adsorption occurs on both CS and SCG surfaces. The overall results support the use of coffee silverskin as a new inexpensive adsorbent material for PTMs from wastewater.

Keywords: potential toxic metals; coffee waste; coffee silverskin; spent coffee grounds; metal adsorption; remediation; Langmuir and Freundlich isotherm models; wastewater

1. Introduction

In the past few years, in the framework of environmental sustainability research, numerous attempts on the reuse and valorization of coffee byproducts and coffee wastewater have been developed [1].

Coffee is the most important food commodity worldwide, being the second-most traded product after oil [2]. Because of this extended market, the coffee industry produces huge volumes of waste. In fact, each stage of the coffee production process (i.e., coffee-cherries processing, dried-beans milling, green-coffee-beans roasting), as well as coffee consumption, each year produces a large volume of biowaste, which contributes to environmental pollution [3]. Among this biowaste, residues with coffee silverskin (CS) and spent coffee grounds (SCG) are the most significantly produced. Since sustainability development has to be prioritized, research devoted to valorizing and reusing this kind of waste should be strongly encouraged in a strategy for the treatment of end-of-life materials [4].

CS is a thin integument of the outer layer of green coffee beans that comes off during the roasting process [5]. Moreover, it is generally produced in high volumes, takes fire very easily, and its disposal requires the use of energy for the necessary compaction process.

SCG are also a well-known residual material obtained from instant-coffee preparation or coffee brewing. SCG are a residue obtained during the treatment of raw coffee powder with hot water or steam for instant-coffee preparation, characterized by fine particle size, high humidity, organic load, and acidity [6]. In order to reduce waste-disposal costs and avoid environmental pollution, coffee companies focus on possible alternative uses of SCG. Due to their composition, SCG possess functional properties, such as water-holding capacity, oil-holding capacity, emulsifing activity, and antioxidant potential, which make the material suitable for manifold applications (e.g., adsorbers, fillers, and additives for polymer composites, supplements in animal feed, soil fertilizers, etc.) [3,7]. Moreover, the suitability of spent coffee grounds as adsorbent for the removal of metal ions from aqueous solutions has been largely explored [6], whereas CS has not received, so far, the same attention. Currently, CS has no commercial value and is used only as fuel and soil fertilizer. Recently, some attempts were made to reuse it in different fields, for example, as a source of some bioactive compounds [7–9]. To this regard, the presence of undesirable products, such as ochratoxin, classified as a possible human carcinogen by the International Agency for Research on Cancer (IARC), strongly limits its use for food, cosmetic, and pharmaceutical purposes [10]. Therefore, alternative applications are desirable. As far as we know, the only example of its possible use was reported by Lavecchia et al., who magnetically modified the CS by contact with an aqueous ferrofluid containing magnetite nanoparticles in order to obtain a magnetic material adsorbent for the removal of xenobiotics from wastewater [11].

Potential toxic metals (PTMs) are commonly distributed in the environment, and their pollution sources stem from numerous industrial activities. Among these, zinc, nickel, and copper are essential elements for life and are micronutrients in trace amounts [12]; however, their untreated and uncontrolled discharge is toxic to ecosystems [13,14]. This toxicity affects humans causing several diseases (e.g., Wilson, Alzheimer's) and gastrointestinal problems [15,16]. Moreover, recent studies reported their negative effect on brain, liver, and carcinogenic diseases [17–20]. Therefore, the removal of Zn^{2+}, Cu^{2+}, and Ni^{2+} ions from wastewater streams is an important research task.

Despite various processes of PTMs elimination that are currently adopted (precipitation, electrocoagulation, membrane cementing or separation, solvent extraction, ion-resin exchange), adsorption is a cost-effective operation for the removal of potential toxic metals from diluted wastewater. Generally, commercially available zeolites or activated carbons are used as adsorbent materials [21]; however, owing to environmental sustainability problems, recent research has been oriented toward inexpensive and sustainable vegetable waste such as that deriving from tea [22], coffee [23], orange [24], and oil [25].

The present study aims to investigate the adsorption potential of selected metal ions (Zn^{2+}, Cu^{2+}, and Ni^{2+}) using the coffee silverskin coming from a local coffee roaster (Caffè Mauro S.p.A). Moreover, the adsorption properties of CS were compared with those of SCG, previously studied for removing metal ions from wastewater.

Thus, this contribution is focused on: (i) exploring the suitability of coffee silverskin for the removal of Zn^{2+}, Cu^{2+}, and Ni^{2+} ions from aqueous solutions by studying the retention profile of the latter under different analytical parameters (contact time, analyte concentration, pH); (ii) comparison of the adsorption properties of CS and SCG; and (iii) predicting the kinetic and isotherm models of metal-ion uptake.

2. Materials and Methods

CS and SCG were obtained from a local coffee roaster (Caffè Mauro S.p.A). All chemicals were purchased from CARLO ERBA and used without further purification. The metal sources used were nickel nitrate hexahydrate (99.9% purity), copper nitrate hemi (pentahydrate) (99.9% purity), and zinc nitrate hexahydrate (98.0% purity).

In order to remove impurities, CS and SCG were washed several times with hot water until the washing solution was colorless. Solids were then rinsed and oven-dried at 80 °C for 24 h. Finally, CS and SCG were ground, sieved to <100 μm, dried again at 150 °C for 24 h, and stored at room temperature until use, as similarly reported elsewhere [26].

Different batch tests were carried out to remove metal ions by single, binary, or ternary aqueous solutions.

The amount of solids used for all experiments was 0.12 g and the volume of aqueous solution was 20 mL, with a metal concentration ranging between 20 and 100 ppm, as previously reported [27].

After contact with metal solutions for a fixed time, the concentrations of metal ions in the filtered suspensions were determined using the ICP-OES instrument (Perkin-Elmer Optima 800, Waltham, MA, USA). Each of the experiments was carried out three times and the average value, having a deviation standard lower than 3%, was considered.

Solution volume was kept constant and adsorption efficiency was evaluated as follows:

$$\text{Adsorption } \% = \frac{C_0 - C_t}{C_0} \times 100 \tag{1}$$

where C_0 (mg/L) is the initial concentration in aqueous solution, and C_t (mg/L) is the concentration at time t.

In order to investigate the morphological features and elemental compositions of adsorbent materials, characterization with the SEM-EDX technique was carried out on a Phenom Pro-X scanning electron microscope (SEM) equipped with an energy-dispersive X-ray (EDX). EDX analysis was used in order to evaluate the content and the dispersion of metals, acquiring, for all samples, at least 20 points for 3 different magnifications.

X-ray spectra were recorded with a Bruker D2 Phaser, using Cu Kα radiation at 30 kV and 20 mA. The peaks attribution was made according to the more known databases. Diffraction angles 2θ were varied between 10° and 80° in steps of 0.02° and a count time of 5 s per step [28,29].

The effects of pH on the adsorption of PTMs were examined to find the adsorption mechanism. The pH value was measured using combined glass electrodes (inoLab pH/Cond 720, WTW, Weilheim Germany). The concentration of metal ions in solutions was in the range 20–200 μM, with a pH value equal to 4.0.

The Langmuir (valid for monolayer sorption onto a surface with a finite number of identical sites and uniform adsorption energies) and Freundlich (applied to ions adsorption on heterogeneous surfaces) isotherm models were applied to study the adsorption behavior and to determine metal-adsorption capacity on CS and SCG.

The Langmuir isotherm is given by Equation (2):

$$q_e = \frac{q_{max}\, b\, C_e}{1 + b\, C_e} \tag{2}$$

where q_{max} is the amount of adsorption corresponding to the monolayer coverage, b is the Langmuir constant that is related to the energy of adsorption, and C_e is the equilibrium liquid-phase concentration. Equation (2) can be linearized to determine Langmuir parameters q_{max} and b.

$$\frac{1}{q_e} = \frac{1}{q_{max}\, b}\frac{1}{C_e} + \frac{1}{q_{max}} \tag{3}$$

The Freundlich equation allows calculating the parameters included in Equation (4) or, better, in linearized Equation (5):

$$q_e = K_f\, C_e{}^n \tag{4}$$

$$ln\, q_e = ln\, K_f + n\, ln\, C_e \tag{5}$$

where K_f (mg/g) is related to adsorption affinity, and $1/n$ is the heterogeneity factor related to the intensity of the adsorption.

Batch experiments on the adsorption of binary and ternary aqueous metal cation species in mixtures were carried out using the same initial total molar amount of cations.

3. Results and Discussion

3.1. Adsorption of Single Metals

The pH of the solution is a significant factor affecting the adsorption of metal ions since competitive adsorption between hydrogen/metal ions occurs. The investigated metal ions were present as Me^{2+}-deriving species (Cu^{2+}, Zn^{2+}, Ni^{2+}) forming, in water, coordinated aqueous complexes. Only at high pH values could hydroxyl species $Me(OH)_3^-$ and $Me(OH)_4{}^{2-}$ be detected. Preliminary tests, carried out by varying the pH, indicate that in order to achieve the dominant species for all metals, aqueous Me^{2+} complexes, the solution's pH value should be maintained in the range 4–5. In this range, no hydroxide precipitation occurs, as previously observed [30–32].

The adsorption isotherm of both materials used, CS and SCG, is shown in Figure 1. It is possible to notice that for both adsorbents the achievement of adsorption equilibrium occurs after about 20 min. To this regard, the adsorption efficiency of Ni^{2+} ions is the same for both materials, with a value of around 50%, whereas the CS adsorbent exhibits great efficiency in removing Cu^{2+} ions. On the other hand, SCG show higher efficiency, with respect to CS, in removing both Cu^{2+} and Zn^{2+} ions. The comparison of adsorption efficiencies, relative to both SCG and CS, may be better understood on the basis of the following consideration.

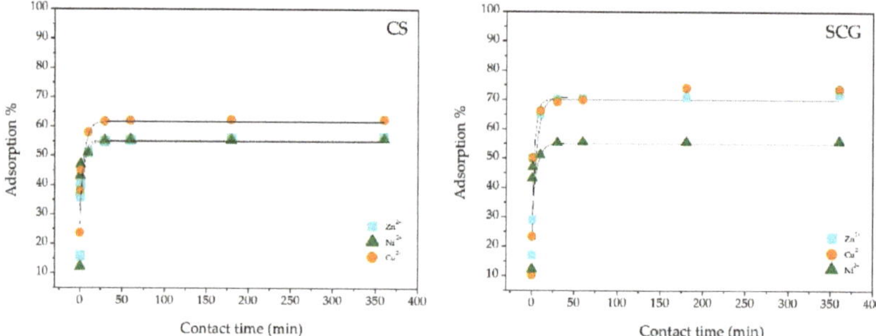

Figure 1. Contact time versus adsorption on coffee silverkin (CS) and spent coffee grounds (SCG) (C_0 = 100 ppm pH = 4 T = 25 °C).

The first stems from the different morphological features related to the external surface of both adsorbents. In principle, in fact, the presence of fissures and holes contributes to metal-ion diffusion through the adsorption surface. From the SEM micrographs shown in Figure 2, it is possible to infer the relevant differences in the CS and SCG morphological surface. SCG exhibit three-dimensional morphology formed by agglomerated grains, having dimensions of about ten microns, whereas CS shows a lamellar structure typical of vegetable teguments with a high surface area, preferentially ordered in two dimensions. Therefore, the bidimensionality of CS can contribute minor adsorption efficiency in comparison to that of SCG.

Another aspect that affects the observed adsorption difference can derive from the different amounts of lignocellulosic components in both structures (Figure 3). In principle, the adsorption ability of materials can be ascribed to favorable electrostatic interactions between positively charged metal ions and the opposite charge of the adsorbing surface. It is worth to underline that lignocellulosic biomasses are characterized by negatively charged surface sites, including hydroxyl and carboxyl groups as well as etheric oxygen atoms [33]. These groups belong to those generally called hard bases in the well-known hard and soft acid-bases theory [34,35]. Therefore, the higher content of lignocellulosic components in the SCG with respect to the CS contributes to explaining the higher adsorption efficiency of the former.

Figure 2. Scanning elecron microscope (SEM) micrographs of (**a**) CS and (**b**) SCG.

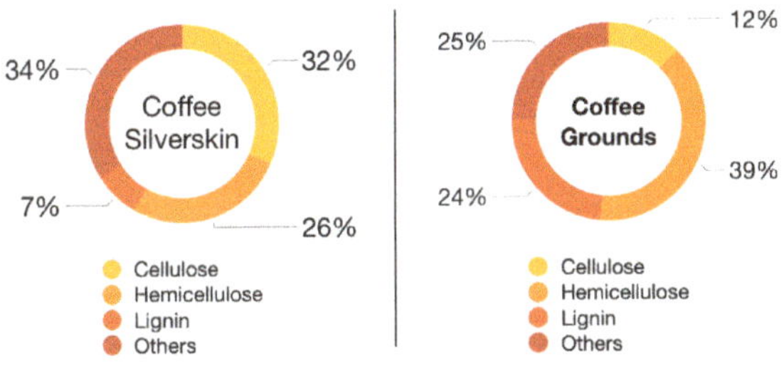

Figure 3. Compound composition of CS and SCG.

It is worth to emphasize that the investigated adsorbing materials were not affected by any contact with metal-ion solutions (Figure 4): the morphologies of CS and SCG were preserved and the mapping of metal ions confirmed the uniform distribution on materials recovered after the batch tests. The retention of metals by CS was also confirmed by X-Ray analysis on the CS recovered after batch adsorption. In Figure 5, it is possible to notice the deposition of nitrate salt precipitated on the adsorbing surface and peaks corresponding to Zn^{2+} species. Similar patterns were also recorded in other metal experiments (not shown). The precipitation of potassium nitrate is due to the leaching of potassium from the CS in the stirred conditions of the experiments. The mineral composition of the CS in fact mainly consists of potassium 5%w/w per 100 g of CS, determined by EDX analysis.

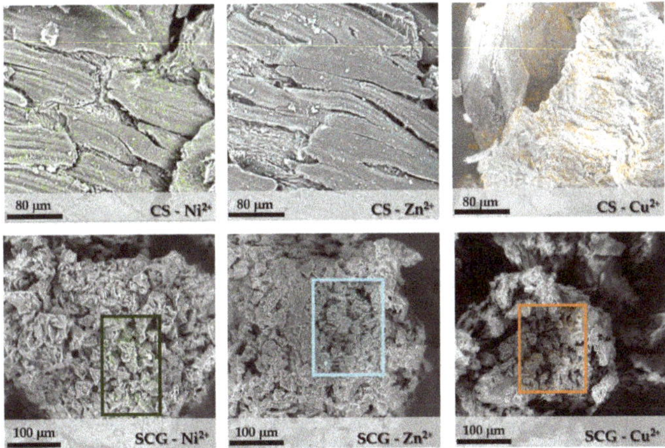

Figure 4. SEM micrographs and energy-dispersive X-ray (EDX) maps of CS and SCG.

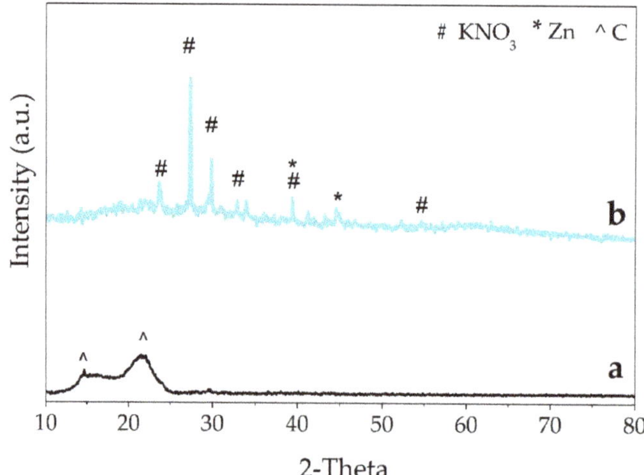

Figure 5. X-ray spectra of (**a**) CS as made and (**b**) CS after Zn^{2+} ion removal.

3.2. Isotherm-Model Application

Adsorption data relative to our systems were fitted by applying the Langmuir equation, and the results are reported in Table 1. Fitting was excellent for all metal-adsorbent systems considered (Figure 6) and the correlation coefficients of the linear regression (r^2) were found to be higher than 0.986 (Table 1).

Table 1. Langmuir and Freundlich isotherm parameters for Cu^{2+}, Zn^{2+}, and Ni^{2+} ions onto CS and SGC using a single-component solution.

PTM	Coffee Silverskin						Spent Coffee Grounds					
	Langmuir			Freundlich			Langmuir			Freundlich		
	q_{max}	b	r^2	K_f	n	r^2	q_{max}	b	r^2	K_f	n	r^2
Cu^{2+}	9.58	2.60	0.996	7.06	2.06	0.947	10.00	3.05	0.986	8.63	2.06	0.903
Ni^{2+}	1.43	15.05	0.994	1.30	13.00	0.827	1.67	20.22	0.999	1.58	13.00	0.981
Zn^{2+}	15.17	0.62	0.989	5.69	1.37	0.987	8.23	4.45	0.991	6.69	1.37	0.990

The maximum adsorption amounts on CS, fitted and obtained by the Langmuir model, were 15.17, 9.58, and 1.43 mg/g, respectively, for Cu^{2+}, Zn^{2+}, and Ni^{2+} ions. Therefore, the adsorption affinity of CS gives the sequence $Cu^{2+} > Zn^{2+} > Ni^{2+}$. A similar trend was observed for SCG.

From Langmuir isotherm coefficients, it is possible to also define dimensionless parameter R_L:

$$R_L = \frac{1}{(1 + bC_0)} \quad (6)$$

This can be considered a separation factor that can determine the type of isotherm [36].

Values of R_L denote if the type of the Langmuir isotherm is irreversible ($R_L = 0$), favorable ($0 < R_L < 1$), linear ($R_L = 1$) or unfavorable ($R_L > 1$); in all our cases, values of R_L ranged between 0 and 1.

On the other hand, by applying the Freundlich isotherm models, adsorption affinity K_f values, relative to Cu^{2+}, Zn^{2+}, and Ni^{2+} ions on CS, were 7.06, 5.69, 1.30, respectively, indicating that Cu^{2+} and Zn^{2+} are preferably adsorbed. The n value indicates the degree of nonlinearity between solution concentrations and adsorption according to the following considerations: if $n = 1$, adsorption is linear;

if $n < 1$, adsorption implies a chemical process; if $n > 1$, adsorption stems from a physical process. The n values on using the Freundlich equation for Cu^{2+} and Zn^{2+} adsorption, both on the CS and on the SCG, are lower than 3 (Table 1). The value of $n > 1$ is very common and it is generally ascribed to dispersed surface sites or to any factor that may generate a decrease in adsorbent–adsorbate interaction. Generally, the increase of surface density and of n values, within the range of 1–10, means that adsorption was good [37]. In the present study, with respect to Cu^{2+} and Zn^{2+} ion adsorption, since n lies between 1 and 10, it is possible to infer that the adsorption of metal ions onto CS and SCG is essentially physical. For nickel ions, the n value, greater than 10, indicates minor mobility and a higher retention of metals in the solid phase [38].

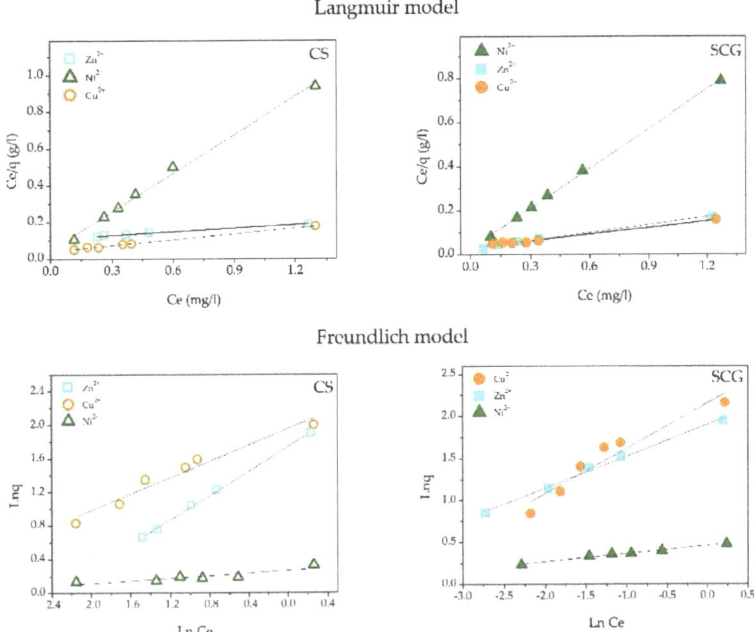

Figure 6. Langmuir and Freundlich isotherms for adsorption of metal ions on CS and SCG.

However, the results demonstrate that the fitting for all adsorbing materials is more suitable with the Langmuir model (r^2 = 0.986–0.999) for PTM adsorption than with the Freundlich one (r^2 = 0.827–0.987), so the adsorption characteristics were mainly of monolayer type.

3.3. Comparison of Metal Adsorption in Competitive Systems

In order to evaluate the affinity of metal ions towards the CS material, we also performed batch-experiment adsorption of binary and ternary aqueous metal cation species in mixtures Cu^{2+}/Zn^{2+}, Cu^{2+}/Ni^{2+}, Zn^{2+}/Ni^{2+} and $Cu^{2+}/Zn^{2+}/Ni^{2+}$ using the same initial total molar amount of cations.

Figure 7 shows the experimental data for binary and ternary mixtures and allows comparing the relative adsorption results with those obtained with separately using single metals.

In principle, in multicomponent system solutions, ions exhibit competitive adsorption. Therefore, the amount of cations adsorbed from binary solutions is lower than that obtained by using a single metal solution, and this is in agreement with other reports [30,39].

In particular, results suggest that the adsorption of metals on CS in a multicomponent system is, for all investigated metals, about 10%–15% lower than that detected in a single-component system.

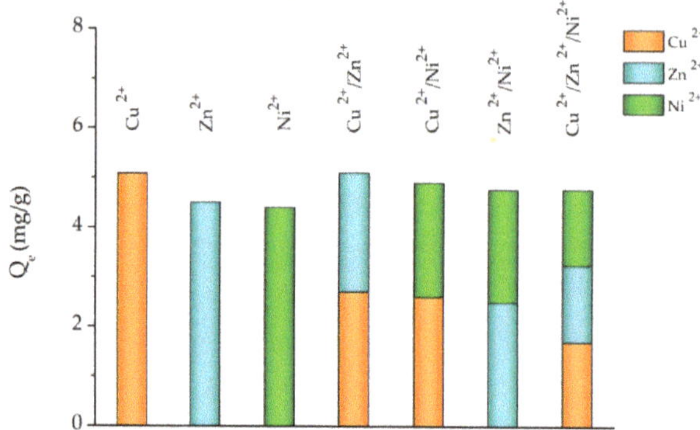

Figure 7. Single and competitive adsorption of Cu^{2+}, Zn^{2+}, and Ni^{2+} on CS: amount of metal ions adsorbed on CS.

However, total adsorption was higher than that observed in using a single component, in agreement with already observed results relative to other adsorbents [40,41]. The affinity order towards metals exploited by the CS material, both in the binary and ternary system, is the same as that found in the single one ($Cu^{2+} > Zn^{2+} > Ni^{2+}$).

The affinity and adsorption capacity of different metals could suggest a correlation between different metal parameters (e.g., electronegativity, coordination-complex radius, and stability constants of the associated metal hydroxide or solvated species) [42]. Electronegativity is regarded as a contributing parameter to metal-ion uptake. In fact, when metal cations are adsorbed on a negatively charged surface, the attraction of a negative charge plays a significant role in the adsorption process. However, this is not a general rule, as verified by different works on adsorption processes [9]. In principle, adsorption of metal cations on hard-containing surface groups depends on an interplay of factors belonging to both ions and surface. Indeed Ni^{2+}, Cu^{2+}, and Zn^{2+} have d^8, d^9, and d^{10} external electronic structure, respectively, and the observed trend towards their relative adsorption mainly depends on the stability of the aqueous coordinated complexes. All of them form octahedral $Me(OH_2)_6{}^{2+}$ complexes, and no information on their stability may be obtained from the literature. Furthermore, all three cations are considered on the borderline of the hard and soft acid and base classification [34,35].

4. Sustainability Considerations/Conclusions

Global coffee production has been extending to include a large industrial system, with a total coffee consumption of 157,858 in thousand 60 kg bags during the 2016/2017 period [42].

Among the huge volume of coffee waste as a result of the coffee production and consumption process, CS is generally considered the most significantly produced and, at the same time, difficult to manage. Indeed, CS represents not only solid waste that contributes to environmental pollution, but it also has high impact on the economy of coffee-roaster companies. In fact, the storage of CS before its disposal in compost or landfill needs specific machinery for compaction and relative high-energy consumption. Furthermore, its biomass releases both CO_2 and methane in landfills, with consequent environmental impact. Basically, the alternative to landfills is to bury or burn CS waste, with the unavoidable production of greenhouse gasses. According to the United States Environmental Protection Agency (EPA) [43], burning garbage constituted by biomass releases more carbon dioxide than burning coal. It is therefore clear that alternative solutions should be considered. Although a 'zero-waste' industry is not easily achievable in the near future, research must strive to be policy-aimed to achieve sustainable results in the reuse of byproduct materials [44].

The coffee waste used in this study derived from the roasting of green coffee beans. In order to perform the adsorption of potential toxic metals like Cu^{2+}, Zn^{2+}, and Ni^{2+} ions, no chemical modification or further treatment was used to obtain the adsorbing materials. This aspect is to be emphasized, since it is also necessary to ensure that waste treatment does not create any extra problems on the environment that could eventually undermine the beneficial effects of utilizing waste products such as CS or SCG [45].

Furthermore, CO_2 emissions should decrease due to the reduced use of incinerators for these organic wastes. Meanwhile, recovery of natural resources can be simultaneously achieved from the viewpoint of waste management.

Accordingly, the results of this preliminary laboratory work provide a possible sustainable route to recover inexpensive biological waste for its use as adsorbent of PTMs from wastewater. Further works will be devoted to identifying the regeneration methods of coffee silverskin, the recovery of the adsorbate, and the subsequent regeneration and reuse of the adsorbent material. Finally, quantitative consideration of sustainability (Life-Cycle Assessment, LCA analysis) on the applicability of CS and SCG in water remediation should be carried out in order to validate the entire process from an economical and environmental point of view.

Author Contributions: Conceptualization, P.F., L.B. and F.M.; Formal analysis, A.M., E.P.; Investigation, A.M., E.P., G.P.; Project administration, P.F., L.B., F.M. and G.P.; Supervision, P.F. and F.M.; Writing—original draft, P.F.; Writing—review & editing, P.F., A.M. and F.M.

Funding: This publication was supported by the Mediterranean University of Reggio Calabria.

Acknowledgments: The authors thank Caffè Mauro S.p.A. for providing coffee silverskin and spent coffee grounds, Ing. Giuseppe Modafferi (Caffè Mauro S.p.A.) and Saverio Festa (Caffè Mauro S.p.A.) for fruitful discussions, and Ing. Letizia Buonsanti for kinetic adsorption experiments.

Conflicts of Interest: The authors declare no conflicts of interest.

References

1. Figueroa, G.A.; Homann, T.; Ravel, H.M. Coffee Production Wastes: Potentials and Perspective. *Austin Food Sci.* **2016**, *1*, 1014.
2. Esquivel, P.; Jimenez, V.M. Functional Properties of coffee and coffee by-products. *Food. Res. Int.* **2012**, *46*, 488–495. [CrossRef]
3. Kovalcik, A.; Obruca, S.; Marova, I. Valorization of Spent Coffee Grounds: A review. *Food Bioprod. Process.* **2018**, *110*, 104–119. [CrossRef]
4. Coppola, L.; Bellezze, T.; Belli, A.; Bignozzi, M.C.; Bolzoni, F.; Brenna, A.; Cabrini, M.; Candamano, S.; Cappai, M.; Caputo, D.; et al. Binders alternative to Portland cement and waste management for sustainable construction-part 1. *J. Appl. Biomater. Funct. Mater.* **2018**, *16*, 186–202. [CrossRef] [PubMed]
5. Alves, R.C.; Rodrigues, F.; Antónia Nunes, M.; Vinha, A.F.; Oliveira, M.B.P.P. State of the art in coffee processing by-products. In *Handbook of Coffee Processing by-Products*, 1st ed.; Galanakis, C.M., Ed.; Academic Press: London, UK, 2017; pp. 1–26.
6. Anastopoulos, I.; Karamesouti, M.; Mitropoulos, A.C.; Kyzas, G.Z. A review for coffee adsorbents. *J. Mol. Liq.* **2017**, *229*, 555–565. [CrossRef]
7. Castro, C.S.; Abreu, A.L.; Silva, C.L.; Guerreiro, M.C. Phenol adsorption by activated carbon produced from spent coffee grounds. *Water Sci. Technol.* **2011**, *64*, 2059–2065. [CrossRef] [PubMed]
8. Costa, A.S.; Alves, R.C.; Vinha, A.F.; Costa, E.; Costa, C.S.; Nunes, M.A.; Almeida, A.A.; Santos-Silva, A.; Oliveira, M.B.P.P. Nutritional, chemical and antioxidant/pro-oxidant profiles of silverskin, a coffee roasting by-product. *Food Chem.* **2018**, *267*, 28–35. [CrossRef] [PubMed]
9. Costa, A.S.; Alves, R.C.; Vinha, A.F.; Barreira, S.V.; Nunes, M.A.; Cunha, L.M.; Oliveira, M.B.P.P. Optimization of antioxidants extraction from coffee silverskin, a roasting by-product, having in view a sustainable process. *Ind. Crops Prod.* **2014**, *53*, 350–357. [CrossRef]
10. Iriondo-DeHond, A.; Haza, A.I.; Ávalos, A.; del Castillo, M.D.; Morales, P. Validation of coffee silverskin extract as a food ingredient by the analysis of cytotoxicity and genotoxicity. *Food Res. Int.* **2017**, *100*, 791–797. [CrossRef] [PubMed]

11. Di Battista, A.; Lavecchia, R. Magnetically modified coffee silverskin for the removal of xenobiotics from wastewater. *Chem. Eng. Trans.* **2013**, *35*, 1375–1380. [CrossRef]
12. Volesky, B. Detoxification of metal-bearing effluents: Biosorption for the next century. *Hydrometallurgy* **2001**, *59*, 203–216. [CrossRef]
13. Meers, E.; Ruttens, A.; Hopgood, M.J.; Samson, D.; Tack, F.M.G. Comparison of EDTA and EDDS as potential soil amendments for enhanced phytoextraction of heavy metals. *Chemosphere* **2005**, *58*, 1011–1022. [CrossRef] [PubMed]
14. Race, M.; Ferraro, A.; Fabbricino, M.; La Marca, A.; Panico, A.; Spasiano, D.; Tognacchini, A.; Pirozzi, F. Ethylenediamine-N,N′-Disuccinic Acid (EDDS)—Enhanced Flushing Optimization for Contaminated Agricultural Soil Remediation and Assessment of Prospective Cu and Zn Transport. *Int. J. Environ. Res. Public Health* **2018**, *15*, 543. [CrossRef] [PubMed]
15. Bandmann, O.; Weiss, K.H.; Kaler, S.G. Wilson's disease and other neurological copper disorders. *Lancet Neurol.* **2015**, *14*, 103–113. [CrossRef]
16. Cho, C.H. Zinc: Absorption and role in gastrointestinal metabolism and disorders. *Dig. Dis.* **1991**, *9*, 49–60. [CrossRef] [PubMed]
17. Portbury, S.D.; Adlard, P.A. Zinc signal in brain diseases. *Int. J. Mol. Sci.* **2017**, *18*, 2506. [CrossRef] [PubMed]
18. Prystupa, A.; Błażewicz, A.; Kiciński, P.; Sak, J.J.; Niedziałek, J.; Załuska, W. Serum concentrations of selected heavy metals in patients with alcoholic liver cirrhosis from the Lublin Region in Eastern Poland. *Int. J. Environ. Res. Public Health* **2016**, *13*, 582. [CrossRef] [PubMed]
19. Kim, H.S.; Yeo, J.K.; Young, R.S. An overview of carcinogenic heavy metal: Molecular toxicity mechanism and prevention. *J. Cancer Prev.* **2015**, *20*, 232. [CrossRef] [PubMed]
20. Pietropaolo, A.; Satriano, C.; Strano, G.; La Mendola, D.; Rizzarelli, E. Different zinc (II) complex species and binding modes at Aβ N-terminus drive distinct long range cross-talks in the Aβ monomers. *J. Inorg. Biochem.* **2015**, *153*, 367–376. [CrossRef] [PubMed]
21. Ince, M.; Ince, O.K. An Overview of Adsorption Technique for Heavy Metal Removal from Water/Wastewater: A Critical Review. *Int. J. Pure Appl. Sci.* **2017**, *3*, 10–19. [CrossRef]
22. Amarasinghe, B.M.W.P.K.; Williams, R.A. Tea waste as low cost adsorbent for the removal of Cu and Pb from wastewater. *Chem. Eng. J.* **2007**, *132*, 299–309. [CrossRef]
23. Kaikake, K.; Hoaki, K.; Sunada, H.; Dhakal, R.P.; Baba, Y. Removal characteristics of metal ions using degreased coffee beans: Adsorption equilibrium of cadmium(II). *Bioresour. Technol.* **2007**, *98*, 2787–2791. [CrossRef] [PubMed]
24. Perez-Marin, A.B.; Zapata, V.M.; Ortuno, J.F.; Aguilar, M.; Saez, J.; Lorens, M. Removal of cadmium from aqueous solutions by adsorption onto orange waste. *J. Hazard. Mater. B* **2007**, *139*, 122–131. [CrossRef] [PubMed]
25. Fiol, N.; Villaescusa, I.; Martinez, M.; Miralles, N.; Poch, J.; Serarols, J. Sorption of Pb(II), Ni(II), Cu(II) and Cd(II) from aqueous solution by olive stone waste. *Sep. Purif. Technol.* **2006**, *50*, 132–140. [CrossRef]
26. Zuorro, A.; Lavecchia, R.; Natali, S. Magnetically modified agro-industrial wastes as efficient and easily recoverable adsorbents for water treatment. *Chem. Eng. Trans.* **2014**, *38*, 349–354. [CrossRef]
27. Chavan, A.A.; Pinto, J.; Liakos, I.; Bayer, I.S.; Lauciello, S.; Athanassiou, A.; Fragouli, D. Spent coffee bioelastomeric composite foams for the removal of Pb^{2+} and Hg^{2+} from water. *ACS Sustain. Chem. Eng.* **2016**, *4*, 5495–5502. [CrossRef]
28. Mauriello, F.; Paone, E.; Pietropaolo, R.; Balu, A.M.; Luque, R. Catalytic Transfer Hydrogenolysis of Lignin-Derived Aromatic Ethers Promoted by Bimetallic Pd/Ni Systems. *ACS Sustain. Chem. Eng.* **2018**, *6*, 9269–9276. [CrossRef]
29. Frontera, P.; Malara, A.; Stelitano, S.; Fazio, E.; Neri, F.; Scarpino, L.; Antonucci, P.L.; Santangelo, S. A new approach to the synthesis of titania nano-powders enriched with very high contents of carbon nanotubes by electro-spinning. *Mater Chem. Phys.* **2015**, *153*, 338–345. [CrossRef]
30. Sud, D.; Mahajan, G.; Kaur, M.P. Agricultural waste material as potential adsorbent for sequestering heavy metal ions from aqueous solutions—A review. *Bioresour. Technol.* **2008**, *99*, 6017–6027. [CrossRef] [PubMed]
31. Sitko, R.; Turek, E.; Zawisza, E.; Malicka, E.; Talik, E.; Heimann, J.; Gagor, A.; Feist, B.; Wrzalik, R. Adsorption of divalent metal ions from aqueous solutions using graphene oxide. *Dalton Trans.* **2013**, *42*, 5682–5689. [CrossRef] [PubMed]

32. Azouaou, N.; Sadaoui, Z.; Djaafri, A.; Mokaddem, H. Adsorption of cadmium from aqueous solution onto untreated coffee grounds: Equilibrium, kinetics and thermodynamics. *J. Hazard. Mater.* **2010**, *184*, 126–134. [CrossRef] [PubMed]
33. Demirbas, A. Heavy metal adsorption onto agro-based waste material: A review. *J. Hazard. Mater* **2008**, *157*, 220–229. [CrossRef] [PubMed]
34. Huheey, J.E. *Inorganic Chemistry: Principles of Structure and Reactivity*; Harper International Edition; Harper & Row: New York, NY, USA, 2006.
35. Shiver, D.F.; Atkins, P.W.; Langford, G.H. *Inorganic Chemistry*, 2nd ed.; Oxford University Press: Oxford, UK, 1993.
36. Hall, K.R.; Eagleton, L.C.; Acrivos, A.; Vermevlem, T. Pore-and solid-diffusion kinetics in fixed-bed adsorption under constant-pattern conditions. *Ind. Eng. Chem. Fundam.* **1966**, *5*, 212–223. [CrossRef]
37. Kumar, Y.P.; King, P.; Prasad, V.S.R.K. Adsorption of zinc from aqueous solution using marine green algae-Ulva fasciata sp. *Chem. Eng. J.* **2007**, *129*, 161–166. [CrossRef]
38. Reed, B.E.; Matsumoto, M.-R. Modeling cadmium adsorption by activated carbon using the Langmuir and Freundlich isotherm expressions. *Sep. Sci. Technol.* **1993**, *28*, 2179–2195. [CrossRef]
39. Morera, M.T.; Echeverria, J.C.; Mazkiaran, C.; Garrido, J.J. Isotherm and sequential extraction procedures for evaluating sorption and distribution of heavy metals in soils. *Environ. Pollut.* **2001**, *113*, 135–144. [CrossRef]
40. Wang, X.S.; Li, Z.Z. Competitive Adsorption of Nickel and Copper Ions from Aqueous Solution Using Nonliving Biomass of the Marine Brown Alga Laminaria japonica. *Clean* **2009**, *37*, 663–668. [CrossRef]
41. Futalan, C.M.; Kan, C.C.; Dalida, M.L.; Hsien, K.J.; Pascua, C.; Wan, M.W. Comparative and competitive adsorption of copper, lead, and nickel using chitosan immobilized on bentonite. *Carbohydr. Polym.* **2011**, *83*, 528–536. [CrossRef]
42. Trade Statistics Tables: Looking for quick facts and figures on the global coffee trade? Available online: http://www.ico.org/trade_statistics.asp?section=Statistics (accessed on 1 October 2018).
43. United States Environmental Protection Agency; Relative Risk Reduction Strategies Committee. *Reducing Risk: Setting Priorities and Strategies for Environmental Protection: The Report of the Science Advisory Board, Relative Risk Reduction Strategies Committee, to William K. Reilly, Administrator, United States Environmental Protection Agency*; US Environmental Protection Agency, Science Advisory Board: Washington, DC, USA, 1990.
44. Reh, L. Process engineering in circular economy. *Particuology* **2013**, *11*, 119–133. [CrossRef]
45. Mirabella, N.; Castellani, V.; Sala, S. Current options for the valorization of food manufacturing waste: A review. *J. Clean. Prod.* **2014**, *65*, 28–41. [CrossRef]

Article

Adsorption of Hexavalent Chromium Using Banana Pseudostem Biochar and Its Mechanism

Shuang Xu, Weiguang Yu, Sen Liu, Congying Xu, Jihui Li * and Yucang Zhang *

Key Laboratory of Ministry of Education for Advanced Materials in Tropical Island Resources, College of Materials and Chemical Engineering, Hainan University, Haikou 570228, China; Shuangxu0216@163.com (S.X.); yuweiguang666@163.com (W.Y.); haidaliusen@163.com (S.L.); xucongying1229@sina.com (C.X.)

* Correspondence: lijihui@hainu.edu.cn (J.L.); yczhang@hainu.edu.cn (Y.Z.)

Received: 1 October 2018; Accepted: 14 November 2018; Published: 17 November 2018

Abstract: A low-cost biochar was prepared through slow pyrolysis of banana pseudostem biowaste at different temperatures, and characterized by surface area and porosity analysis, scanning electron microscopy (SEM), Fourier-transform infrared (FTIR) spectroscopy, and X-ray photoelectron spectroscopy (XPS). It was shown that the biochar prepared at low pyrolysis temperature was rich in oxygen-containing groups on the surface. Adsorption experiments revealed that the biochar prepared at 300 °C (BB300) was the best adsorbent for Cr(VI) with 125.44 mg/g maximum adsorption capacity at pH 2 and 25 °C. All the adsorption processes were well described by pseudo-second-order and Langmuir models, indicating a monolayer chemiadsorption. Furthermore, it was demonstrated that adsorption of Cr(VI) was mainly attributed to reduction of Cr(VI) to Cr(III) followed by ion exchange and complexation with the biochar.

Keywords: biochar; banana pseudostem; chromium; adsorption mechanism

---

1. Introduction

Chromium, a toxic heavy metal, is widely discharged into the aquatic environment from various industries [1,2], causing serious harm to human health directly or indirectly [3–5]. Usually, chromium exists in hexavalent and trivalent forms in polluted water. Hexavalent Cr is highly toxic, carcinogenic and allergenic, while trivalent Cr is an essential trace element for mammals and less toxic [2,6–8]. Many treatment methods have been developed for removal of hexavalent chromium from water in the past decades, such as chemical precipitation [9], sedimentation [10], flocculation [11], adsorption [12], ultrafiltration [13], ion exchange [14], chemical coagulation [15], and so on. Being effective and operationally simple, adsorption is a useful method for removal of hexavalent chromium, solving the problems of sludge disposal [16,17].

Biochar, a cost-effective and green carbon material prepared from biomass through thermal conversion in an oxygen deficient environment [18,19], has been widely used to remove or immobilize Cr(III) and Cr(VI) due to their large specific surface area, high porosity and abundant functional groups on surface [20,21]. As the adsorption capabilities of biochar are significantly affected by biomass types [20,21], a series of biomass including oak wood, oak bark [22], ramie [23], cotton stalk [24], beet tailing [25], coconut coir [26], leaf of *Leersia hexandra* Swartz [27] and so on have been employed to make biochars by pyrolysis at different temperatures for removal of hexavalent chromium in the past decades. The adsorption capability of these biochars ranged from 3.03 mg/g to 349.81 mg/g, depending on the biomass feedstock. However, the application of biochars for removal of Cr(VI) was limited by either low adsorption efficiencies or less production of biomass feedstocks. For example, the adsorption capability of *Leersia hexandra* Swartz leaf biochar was up to 349.81 mg/g, but the production of *Leersia hexandra* Swartz leaf was less; even oak wood was available in large quantities, the adsorption capability of oak wood biochar was only 3.03 mg/g. As a result, biochars with high

Cr(VI) adsorption capability from large-scale and low-cost biowaste are still highly necessary for remediation of Cr(VI) containing waste water.

Banana pseudostem, a major residual waste of banana which is one of most widely distributed consumed food crops, is massively generated every year [28]. Banana pseudostem consists mainly of cellulose, hemicellulose, pectin and lignin [29]. Here, banana pseudostem was used as material to prepare biochar through direct pyrolysis at different temperatures and both absorption capability and adsorption mechanism of the biochar for Cr(VI) were investigated in aqueous solution.

2. Materials and Methods

2.1. Materials

All chemicals including HCl, KOH, H_2SO_4, H_3PO_4, CH_3COCH_3, $C_{13}H_{14}N_4O$ and $K_2Cr_2O_7$ were analytical grade reagents and purchased from Aladdin or Macklin in Shanghai of China. A stock solution (1000 mg/L) of Cr(VI) was prepared by dissolving $K_2Cr_2O_7$ in deionized water (Water treatment system, LD-UPW-20, Leading, Shanghai, China). The stock solution was then diluted to desired concentrations. The pH of Cr(VI) solution was adjusted by suitable concentration solution of HCl or KOH.

2.2. Preparation of Banana Pseudostem Biochars

Banana pseudostem, the layered pseudostem wrapped in the trunk of banana trees, was supplied by the farms around Haikou City, Hainan Province, China. The sample was air-dried at 60 °C by blast drying oven and then smashed to pass through a 60 mesh sieve (0.3 mm). The banana pseudostem powder was heated with a rate of 5 °C min^{-1} until it reached target temperature (200, 300, 400, 500, 600 °C) and then maintained the temperature for 1 h in a muffle furnace with tubular reactor under suitable nitrogen flow rate. The resulting biochars were cooled down to room temperature under nitrogen flow. Then the biochar was washed with deionized water for several times, dried at 60 °C for 24 h. The dried biochars were stored in an airtight desiccator prior to use and were abbreviated as BB200, BB300, BB400, BB500 and BB600 respectively, according to the pyrolysis temperature.

2.3. Characterization of Banana Pseudostem Biochars

The textural property was analyzed using N_2 sorption at 77 K on an ASAP 2460 surface area and porosity analyzer (Micromeritics, Norcross, GA, USA). The biochars (0.1–0.2 g) were degassed at 100 °C in vacuum before test. The surface area, pore volume and average pore size were all calculated by the Brunauer-Emmett-Teller (BET) method. The microscopic properties and surface morphologies were characterized using a S-3000N scanning electron microscope (Hitachi, Tokyo, Japan) operated at 10 kV. The surface functional groups of the biochars were characterized by Fourier-transform infrared (FTIR) spectrometer (Bruker Tensor 27, Ettlingen, Germany). The biochars were grinded with KBr and rolled into sheets, then recorded between 400 and 4000 cm^{-1}. The surface elemental content (C, O, N and Cr) and surface functional groups were determined by X-ray photo-electron spectra (XPS) with an Escalab 250Xi spectrometer (Thermo Scientific, Waltham, MA, USA).

2.4. Adsorption Experiments

Batch adsorption experiments of Cr(VI) were performed by adding biochars (50 mg) and aqueous K_2CrO_4 solution (50 mL) into 100 mL sealed conical flasks and shaking at 180 rpm on a thermostat shaker for desired time. For effect of pH on adsorption efficiency, the adsorptions were tested at pH ranged from 1.0 to 5.0 using a 200 mg/L Cr(VI) concentration. For kinetic experiments, BB300, BB400, BB500 and BB600 were mixed with Cr(VI) solution at pH 2, respectively, then the mixtures were shaken for different time interval (0.5, 1, 2, 3, 5, 9, 13, 18, 24, 30, 36, 42 and 48 h). Adsorption isotherms were carried out for 48 h by maintaining the temperature at 25 °C using different Cr(VI) concentrations (25, 50, 100, 200, 300, 400 and 600 mg/L) with initial pH 2.

The Cr(VI) concentrations before and after adsorption were determined on a UV-vis spectrophotometer (MAPADA UV-3300PC, Shanghai, China) at wavelength of 540 nm using 1.5-diphenylcarbazide as indicator. The adsorption amount of Cr onto banana pseudostem biochars was calculated by the difference of concentrations before and after adsorption. The adsorption capacities of biochars towards Cr(VI) were determined as follows:

$$q_e = \frac{(C_0 - C_e)V}{M} \quad (1)$$

where q_e (mg/g) is the adsorption capabilities of biochars; C_0 (mg/L) and C_e (mg/L) are the Cr(VI) concentrations before and after adsorption, respectively; V (L) is the adsorbate solution volume; M (g) is the dosage of biochar.

3. Results and Discussion

3.1. Characterization of Banana Pseudostem Biochars

The FTIR spectra of banana pseudostem biochars are shown in Figure 1. The peak at 3448 cm^{-1} was assigned to hydroxyl group, and the peaks at about 2928 and 2860 cm^{-1} were assigned to aliphatic –CH and $-CH_2$ stretching vibrations, respectively [30]. The peaks at 1630 cm^{-1} resulted from both the C=O stretching vibration and C=C stretching vibration of aromatic ring. The peak at about 1433 cm^{-1} also corresponded to aromatic C=C stretching vibration of aromatic ring [31]. Obviously, the peak at 1630 cm^{-1} reduced and the peak at 1433 cm^{-1} increased while increasing the pyrolysis temperature to 500 and 600 °C, as decarbonylation and aromatization intensively took place at high temperature [32]. The peak at 1385 cm^{-1} was symmetrical stretching vibration of COO^- [33,34]. The peak at 1317 cm^{-1} was assigned to aliphatic CH_2 deformation vibration [31]. The peaks at 1117 cm^{-1} and 1067 cm^{-1} represented alkoxy C–O and aromatic C–O stretching vibrations, respectively. The shift of C–O peaks also confirmed that aromatization took place as increasing pyrolysis temperature [35]. The peaks at 876 cm^{-1} and 781 cm^{-1} were assigned to Si–O–Si symmetric stretching vibrations [36].

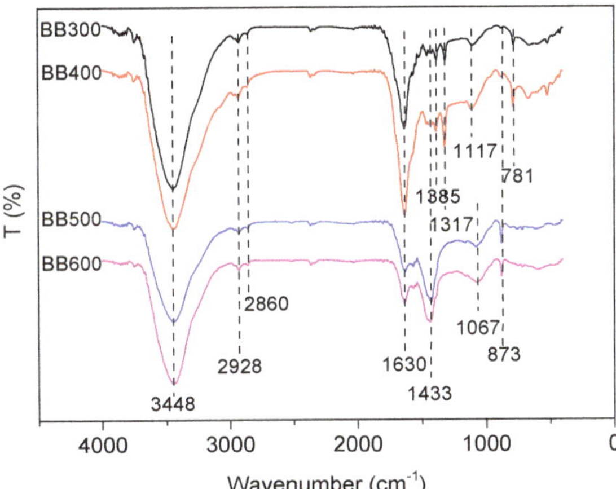

Figure 1. The Fourier-transform infrared (FTIR) spectra of banana pseudostem biochars.

The surface area, pore volume and average pore size of banana pseudostem biochars are given in Table 1. The biochars were of mesoporous structure [37], the average pore diameter (PD) and pore volume (PV) increased with pyrolysis temperature. The BET surface area (SA) slightly increased as pyrolysis temperature increased from 300 to 500 °C, then decreased as reaching 600 °C. This was probably because of formation of more mesopores caused by escape of volatile substances with increasing temperature and destruction of pore structures at high temperature [38]. The same correlation between surface area and pyrolysis temperature was also observed in previous biochar adsorption study [39].

Table 1. The surface area, pore volume and pore diameter of banana pseudostem biochars.

BC	BB300	BB400	BB500	BB600
SA (m^2/g)	4.98	6.62	11.27	8.53
PV (cm^3/g)	0.00959	0.01458	0.03932	0.04389
PD (nm)	10.39	11.85	15.60	19.45

The C 1s XPS peaks of banana pseudostem biochars are shown in Figure 2, and the ratios of C–O, C=O, COO and CO_3^{2-} bonds to the C–C bonds based on peak area are presented in Table 2. The total content of O-containing groups (TCOG) including C–O, C=O, COO and CO_3^{2-}, an important parameter affecting the adsorption capability of biochars [40], generally decreased as pyrolysis temperature increased from 300 to 600 °C, as previously reported by Li [35]. Both C–O and C=O bonds sharply decreased, and COO and CO_3^{2-} basically remained unchanged, as the pyrolysis temperature increased from 300 to 400 °C. This might be due to the intense decomposition of cellulose and hemicellulose by breaking the C–O and C=O groups [41–43]. While the pyrolysis temperature increased to 500 °C, the C=O group sharply reduced with slight reduction of CO_3^{2-}, and the COO obviously increased, indicating C=O and CO_3^{2-} groups generated at low temperature were broken and COO groups were created at high temperature. With further increase of pyrolysis temperature to 600 °C, there were not significant changes of O-containing groups, as the degradation became relatively gradual when the temperature was up to 500 °C [26,32].

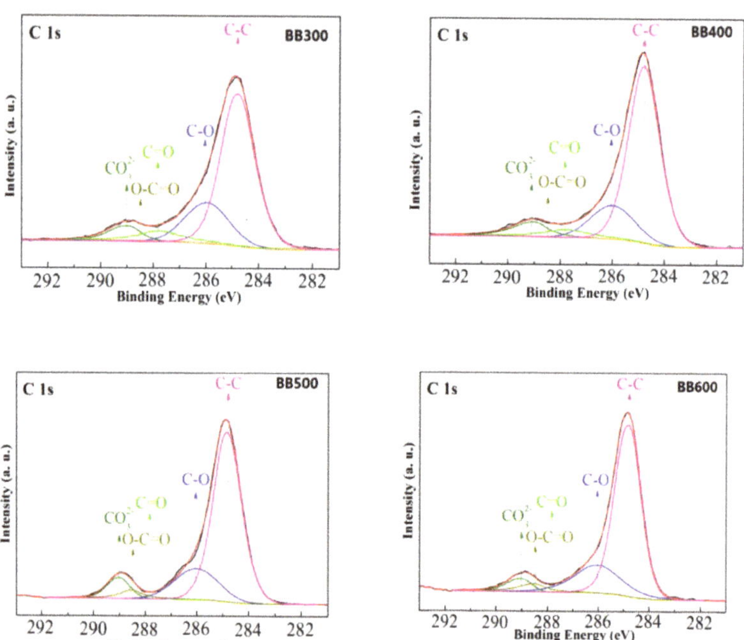

Figure 2. The XPS C 1s scans and peak fitting for banana pseudostem biochars.

Table 2. The peak area ratios of C–O, C=O, O–C=O and CO_3^{2-} bonds to the C–C bonds of C 1s XPS spectra of biochars.

B.E. (eV)	Ass.	BB300	BB400	BB500	BB600	BB300+Cr
C1 (286.0)	C–O (%)	34.14	26.25	26.94	29.56	43.52
C2 (287.8)	C=O (%)	12.62	8.50	0.11	0.13	0.52
C3 (288.5)	COO (%)	0.74	0.85	5.20	5.73	8.81
C4 (289.0)	CO_3^{2-} (%)	10.47	10.39	9.85	7.19	5.35
	TCOG	57.97	45.99	42.10	42.61	58.20

The surface morphologies of biochars are shown in Figure 3. The SEM image of BB300 revealed rod bundles with plicated surface and irregular small pieces, and BB400 also showed similar morphology. The SEM image of BB500 indicated that the rod bundles were partly destructed and became incompact. While the pyrolysis temperature was further raised to 600 °C, the rod bundles were almost destructed to small pieces and irregular sheets with some small strips.

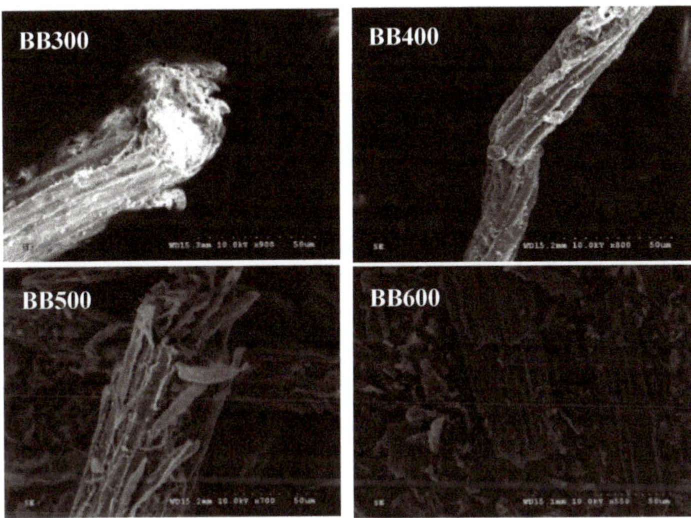

Figure 3. The SEM images of banana pseudostem biochars.

3.2. Effects of pH and Pyrolysis Temperature

The adsorption of Cr(VI) can be significantly influenced by pH, due to its effects on (1) distribution of surface charge and ionic state of functional groups on the biochar surface [23,44,45], (2) formation species and oxidizing ability of chromates [22,44,46]. As seen in Figure 4, the Cr(VI) adsorption efficiencies of all banana pseudostem biochars generally increased with decreasing initial pH of Cr(VI) solution, and the maximum adsorption capacity was obtained at pH 1, which fully indicated that the biochars were more active under acidic conditions for Cr(VI) adsorption. For example, the adsorption of BB300 increased gradually from 6.94 mg/g to 18.36 mg/g with the decrease of pH from 5.0 to 3.0, then increased sharply to 189.89 mg/g as the pH decreased to 1.0. This was caused by the formation of massive positive charges on biochar surface and HCr_2O^{7-} in the solution at lower pH, which was favourable for the electron interaction between the biochars and chromate anions. Moreover, the oxidizing ability of chromate was significantly improved at low pH, leading to remove more Cr(VI) by reduction [22,23,42].

As a critical parameter influencing the physicochemical properties of biochars, pyrolysis temperature was found to play an important role in Cr(VI) adsorption capabilities of banana pseudostem biochars. The adsorption capability of the biochars remarkably increased with increase of pyrolysis temperature from 200 to 300 °C, then decreased with further increase of pyrolysis temperature. The best adsorption capacity of BB300 was up to 189.89 mg/g at pH 1 and 25 °C (Figure 4). There were not significant changes in the surface area and microporosity of biochars altering pyrolysis temperature (Table 1). Whereas the content of O-containing groups which can be efficient factors for improving the redox, ion exchange and electron interaction abilities of biochars obviously reduced with the increase of pyrolysis temperature (Table 2). Thus, the decrease of adsorption capability with increase of temperature from 300 to 600 °C might be resulted from loss of surface O-containing groups. Similar results were also observed by Zhou et al. and Chen et al. [23,42]. As the adsorption capabilities of BB200 and BB600 were almost the same, the biochars prepared at different temperatures ranged from 300 to 600 °C were selected to study adsorption kinetics and isotherms.

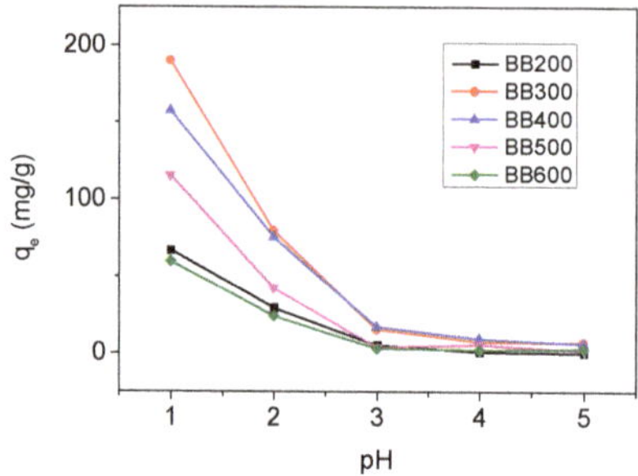

Figure 4. The effect of pH and pyrolysis temperature on adsorption capability (initial Cr(VI) concentration = 200 mg/L; adsorbent dose = 1 g/L; temperature = 25 °C; contact time = 24 h).

3.3. Adsorption Kinetics

The adsorption kinetics were studied at pH 2 and described by the pseudo-first-order (Equation (2)) and pseudo-second-order (Equation (3)) models [47]:

$$\ln(q_e - q_t) = \ln q_e - k_1 t \tag{2}$$

$$\frac{t}{q_t} = \frac{1}{k_2 q_e^2} + \frac{1}{q_e} t \tag{3}$$

where q_e and q_t (mg/g) are the equilibrium and instant adsorption capacities, respectively. k_1 (h^{-1}) and k_2 ($g \cdot mg^{-1} \cdot h^{-1}$) are the corresponding adsorption rate constants.

The adsorption kinetics of banana pseudostem biochars are presented in Figure 5. The adsorption of Cr(VI) gently increased with contact time and almost reached equilibrium after 48 h. The calculated results of pseudo-first-order and pseudo-second-order rate equations are illustrated in Table 3. The correlation coefficients (R^2) of the pseudo-second-order kinetic model were generally better than that of pseudo-first-order kinetic model. Meanwhile, the q_e values calculated from pseudo-second-order equation were more consistent with the experimental results. These supported that the adsorption of Cr(VI) by banana pseudostem biochars was a chemiadsorption process. Adsorption kinetic studies of BB300 at different temperatures showed that Cr(VI) adsorption efficiences of the biochars increased greatly with temperature (Figure 6), suggesting the process was endothermic and high temperature was favorable for chromium removal, which might be due to more facile reduction of Cr(VI) by biochars at higher temperature [48,49]. The results also revealed that the adsorption kinetics were better described by pseudo-second-order model than pseudo-first-order model (Table 4). Additionally, comparing with the experimental results, the higher calculated q_e results by pseudo-second-order model seemed more reasonable, as the adsorptions might not reach equilibrium within 48 h.

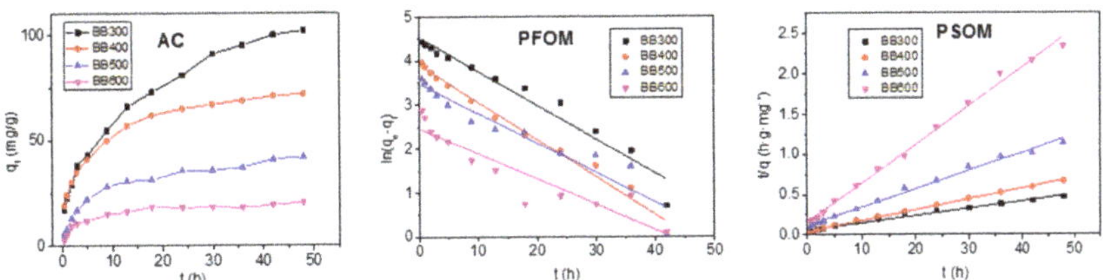

Figure 5. The adsorption kinetics of biochar for Cr(VI). AC is the adsorption curve of Cr(VI); PFOM is the pseudo-first-order model; PSOM is the pseudo-second-order model (initial Cr(VI) concentration = 200 mg/L; adsorbent dose = 1 g/L; pH = 2; temperature = 25 °C).

Table 3. The fitting results of Cr(VI) adsorption using pseudo-first-order and pseudo-second-order models.

	Pseudo-First-Order Model			
	BB300	**BB400**	**BB500**	**BB600**
$q_{e,exp}$	102.45	72.21	42.01	20.58
$q_{e,cal}$	93.03	50.25	32.70	11.35
k_1	7.72×10^{-2}	8.51×10^{-2}	6.74×10^{-2}	5.63×10^{-2}
R^2	0.96073	0.98212	0.90456	0.86206
	Pseudo-Second-Order Model			
	BB300	**BB400**	**BB500**	**BB600**
$q_{e,exp}$	102.45	72.21	42.01	20.58
$q_{e,cal}$	111.35	75.20	44.40	20.88
k_2	1.35×10^{-3}	3.92×10^{-3}	4.29×10^{-3}	1.45×10^{-2}
R^2	0.98054	0.99660	0.99096	0.99493

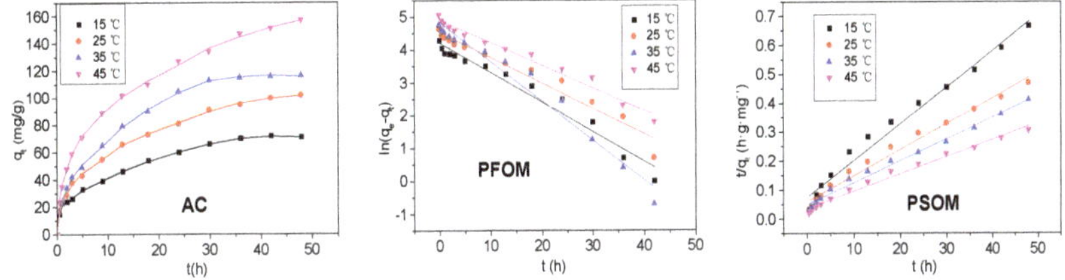

Figure 6. The adsorption kinetics of BB300 for Cr(VI) at different temperatures. AC is the adsorption curve of Cr(VI); PFOM is the pseudo-first-order model; PSOM is the pseudo-second-order model (initial Cr(VI) concentration = 200 mg/L; adsorbent dose = 1 g/L; pH = 2).

Table 4. The fitting results of Cr(VI) adsorption on BB300 at different temperatures using pseudo-first-order and pseudo-second-order models.

	Pseudo-First-Order Model			
	15 °C	**25 °C**	**35 °C**	**45 °C**
$q_{e,exp}$	72.04	102.45	116.50	156.56
$q_{e,cal}$	67.63	94.36	128.17	135.94
k_1	9.04×10^{-2}	7.78×10^{-2}	1.20×10^{-1}	6.90×10^{-2}
R^2	0.96007	0.95519	0.96748	0.97669
	Pseudo-Second-Order Model			
	15 °C	**25 °C**	**35 °C**	**45 °C**
$q_{e,exp}$	72.04	102.45	116.50	156.56
$q_{e,cal}$	61.46	111.86	132.98	169.20
k_2	3.56×10^{-3}	1.35×10^{-3}	1.13×10^{-3}	9.50×10^{-4}
R^2	0.97909	0.98054	0.98770	0.98285

3.4. Adsorption Isotherms

The Langmuir [50] and Freundlich models [51] were used to fit the isothermal adsorption data. These two models were described as Equations (4) and (5) respectively:

$$q_e = \frac{K_L q_{max} C_e}{1 + K_L C_e}, \ (R_L = \frac{1}{1 + K_L C_0}) \quad (4)$$

$$q_e = K_F C_e^n \quad (5)$$

where C_e is the equilibrium concentrations of Cr(VI) in solutions (mg/L); q_e and q_{max} (mg/g) are the equilibrium and maximum adsorption capacities of Cr(VI), respectively; The R_L parameter is the dimensionless adsorption factor; C_0 (mg/L) is the initial concentration of Cr(VI) in the solution; K_L (L/mg) is the Langmuir constant related to binding energy. K_F ((mg/g) $(mg/L)^{-n}$) is the Freundlich constant related to adsorption capacity and n is an indicator of the adsorption intensity.

Batch adsorption isotherms are shown as Figure 7. The equilibrium adsorption capacities of all biochars generally increased with concentration of Cr(VI) and reached maximum, the maximum adsorption capacities of BB300, BB400, BB500 and BB600 were 125.44 mg/g, 80.33 mg/g, 43.47 mg/g and 21.53 mg/g respectively. The adsorption isotherms was better fitted by Langmuir model than Freundlich model (Table 5), indicating monolayer adsorption onto homogeneous surface with limited active sites was a dominating process for Cr(VI) removal by banana pseudostem biochars [52,53]. The average value of adsorption factor (R_L) of Langmuir isotherm was less than 1, suggesting that the removal of Cr(VI) by the biochars was a highly efficient process.

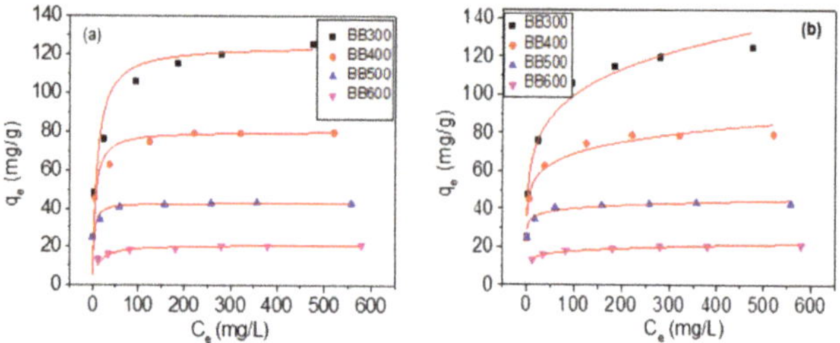

Figure 7. The adsorption isotherms of biochars for Cr(VI). (**a**) is the Langmuir model; (**b**) is the Freundlich model (initial Cr(VI) concentration = 25–600 mg/L; adsorbent dose = 1 g/L; pH = 2; contact time = 48 h).

Table 5. The parameters of Langmuir and Freundlich isotherms of Cr(VI) adsorption.

	LM			
	BB300	**BB400**	**BB500**	**BB600**
q_{max} (mg/g)	125.44	80.33	43.47	21.53
K_L (L/mg)	0.1024	0.1910	0.3929	0.0796
R^2	0.9977	0.9996	0.9999	0.9997
	FM			
	BB300	**BB400**	**BB500**	**BB600**
KF (mg/g)·$(mg/L)^{-n}$	42.64	39.07	30.25	10.67
n	0.185	0.124	0.062	0.115
R^2	0.9813	0.9609	0.8030	0.9554

3.5. Adsorption Mechanism

The XPS and FTIR spectrometers were also performed after adsorption for exploring the Cr(VI) adsorption mechanism, as shown in Figures 8 and 9 respectively. The XPS results of BB300 after adsorption revealed that Cr(III) was primarily immobilized on the surface of BB300 with a small amount of Cr(VI) (Cr(III)/Cr(VI) = 97.51/2.49). The content of O-containing groups on the surface slightly increased and both C–O and COO obviously increased (Table 2). In contrast, the content of C=O and CO_3^{2-} sharply reduced. These results indicated that reduction of Cr(VI) to Cr(III) [54–56] and adsorption of both Cr(III) and Cr(VI) onto the biochar surface were involved during the process. Additionally, the FTIR showed that the COO peak at 1686 cm^{-1}, aromatic C=C peak at 1429 cm^{-1} and C–O peak at 1109 cm^{-1} obviously increased for BB300 and BB400 after adsorption, implying that the biochars were oxidized by Cr(VI) and COO, C–O and aromatic ring were formed. Hence, the reduction of Cr(VI) by BB300 and BB400 during the adsorption was caused by oxidation and aromatization of the biochars. Differently, the aromatic C=C peak at 1421 cm^{-1} diminished and C=O peak at 1620 cm^{-1} increased for BB500 and BB600 after adsorption, indicating the reduction of Cr(VI) by BB500 and BB600 resulted from oxidation and dearomatization of the biochars. Moreover, adsorption kinetics were described well by pseudo-second-order model, suggesting the removal of Cr(VI) by the biochars was a chemical process. As a result, the decrease of adsorption capability of banana pesudostem biochar with increasing pyrolysis temperature was explained by the lost of surface O-containing groups which played an important role in reduction of Cr(VI). In general, the adsorption of Cr(VI) was mainly attributed to reduction of Cr(VI) to Cr(III) followed by complexation and ion exchange rather than electron interaction with biochar, as the biochar surface was probably positive at pH 2, hindering the electron interaction of Cr(III) cation with biochar. The precipitation of $Cr(OH)_3$ on the surface of biochar also could be ruled out, as the predominant species of trivalent chromium was Cr(III) cation at low pH [57].

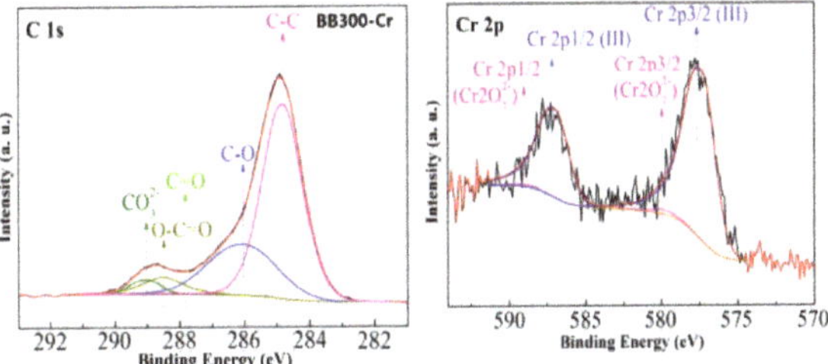

Figure 8. The XPS C 1s and Cr 2p scans and peaks fitting for BB300 after adsorption.

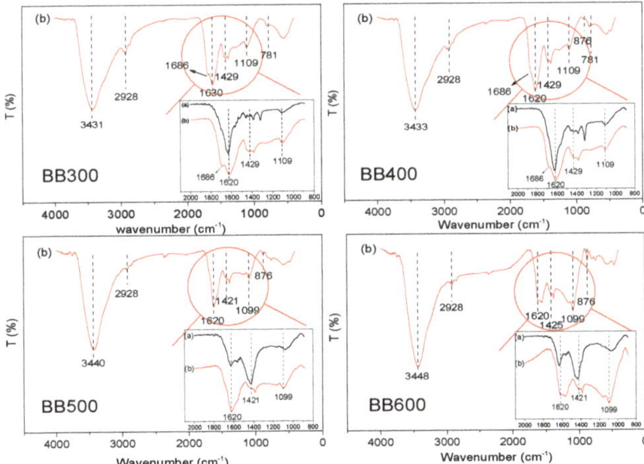

Figure 9. The FTIR spectra of biochars before and after adsorption of Cr(VI). Spectra (**a**) were before adsorption; Spectra (**b**) were after adsorption.

4. Conclusions

The Cr(VI) adsorption capabilities of banana pseudostem biochar were highly affected by pyrolysis temperature. Low pH and high adsorption temperature were favourable for adsorption of Cr(VI). The biochar prepared at 300 °C exhibited best adsorption capability with 125.44 mg/g maximum adsorption capacity at pH 2 and 25 °C. The high adsorption capability of the biochar might be attributed to high content of oxygen-containing functional groups on surface. The Cr(VI) adsorption by BB300 and BB400 resulted from their oxidation and aromatization with ion exchange and complexation, whereas the Cr(VI) adsorption by BB500 and BB600 was ascribed to their oxidation and dearomatization followed by ion exchange and complexation. As the feedstock is large-scale available and highly renewable biowaste, banana pseudostem biochar is a low-cost adsorbent and expected to have a practical application for remediation of chromium-polluted water in the future.

Author Contributions: Conceptualization, J.L. and Y.Z.; Methodology, J.L.; Software, S.X. and W.Y.; Validation, S.X. and S.L.; Formal Analysis, J.L. and Y.Z.; Investigation, S.X., W.Y. and S.L.; Resources, C.X.; Data Curation, S.X.; Writing-Original Draft Preparation, S.X.; Writing-Review & Editing, J.L.; Visualization, J.L.; Supervision, J.L. and Y.Z.; Project Administration, C.X.; Funding Acquisition, Y.Z."

Funding: This research was funded by the key Scientific Research Project Funding of Hainan Province [ZDYF2017005], the National Natural Science Foundation of China [21801053] and the Innovation Project of Science and Technology Association of Hainan Province for Distinguished Young Scholar [HAST201625]; and the APC was funded by the key Scientific Research Project Funding of Hainan Province [ZDYF2017005].

Conflicts of Interest: The authors declare no conflicts of interest.

References

1. Bayazit, Ş.S.; Kerkez, Ö. Hexavalent chromium adsorption on superparamagnetic multi-wall carbon nanotubes and activated carbon composites. *Chem. Eng. Res. Des.* **2014**, *92*, 2725–2733. [CrossRef]
2. Pradhan, D.; Sukla, L.B.; Sawyer, M.; Rahman, P.K.S.M. Recent bioreduction of hexavalent chromium in wastewater treatment: A review. *J. Ind. Eng. Chem.* **2017**, *55*, 1–20. [CrossRef]
3. Costa, M.; Klein, C.B. Toxicity and carcinogenicity of chromium compounds in humans. *Crit. Rev. Toxicol.* **2006**, *36*, 155–163. [CrossRef] [PubMed]
4. Li, Q.; Liu, H.; Alattar, M.; Jiang, S.; Han, J.; Ma, Y.; Jiang, C. The preferential accumulation of heavy metals in different tissues following frequent respiratory exposure to PM2.5 in rats. *Sci. Rep.* **2015**, *5*, 16936–16944. [CrossRef] [PubMed]
5. Niu, Y.; Jiao, W.; Yu, H.; Niu, Y.; Pang, Y.; Xu, X.; Guo, X. Spatial evaluation of heavy metals concentrations in the surface sediment of Taihu Lake. *Int. J. Environ. Res. Public Health* **2015**, *12*, 15028–15039. [CrossRef] [PubMed]
6. Kotaś, J.; Stasicka, Z. Chromium occurrence in the environment and methods of its speciation. *Environ. Pollut.* **2000**, *107*, 263–283. [CrossRef]
7. Bagchi, D.; Stohs, S.J.; Downs, B.W.; Bagchi, M.; Preuss, H.G. Cytotoxicity and oxidative mechanisms of different forms of chromium. *Toxicology* **2002**, *180*, 5–22. [PubMed]
8. Tu, W.; Li, K.; Shu, X.; Yu, W.W. Reduction of hexavalent chromium with colloidal and supported palladium nanocatalysts. *J. Nanopart. Res.* **2013**, *15*, 1593–1602.
9. Kongsricharoem, N.; Polprasert, C. Chromium removal by a bipolar electro-chemical precipitation process. *Water Sci. Technol.* **1996**, *34*, 109–116.
10. Song, Z.; Williams, C.J.; Edyvean, R.G.J. Sedimentation of tannery wastewater. *Water Res.* **2000**, *34*, 2171–2176. [CrossRef]
11. Wang, G.; Chang, Q.; Han, X.; Zhang, M. Removal of Cr(VI) from aqueous solution by flocculant with the capacity of reduction and chelation. *J. Hazard. Mater.* **2013**, *248–249*, 115–121. [CrossRef] [PubMed]
12. Karthikeyan, T.; Rajgopal, S.; Miranda, L.R. Chromium(VI) adsorption from aqueous solution by *Hevea brasilinesis* sawdust activated carbon. *J. Hazard. Mater.* **2005**, *124*, 192–199. [CrossRef] [PubMed]
13. Ghosh, G.; Bhattacharya, P.K. Hexavalent chromium ion removal through micellar enhanced ultrafiltration. *Chem. Eng. J.* **2006**, *119*, 45–53. [CrossRef]

14. Rengaraj, S.; Yeon, K.-H.; Moon, S.-H. Removal of chromium from water and wastewater by ion exchange resins. *J. Hazard. Mater.* **2001**, *B87*, 273–287. [CrossRef]
15. Achouri, O.; Panico, A.; Bencheikh-Lehocine, M.; Derbal, K.; Pirozzi, F. Effect of chemical coagulation pretreatment on anaerobic digestion of tannery wastewater. *J. Environ. Eng.* **2017**, *143*, 04017039. [CrossRef]
16. Mohan, D.; Singh, K.P.; Singh, V.K. Trivalent chromium removal from wastewater using low cost activated carbon derived from agricultural waste material and activated carbon fabric cloth. *J. Hazard. Mater.* **2006**, *135*, 280–295. [CrossRef] [PubMed]
17. Yan, L.; Kong, L.; Qu, Z.; Li, L.; Shen, G. Magnetic biochar decorated with ZnS nanocrytals for Pb (II) removal. *ACS Sustain. Chem. Eng.* **2014**, *3*, 125–132. [CrossRef]
18. Ahmad, M.; Rajapaksha, A.U.; Lim, J.E.; Zhang, M.; Bolan, N.; Mohan, D.; Vithanage, M.; Lee, S.S.; Ok, Y.S. Biochar as a sorbent for contaminant management in soil and water: A review. *Chemosphere* **2014**, *99*, 19–33. [CrossRef] [PubMed]
19. Cheng, B.H.; Zeng, R.J.; Jiang, H. Recent developments of post-modification of biochar for electrochemical energy storage. *Bioresour. Technol.* **2017**, *246*, 224–233. [CrossRef] [PubMed]
20. Oliveira, F.R.; Patel, A.K.; Jaisi, D.P.; Adhikari, S.; Lu, H.; Khanal, S.K. Environmental application of biochar: Current status and perspectives. *Bioresour. Technol.* **2017**, *246*, 110–122. [CrossRef] [PubMed]
21. Mohan, D.; Sarswat, A.; Ok, Y.S.; Pittman, C.U., Jr. Organic and inorganic contaminants removal from water with biochar, a renewable, low cost and sustainable adsorbent—A critical review. *Bioresour. Technol.* **2014**, *160*, 191–202. [CrossRef] [PubMed]
22. Mohan, D.; Rajput, S.; Singh, V.K.; Steele, P.H.; Pittman, C.U., Jr. Modeling and evaluation of chromium remediation from water using low cost bio-char, a green adsorbent. *J. Hazard. Mater.* **2011**, *188*, 319–333. [PubMed]
23. Zhou, L.; Liu, Y.; Liu, S.; Yin, Y.; Zeng, G.; Tan, X.; Hu, X.; Hu, X.; Jiang, L.; Ding, Y.; et al. Investigation of the adsorption-reduction mechanisms of hexavalent chromium by ramie biochars of different pyrolytic temperatures. *Bioresour. Technol.* **2016**, *218*, 351–359. [CrossRef] [PubMed]
24. Duan, S.; Ma, W.; Pan, Y.; Meng, F.; Yu, S.; Wu, L. Synthesis of magnetic biochar from iron sludge for the enhancement of Cr (VI) removal from solution. *J. Taiwan Inst. Chem. Eng.* **2017**, *80*, 835–841. [CrossRef]
25. Dong, X.; Ma, L.Q.; Li, Y. Characteristics and mechanisms of hexavalent chromium removal by biochar from sugar beet tailing. *J. Hazard. Mater.* **2011**, *190*, 909–915. [CrossRef] [PubMed]
26. Shen, Y.S.; Wang, S.L.; Tzou, Y.M.; Yan, Y.Y.; Kuan, W.H. Removal of hexavalent Cr by coconut coir and derived chars—The effect of surface functionality. *Bioresour. Technol.* **2012**, *104*, 165–172. [CrossRef] [PubMed]
27. Zhang, X.; Zhang, X.; Chen, Z. Biosorption of Cr(VI) from aqueous solution by biochar derived from the leaf of *Leersia hexandra* Swartz. *Environ. Earth Sci.* **2017**, *76*, 67–74. [CrossRef]
28. Shah, M.P.; Reddy, G.V.; Banerjee, R.; Ravindra Babu, P.; Kothari, I.L. Microbial degradation of banana waste under solid state bioprocessing using two lignocellulolytic fungi (*Phylosticta* spp. MPS-001 and *Aspergillus* spp. MPS-002). *Process. Biochem.* **2005**, *40*, 445–451. [CrossRef]
29. Jayaprabha, J.S.; Brahmakumar, M.; Manilal, V.B. Banana pseudostem characterization and its fiber property evaluation on physical and bioextraction. *J. Nat. Fibers* **2011**, *8*, 149–160. [CrossRef]
30. Li, D.-C.; Ding, J.-W.; Qian, T.-T.; Zhang, S.; Jiang, H. Preparation of high adsorption performance and stable biochar granules by $FeCl_3$-catalyzed fast pyrolysis. *RSC Adv.* **2016**, *6*, 12226–12234. [CrossRef]
31. Özçimen, D.; Ersoy-Meriçboyu, A. Characterization of biochar and bio-oil samples obtained from carbonization of various biomass materials. *Renew. Energy* **2010**, *35*, 1319–1324. [CrossRef]
32. Chen, B.; Chen, Z. Sorption of naphthalene and 1-naphthol by biochars of orange peels with different pyrolytic temperatures. *Chemosphere* **2009**, *76*, 127–133. [CrossRef] [PubMed]
33. Lian, F.; Xing, B.; Zhu, L. Comparative study on composition, structure, and adsorption behavior of activated carbons derived from different synthetic waste polymers. *J. Colloid Interface Sci.* **2011**, *360*, 725–730. [CrossRef] [PubMed]
34. Song, Z.; Lian, F.; Yu, Z.; Zhu, L.; Xing, B.; Qiu, W. Synthesis and characterization of a novel MnO_x-loaded biochar and its adsorption properties for Cu^{2+} in aqueous solution. *Chem. Eng. J.* **2014**, *242*, 36–42.
35. Li, G.; Zhu, W.; Zhang, C.; Zhang, S.; Liu, L.; Zhu, L.; Zhao, W. Effect of a magnetic field on the adsorptive removal of methylene blue onto wheat straw biochar. *Bioresour. Technol.* **2016**, *206*, 16–22. [CrossRef] [PubMed]

36. Qian, L.; Zhang, W.; Yan, J.; Han, L.; Chen, Y.; Ouyang, D.; Chen, M. Nanoscale zero-valent iron supported by biochars produced at different temperatures: Synthesis mechanism and effect on Cr(VI) removal. *Environ. Pollut.* **2017**, *223*, 153–160. [PubMed]
37. Sun, L.; Wan, S.; Luo, W. Biochars prepared from anaerobic digestion residue, palm bark, and eucalyptus for adsorption of cationic methylene blue dye: Characterization, equilibrium, and kinetic studies. *Bioresour. Technol.* **2013**, *140*, 406–413. [PubMed]
38. Tan, X.; Liu, Y.; Zeng, G.; Wang, X.; Hu, X.; Gu, Y.; Yang, Z. Application of biochar for the removal of pollutants from aqueous solutions. *Chemosphere* **2015**, *125*, 70–85. [PubMed]
39. Zeng, Z.-W.; Tian, S.-R.; Liu, Y.-G.; Tan, X.-F.; Zeng, G.-M.; Jiang, L.-H.; Yin, Z.-H.; Liu, N.; Liu, S.-B.; Li, J. Comparative study of rice husk biochars for aqueous antibiotics removal. *J. Chem. Technol. Biotechnol.* **2017**, *93*, 1075–1084. [CrossRef]
40. Chen, B.; Zhou, D.; Zhu, L. Transitional adsorption and partition on nonpolar and polar aromatic contaminants by biochars of pine needles with different pyrolytic temperatures. *Environ. Sci. Technol.* **2008**, *42*, 5137–5143. [CrossRef] [PubMed]
41. Abdullah, N.; Sulaiman, F.; Miskam, M.A.; Taib, R.M. Characterization of banana (*Musa* spp.) pseudo-stem and fruit-bunch-Stem as a potential renewable energy resource. *Int. J. Biol. Veter. Agric. Food Eng.* **2014**, *8*, 712–716.
42. Chen, D.; Yu, X.; Song, C.; Pang, X.; Huang, J.; Li, Y. Effect of pyrolysis temperature on the chemical oxidation stability of bamboo biochar. *Bioresour. Technol.* **2016**, *218*, 1303–1306. [CrossRef] [PubMed]
43. Zhou, W.; Li, W.; Li, J.; Zhang, Y. Characterization of cellulose from banana pseudo-stem by polyhydric alcohols liquefaction. *Renew. Energy Resour.* **2016**, *34*, 285–291.
44. Chen, T.; Zhou, Z.; Xu, S.; Wang, H.; Lu, W. Adsorption behavior comparison of trivalent and hexavalent chromium on biochar derived from municipal sludge. *Bioresour. Technol.* **2015**, *190*, 388–394. [CrossRef] [PubMed]
45. Tytlak, A.; Oleszczuk, P.; Dobrowolski, R. Sorption and desorption of Cr(VI) ions from water by biochars in different environmental conditions. *Environ. Sci. Pollut. Res. Int.* **2015**, *22*, 5985–5994. [CrossRef] [PubMed]
46. Di Natale, F.; Erto, A.; Lancia, A.; Musmarra, D. Equilibrium and dynamic study on hexavalent chromium adsorption onto activated carbon. *J. Hazard. Mater.* **2015**, *281*, 47–55. [CrossRef] [PubMed]
47. Nityanandi, D.; Subbhuraam, C.V. Kinetics and thermodynamic of adsorption of chromium(VI) from aqueous solution using puresorbe. *J. Hazard. Mater.* **2009**, *170*, 876–882. [CrossRef] [PubMed]
48. Patel, S.; Mishra, B.K. Oxidation of alcohol by lipopathic Cr(VI): A mechanistic study. *J. Org. Chem.* **2006**, *71*, 6759–6766. [CrossRef] [PubMed]
49. Xu, F.; Ma, T.; Zhou, L.; Hu, Z.; Shi, L. Chromium isotopic fractionation during Cr(VI) reduction by Bacillus sp. under aerobic conditions. *Chemosphere* **2015**, *130*, 46–51. [CrossRef] [PubMed]
50. Langmuir, I. The adsorption of gases on plane surfaces of glass, mica and platinum. *J. Am. Chem. Soc.* **1918**, *40*, 1361–1403. [CrossRef]
51. Albadarin, A.B.; Mo, J.; Glocheux, Y.; Allen, S.; Walker, G.; Mangwandi, C. Preliminary investigation of mixed adsorbents for the removal of copper and methylene blue from aqueous solutions. *Chem. Eng. J.* **2014**, *255*, 525–534. [CrossRef]
52. Langmuir, I. The constitution and fundamental properties of solids and liquids. *J. Am. Chem. Soc.* **1916**, *38*, 2221–2295. [CrossRef]
53. Özcan, A.S.; Erdem, B.; Özcan, A. Adsorption of acid blue 193 from aqueous solutions onto BTMA-bentonite. *Colloids Surf. Physicochem. Eng. Asp.* **2005**, *266*, 73–81. [CrossRef]
54. Wiberg, K.B.; Schaferlb, H. Chromic acid oxidation of isopropyl alcohol. preoxidation equilibria. *J. Am. Chem. Soc.* **1969**, *91*, 927–932. [CrossRef]
55. Hiran, B.L.; Chaplot, S.L.; Joshi, V.; Chaturvedi, G. Kinetics of the effect of some bidentate amino acid ligands in the oxidation of lactic acid by chromium(VI). *J. Am. Chem. Soc.* **2002**, *43*, 657–661.
56. Rorek, J.; Radkowsky, A.E. Mechanism of the chromic acid oxidation of cyclobutanol. *J. Am. Chem. Soc.* **1973**, *95*, 7123–7132.
57. Fahim, N.F.; Barsoum, B.N.; Eid, A.E.; Khalil, M.S. Removal of chromium (III) from tannery wastewater using activated carbon from sugar industrial waste. *J. Hazard. Mater.* **2006**, *136*, 303–309. [CrossRef] [PubMed]

Article

Optimal Management of a Hybrid Renewable Energy System Coupled with a Membrane Bioreactor Using Enviro-Economic and Power Pinch Analyses for Sustainable Climate Change Adaption

Tuan-Viet Hoang [1,†], Pouya Ifaei [1,†], Kijeon Nam [1], Jouan Rashidi [1], Soonho Hwangbo [1,2,*], Jong-Min Oh [1] and ChangKyoo Yoo [1,*]

1 Deptartment of Environmental Science and Engineering, College of Engineering, Center for Environmental Studies, Kyung Hee University, Seocheon-dong 1, Giheung-gu, Yongin-si, Gyeonggi-do 446-701, Korea; tuanvietxda@gmail.com (T.-V.H.); pooya_if@hotmail.com (P.I.); spirit1058@gmail.com (K.N.); Jouanra@gmail.com (J.R.); jmoh@khu.ac.kr (J.-M.O.)

2 Process and Systems Engineering Center (PROSYS), Department of Chemical and Biochemical Engineering, Technical University of Denmark, Soltofts Plads 229, 2800 Kgs. Lyngby, Denmark

* Correspondence: hwsoonho@gmail.com or soohw@kt.dtu.dk (S.H.); ckyoo@khu.ac.kr (C.Y.); Tel.: +45-4525-2800 (S.H.); +82-31-201-3824 (C.Y.)

† The first and second authors contributed equally to this paper.

Received: 31 October 2018; Accepted: 11 December 2018; Published: 22 December 2018

Abstract: This study proposed an optimal hybrid renewable energy system (HRES) to sustainably meet the dynamic electricity demand of a membrane bioreactor. The model-based HRES consists of solar photovoltaic panels, wind turbines, and battery banks with grid connectivity. Three scenarios, 101 sub-scenarios, and three management cases were defined to optimally design the system using a novel dual-scale optimization approach. At the system scale, the power-pinch analysis was applied to minimize both the size of components and the outsourced needed electricity (NE) from Vietnam's electrical grid. At a local-scale, economic and environmental models were integrated, and the system was graphically optimized using a novel objective function, combined enviro-economic costs (CEECs). The results showed that the optimal CEECs were $850,710/year, $1,030,628/year, and $1,693,476/year for the management cases under good, moderate, and unhealthy air qualities, respectively. The smallest CEEC was obtained when 47% of the demand load of the membrane bioreactor was met using the HRES and the rest was supplied by the grid, resulting in 6,800,769 kg/year of CO_2 emissions.

Keywords: climate change; enviro-economic analysis; membrane bioreactor; optimization model; power pinch analysis; renewable energy

1. Introduction

Much research has been devoted to tackling contemporary problems including global warming, climate change, environmental pollution, and sustainable development. Sustainable development establishes a basis on which the future world can be built. Hence, a sustainable system may be regarded as a cost-efficient, reliable, and environment-friendly system, which effectively utilizes local resources [1]. Many countries, including developing nations such as Vietnam, should set sustainable development strategies such as an optimal energy management to mitigate climate change impacts, and minimize environmental pollution by application of green energy sources [2,3].

Vietnam enjoys a variety of renewable energy sources such as hydro, solar, wind, biomass, geothermal, and wave energies, along with other Southeast Asian countries. However, few RES projects have been implemented to date, and less than 5.8% of the country's annual electrical demand was met using small hydropower, and renewables in 2017 [4,5]. Due to rapid economic development, urbanization, industrialization, and population growth, Vietnam's energy demand is anticipated to increase by 11–16% in the near future. It is also predicted that the rate of increase in energy demand will triple by 2020 [6]. Sustainable energy production is therefore unlikely without the efficient use of renewable energies [7].

Solar and wind energies are two robust RES in developing countries [8,9]. The intermittency problems associated with weather-driven energies could be solved using appropriate storage facilities to minimize the need for grid-sourced electricity [10]. Vietnam enjoys considerable solar radiation and permeant wind flow [4,11]. Several studies have shown that there is a great wind potential in Vietnam, especially in Ninh Thuan and Binh Thuan provinces [5,12–14]. The wind speed data used in this study were obtained from the Da Loan station in Lam Dong province at 80 m above sea level, as shown in Figure 1a. Solar energy is the other promising RES in Vietnam. According to the National Center for Hydro-Meteorological Forecasting (NHC), Vietnam has been clustered into three solar regions. The northwestern, and southern regions are mainly and dominantly solar because of a high number of sunshine hours, 1897–2102 h/year, and 1900–2900 h/year, respectively. The central region is moderately solar with fewer sunshine hours, 1400–1700 h/year. Regions with at least 1800 h/year of sunny skies are determined to be appropriate for solar power harnessing [4]. In this study, data from the Song Binh station in Binh Thuan province were used, as shown in Figure 1b.

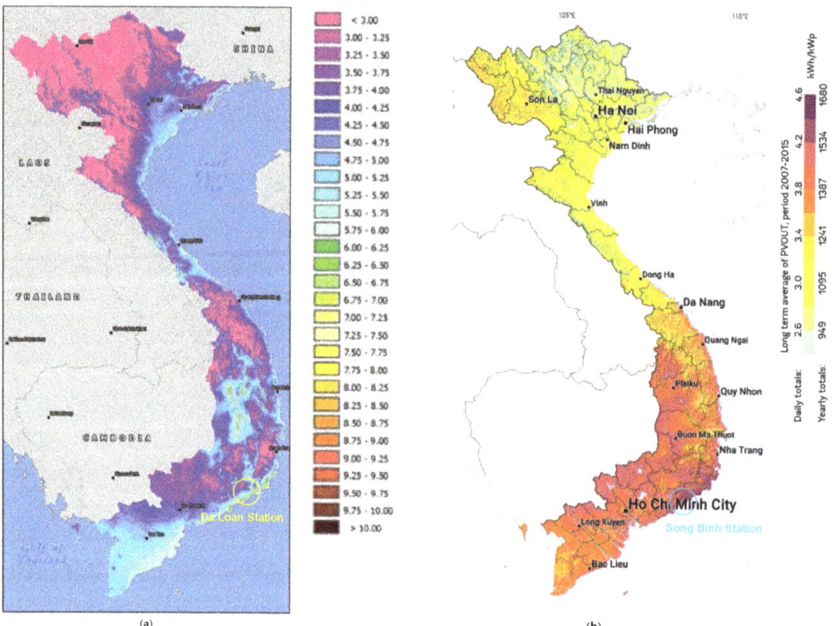

Figure 1. RES map of Vietnam with varying: (**a**) wind speed at 80 m above sea level; and (**b**) solar power [13,14].

Vietnam has undergone rapid economic development, abrupt urbanization, extensive industrialization, and fast population growth compared to other developing countries. Thus, environmental pollution including water and air sectors have become a serious problem in this country [15,16]. According to the Environmental Performance Index and the report of the World Bank, the environmental situation in Vietnam was highlighted by several points in 2018. Vietnam ranked 132 among 180 countries in the overall environmental assessment where air pollution index ranked

161. It should be noted that Vietnam was in the top ten countries in the world with the most polluted air [17]. The main reason comes from industrial emissions, coal combustion and increasing numbers of vehicles using non-environmentally friendly fuels [18,19]. Aquatic ecosystems are dangerously threatened by the high amounts of untreated municipal and industrial wastewater discharge from industrial zones [18–20]. Thus, effective strategies for environmental protection especially wastewater treatment prior to discharge is vital.

Membrane bioreactor (MBR) technology has many advantages over conventional wastewater treatment plants. MBR technology has less sludge, lower environmental footprint, superior effluent quality, and better intensification, with higher retention time [21]. Properties of the membrane are mainly represented by the membrane module manufacturer. The material, internal and external diameter, surface area, nominal pore size, and tensile strength are described in membrane module properties, and the configuration of modules and available filtration area are detailed in membrane frame properties [22]. In addition, a MBR can be applied in a tropical climate and under the technical and socioeconomic conditions of Vietnam [20]. However, the application of MBRs has been extremely limited in Vietnam because of the high economic costs associated with fouling on the membrane surface. The physicochemical interactions between the membrane and components lead to a decrease in flux and an increase in trans-membrane pressure. This phenomenon increases energy consumption during the pumping process [22–24]. Extensive application of MBRs in Vietnam would be both cost-prohibitive and suffer from operational problems. A combination of RES and MBR systems may be a promising sustainable answer to the dilemma [25]. In an RES, selection of an inappropriate size and energy mismanagement in the storage components leads to energy loss and much unmet electrical demand [26]. Hence, an energy management strategy has strongly been recommended for effective utilization and energy loss reduction while meeting the electrical demand [27,28]. However, an optimal management methodology for efficient process integration, one that can meet the dynamic demand load of a MBR using intermittent RES, has yet to be identified.

The aim of the current study was to propose a novel HRES to meet a dynamic demand load of a MBR system under optimum conditions using a dual-scale model. In the first model optimization layer, a power-pinch analysis (PoPA) was used to minimize reliance on the fossil-fuel-dependent power grid of Vietnam. In the second layer, economic and environmental models of the technically optimal energy system were developed to evaluate the performance of the optimal integrated system under three scenarios and 101 sub-scenarios for an optimal management. Subsequently, an optimal management scenario was determined using an enviro-economic model, and the technical and enviro-economic models were employed at two optimization scales.

2. Materials and Methods

2.1. System Configuration and Energy Management Strategy

A schematic representation of the system under study is shown in Figure 2. Wind turbines generate an alternating current (AC), while photovoltaic panels (PV) generate direct current (DC). Because a MBR with an external supply source operates on AC, an inverter is used to convert electricity from DC (generated by PV panels or stored in a battery bank) to AC before delivering it to the MBR. A rectifier can convert AC electricity generated by wind turbines (WTs) to DC for storage in a battery bank (BB). Three electrical losses were considered: during energy conversion in the PV panels and WTs; during charging and discharging of the battery; and during net AC-to-DC conversion [29].

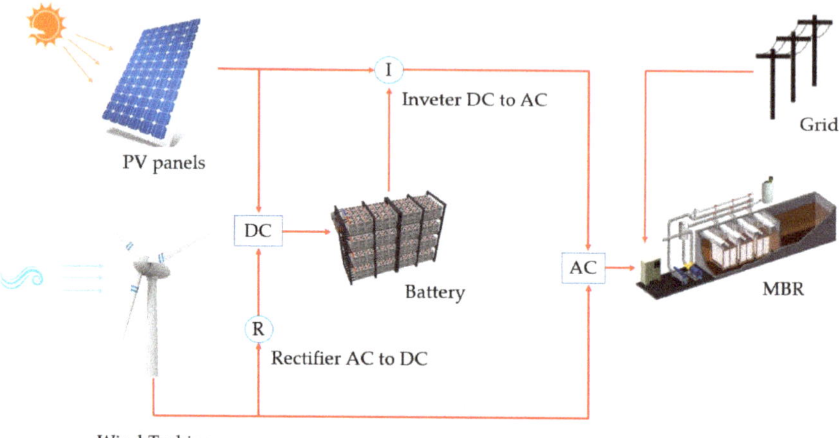

Figure 2. A membrane bioreactor system with the hybrid solar and wind energy sources with a battery.

Figure 3 is a schematic representation of a MBR system. The demand load was obtained from an operating full-scale MBR plant in southern Vietnam. The plant consists of five main basins, including anaerobic, stabilizing, anoxic, aerobic, and a submerged reactor. In the submerged reactor, the membrane module is immerged in the bioreactor. There is no recirculation loop [30]. Thus, a MBR using a submerged reactor has lower energy consumption because of no high pressurizing pump for recirculation [31]. Here, polyvinylidene (PVDF) membrane material was used with the hollow fiber of 0.01 µm diameter because of its acceptable physical and chemical resistance. Volumes of the bioreactors are shown in the figure. The plant's operating temperature was maintained at 20 °C for cycles of 15 min (14.5 min for filtration, with the remaining time for back-washing).

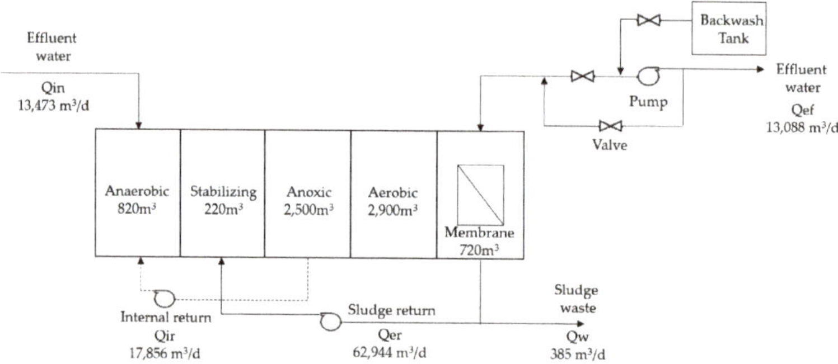

Figure 3. Schematic representation of a full-scale MBR plant in Vietnam.

An optimal energy management approach was proposed to minimize the components' size in the integrated system as well as the power losses, as illustrated in Figure 4 [28]. According to the figure, AC electricity is generated by the WTs, and consumed to meet the demand load of the MBR. If the generated AC is greater than the MBR demand load in any time interval, excess AC is converted to DC voltage via converters, and then stored in batteries. On the other hand, the DC generated by the PV panels is converted into AC, and delivered to the demand if the demand load is greater than the power generated by AC source. The battery is charged when the generated DC is greater than the demand load and vice versa. Outsourced electricity is imported from the grid if the total electricity in the HRES is not adequate to satisfy the demand load. Power losses were considered in the model to obtain realistic results. The conversion efficiency, charge and discharge efficiency in the lead-acid batteries, and self-discharge ratio were assumed to be 95%, 90%, and 0.004%/h, respectively [28]. However, these values could vary according to operating devices in an in-situ application.

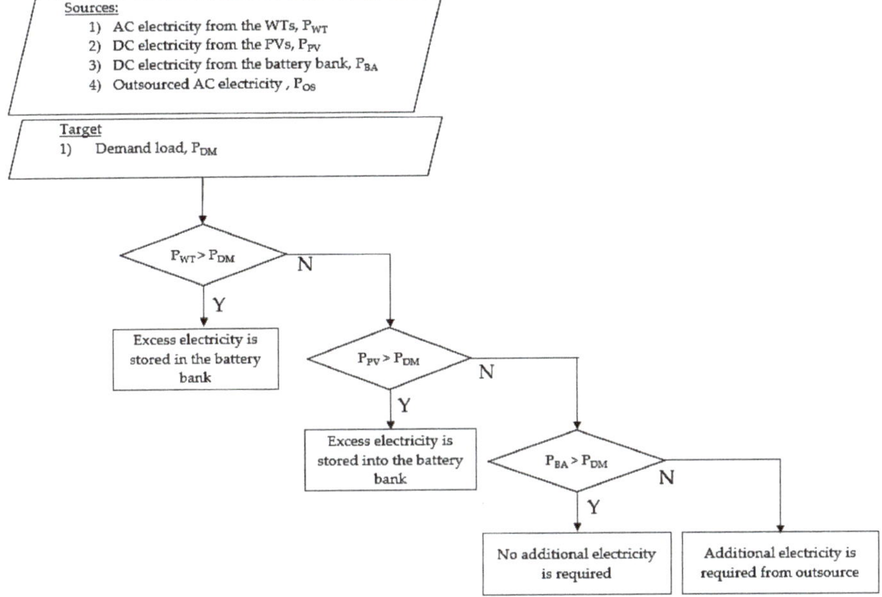

Figure 4. Proposed energy management approach to minimize the system size and power loss.

The research framework used to evaluate the optimal configuration is shown in Figure 5. The raw meteorological data were collected, and the optimal configuration was obtained using the analyzed data in various sub-scenarios. An economic assessment was performed on an optimal integrated system to determine the total annual cost (TAC) and the environmental performance of the integrated system considered total annual carbon dioxide emissions ($TACO_2$) by the grid for all sub-scenarios. Finally, an optimal scenario was determined graphically using the enviro-economic penalty cost. A step-by-step procedure to determine the optimal configuration is detailed as follows:

1. The optimal numbers of PVs and WTs were determined according to sub-scenarios. This study considered three scenarios and 101 sub-scenarios. The sub-scenarios accounted for the percentage of RES contribution in satisfying the electricity in the MBR so that k% RES contribution was used in the $(k + 1)$th sub-scenario. The MBR operated without any contribution from the HRES in the first sub-scenario. To consider the individual contributions of each power source in one HRES configuration, three scenarios were assumed:

    - Scenario 1: The solar (DC) source accounts for 90% of the total amount provided by the RES, while the wind energy (AC) source makes up the remaining 10%. This scenario is suitable for locations where the potential of solar energy source is dominant compared to that of wind energy source.
    - Scenario 2: The wind (AC) source accounts for 90% of the total amount provided by RES, whereas the solar (DC) source makes up the remaining 10%. This scenario should be considered in regions where the wind energy source has higher priority compared to solar energy sources.

- Scenario 3: The wind (AC) and solar (DC) sources contribute equally. In other words, each supplies 50% of the total amount provided by the RES. Here, the potential of both wind and solar energies is highly recommended.

The representative of these three scenarios has the individual contribution in each RES source. This could give us an idea to select an appropriate source based on a specific location.

2. An appropriate battery capacity (BC) and the outsourced needed electricity (NE) for one operational year were determined for each sub-scenario using the storage cascade table (SCT) of PoPA [29].
3. Curves of the combined economic and environmental penalty costs (CEEPC) were plotted using economic and environmental models for all sub-scenarios in three management cases, and the optimal sub-scenario was determined graphically.

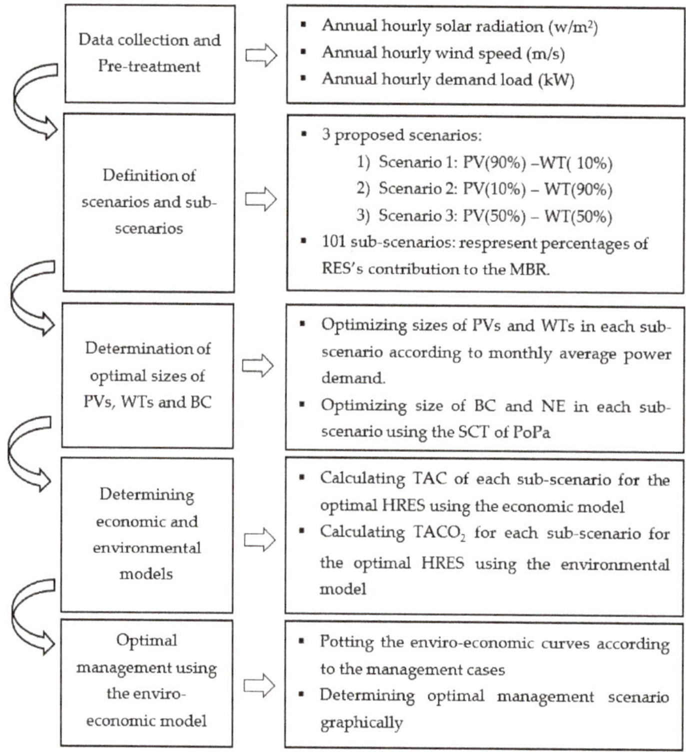

Figure 5. Research framework.

2.2. Data Collection

Solar and wind are known as weather-driven energy sources because of their dependency on the structure of raw meteorological data [9]. Furthermore, the demand load has a dynamic pattern determined by the performance of the MBR.

2.2.1. Solar Radiation and Wind Speed Data

Solar radiation and wind speed data were extracted from Energy Data website, an open platform providing access to energy datasets and data analytics [32]. Wind speed measurements were performed at the Da Loan station in Lam Dong Province in Vietnam using the "Wind measurement for developing wind power plan and wind power project", which is financed by the German Federal Ministry of Environment, Nature Conservation and Nuclear Safety, and the Vietnam Ministry of Industry and Trade. Wind speed data were collected at an anemometer height of 80 m above sea level for one full year. The data were processed by high-quality sensor technology, based on the current international standard IEC 61400-121 [33]. Solar radiation data were obtained from the Song Binh station in Binh Thuan Province, Vietnam, from a project funded by the Energy Sector Management Assistance Program. Solar radiation and wind speed hourly data are presented in the Supplementary Materials and shown graphically in Figure 6. Solar irradiance meets 1 kWh/m^2 at peak time intervals and varies mainly between 400 and 500 W/m^2 in sunny hours. Wind speed data fluctuate because of the site's topography. Wind speeds rarely exceed 12 m/s and usually fluctuate between 2 and 8 m/s.

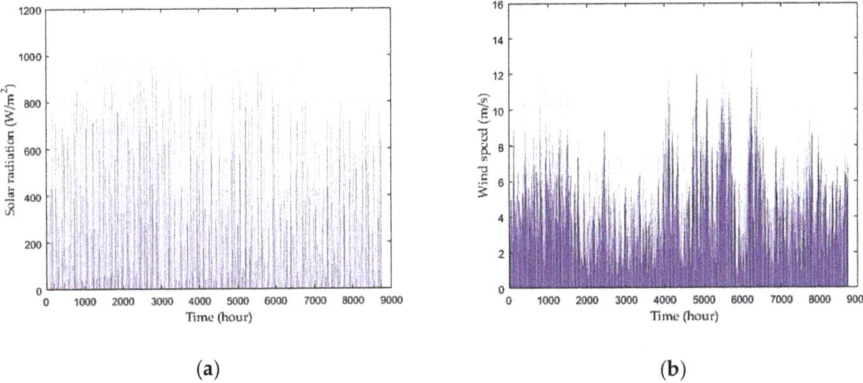

Figure 6. Hourly data of: (**a**) solar radiation at Song Binh station site; and (**b**) wind speed at Da Loan station site.

2.2.2. Demand Load of MBR

The energy consumption of the MBR was calculated according to influent and effluent qualities, while flow rates were obtained from a real MBR plant in Vietnam. The demand load of the MBR plant was the sum of the energy consumption in all processes. Only aerating (AeE), pumping sludge (PuE_{slu}), pumping permeate (PuE_{per}), and mixing was assumed in required bioreactors (MiE). According to the Benchmark Model No. 1, the electricity demand of aeration, pumping, and mixing processes can be calculated using Equations (1)–(4) [34].

$$AeE = \frac{S_O^{Sat}}{(t_1 - t_0) \times 1.8 \times 1000} \int_{t_0}^{t_1} \sum_{i=1}^{5} V_i \times K_L a_i(t) \times dt, \tag{1}$$

$$PuE_{slu} = \frac{1}{t_1 - t_0} \int_{t_0}^{t_1} (0.004 \times Q_{ir}(t) + 0.008 \times Q_{er}(t) + 0.05 \times Q_w(t)) \times dt, \tag{2}$$

$$PuE_{per} = \frac{1}{t_1 - t_0} \int_{t_0}^{t_1} \frac{\Delta P Q_{ef}(t)}{3600\eta} dt, \tag{3}$$

$$MiE = \frac{24}{t_o} \int_{t_0}^{t_1} \sum_{i=1}^{5} (0.005 \times V_i) \times dt \text{ if } K_L a_i(t) < 20\text{d}^{-1}, \quad MiE = 0 \text{ if otherwise}, \tag{4}$$

where t_0 and t_1 are the first and last hours of observation (h), respectively; S_O^{Sat} is the saturated concentration of the dissolved oxygen in water (g/m^3); V_i is the volume of the ith bioreactor (m^3); $K_La_i(t)$ is the oxygen transfer coefficient of the ith bioreactor at time t (h^{-1}); $Q_{ir}(t)$, $Q_{er}(t)$, $Q_w(t)$, and $Q_{ef}(t)$ are the flow rates of the returned sludge from the anoxic to anaerobic bioreactor, the returned sludge from the membrane to stabilizing bioreactor, waste sludge from the membrane bioreactor to a target waste treatment system, and effluent discharging from MBR to sinks in m^3/h, respectively; ΔP is the trans-membrane pressure; and η is the permeate pump efficiency.

The hourly demand load of the MBR plant is given in the Supplementary Materials and displayed graphically in Figure 7. The plant's electrical demand load varies between 300 and 500 kW and reaches 1.4 MW at peak demand.

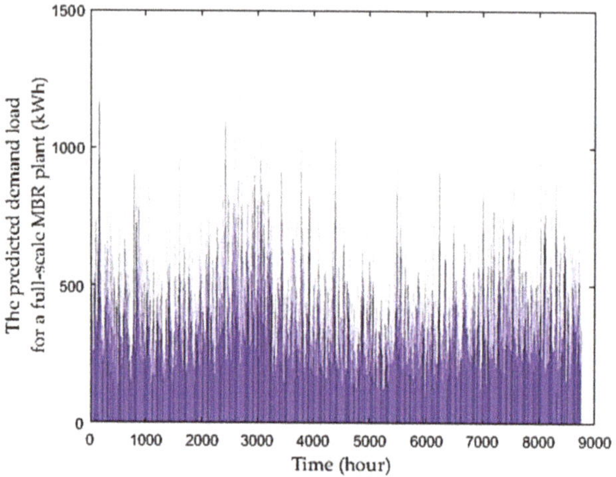

Figure 7. Predicted power demand load for a full-scale MBR plant.

2.3. Optimal Configuration

A numerical tool of PoPA, known as a storage cascade table (SCT), was employed to determine two energy targets, the minimum outsourced electricity supply (MOES) and the available excess electricity for the next day (AEEND) in each operating day during one year [29]. The minimum NE and the optimal BC were then determined. The following simplifying assumptions were considered for the in-silico model development [35,36].

- The intensity of diffuse sky radiation is uniform over the sky dome, and dust and dirt accumulation is neglected on photovoltaic panels.
- The effects of the wind velocity on the PV cell performance have not been considered.
- The power losses in the system were considered as reduction factors in the models.
- The uncertainties in the power market has been neglected.

The SCT was constructed and applied to determine the energy targets. The sub-scenario in which the RES supplied 30% of the total demand load on the second day of operation was used to detail SCT construction. This detailed sub-scenario (the 31st) is categorized under Scenario 2. Wind accounted for 90% of the RES, and solar supplied the rest in the detailed sub-scenario [29]. The SCT table (Table 1) was constructed as detailed as follows.

- Operation hours are given in Column 1 that are listed from "1 h" to "24 h" in a daily pattern. Time intervals are the durations between two adjacent operation hours given in Column 2.

- Hourly DC and AC electricity are measured in kWh, and given in Columns 3 and 4, respectively, calculated using Equations (5) and (6) [37,38].

$$P_{PV} = A_1 \times \eta_1 \times P_f \times \eta_{pc} \times I \times n_1, \quad (5)$$

$$P_{WT} = 0.5 \times C_p \times \rho \times A_2 \times v^3 \times n_2, \quad (6)$$

where P_{PV} is the direct current produced by PV panels, A_1 is the area of the PV panel with a unit area of 1.63 m^2/kWh, η_1 is the efficiency of the module (0.11), P_f is the packing factor assumed to be 0.9, η_{pc} is the power conditioning efficiency assumed to be 0.86, I is the solar radiation (W/m^2), n_1 is the number of PV panels in a solar module, P_{WT} is the produced power by the WTs (kWh), C_p is the wind turbine's power coefficient assumed to be 0.5, ρ is the ambient air density considered as 1.225 kg/m^3, v is the wind speed in m/s, n_2 is the number of WTs in a wind farm, and A_2 is the swept area by the turbine blades that have rotor radius r, which is obtained via Equation (7) [38].

$$A_2 = \pi \times r^2, \quad (7)$$

- The hourly AC electricity demand of the MBR system (P_{DM}) is given in Column 5 in kWh.
- If the AC electricity is not adequate to meet the demand load at any time intervals, the DC electricity is delivered. The required power is obtained using Equation (8) that is given in Column 6 [29].

$$Additional\ amount\ from\ the\ DC\ source = P_{DM} - P_{WT}, \quad (8)$$

- The battery is discharged when the total AC and DC electricity is not enough to meet the demand load, the required amount of battery power is obtained using Equation (9). The calculated quantities are given in Column 7 [29].

$$Second\ additional\ amount\ required\ from\ the\ battery = P_{DM} - P_{WT} - P_{PV} \times c, \quad (9)$$

where c is the conversion efficiency (0.95).

- The battery is charged when the generated DC and AC electricity is greater than the MBR demand load. The amounts of surplus AC and DC power are, respectively, summarized in Columns 8 and 9. The charged (IN) and discharged (OUT) power quantities are, respectively, calculated according to Columns 8 and 9, and Column 7. The IN and OUT are obtained using Equations (10) and (11) and listed in Columns 10 and 11, respectively [29].

$$IN = ACsurplus \times ch + DCsurplus, \quad (10)$$

$$OUT = Second\ additional\ amount\ from\ the\ battery/ch, \quad (11)$$

where ch is charge/discharge efficiency (0.95).

- The maximum electricity storage capacity for the battery (BC) and the outsourced electricity requirement at various time intervals were obtained by cascading the quantities of storage capacity and outsourced electricity columns, respectively. Infeasible storage capacity (ISC) and infeasible outsourced electricity (IOE) that represented the first day of operation are obtained using Equations (12) and (13) and given in Columns 12 and 13, respectively [29]. Feasible storage capacity (FSC) and feasible outsourced electricity (FOE) that represented a normal day of operation are obtained using Equations (14) and (15) and given in Columns 14 and 15, respectively [29]. Here, only the second day of operation was described as a normal day of operation for simplicity.

$$ISC_j^1 = ISC_{j-1}^1 \times (1-s) + IN_j^1 \times c - OUT_j^1/c, \quad (12)$$

$$IOE_j^1 = -ISC_j^1 \times (ch \times c), \quad (13)$$

$$\begin{array}{l} FSC_j^i = ISC_{24}^1 + FSC_{j-1}^i \times (1-s) + IN_j^i \times c - OUT_j^i/c \text{ if } i = 2 \\ FSC_j^i = FSC_{24}^{i-1} + FSC_{j-1}^i \times (1-s) + IN_j^i \times c - OUT_j^i/c \text{ if } i > 2 \end{array} \quad (14)$$

$$FOE_i^j = -FSC_i^j \times (ch \times c)\,, \quad (15)$$

where i represents normal days ($i = 2,3,\ldots,365$), j represents hours ($j = 1,2,\ldots,24$), s is the self-discharged energy in the battery (0.004%/h), and ch is the charge/discharge efficiency (0.95) [39]. The available excess electricity for the next day ($AEEND$) is obtained via Equation (16) [29], and used to determine minimum outsourced electricity supply ($MOES$) as formulated in Equation (17) [29].

$$\begin{array}{l} AEEND = ISC_{24} \text{ for the first day of operation} \\ AEEND = FSC_{24}^i \text{ for the normal days of operation,} \end{array} \quad (16)$$

$$MOES = \sum_{i=1}^{24} FOE_j^i/(1-s)^j \text{ for the normal days of operation,} \quad (17)$$

where ISC_{24}, and FSC_{24}^i represent excess electricity cascaded at 24 h in the first and normal days of operation, respectively.

- It is necessary to calculate BoC and NoE in each single day to determine BC and NE in one operational year. The $BoCs$ are the maximum quantities of ISC given in Column 12 from 0 to 24 h for the first day of operation. The BC and NE are the maximum quantity of $BoCs$ and the sum of all $NoEs$ in an operational year, respectively. The BC and NE are given in Equations (18)–(21) [26,29].

$$BoC_{present} = Max_{j=1}^{24}(FSC_j), \quad (18)$$

$$BC = Max_{i=1}^{365}(BoC_i), \quad (19)$$

$$NoE_{present} = \begin{cases} 0 & if AEEND_{previous} \geq MOES_{present} \\ MOES_{present} - AEEND_{previous} & if AEEND_{previous} < MOES_{present} \end{cases}, \quad (20)$$

$$NE = \sum_{i=1}^{364} NoE_{\mathrm{i}}, \quad (21)$$

where *present* and *previous* represent the present day and the previous day, respectively. NoE is the needed electricity from the grid in a single day (kWh), BoC is the maximum battery capacity in a single day (kWh), NE is the sum of needed electricity from the grid for one operational year, and BC is the maximum battery capacity for one operational year (kWh).

Table 1. Storage cascade table.

1	2	3	4	5	6	7	8	9	10	11	12	13	14	15
Time (h)	Time Interval (h)	DC Source, P_{PV} (kWh)	AC Source, P_{WT} (kWh)	AC Demand, P_{DM} (kWh)	Additional Amount from DC Source (kWh)	Second Additional Amount Required from the Battery (kWh)	AC Surplus (kWh)	DC Surplus (kWh)	Charging into Battery, *IN* (kWh)	Discharging from the Battery, *OUT* (kWh)	Infeasible Storage Capacity, *ISC* (kWh)	Infeasible Outsource Electricity, *IOE* (kWh)	Feasible Storage Capacity, *FSC* (kWh)	Feasible Outsource Electricity, *FOE* (kWh)
													1184.19	2267.37
1	1	0	226.17	631.06	404.88	404.88	0	0	0	426.19	0.00	404.88	710.59	0
2	1	0	388.52	350.10	0	0	388.52	0	369.09	0	332.18	0	1042.75	0
3	1	0	424.17	463.54	39.36	39.36	0	0	0	41.44	286.13	0	996.66	0
4	1	0	424.17	461.18	37.01	37.01	0	0	0	38.96	242.83	0	953.34	0
5	1	0	380.87	395.44	14.57	14.57	0	0	0	15.34	225.78	0	936.26	0
6	1	0.08	309.88	291.87	0	0	309.88	0.08	294.47	0	490.79	0	1201.25	0
7	1	0.32	293.63	270.41	0	0	293.63	0.32	279.27	0	742.12	0	1452.54	0
8	1	0.66	337.11	391.88	54.77	54.14	0	0	0	56.99	678.77	0	1389.17	0
9	1	1.19	54.05	343.57	289.52	288.39	0	0	0	303.56	341.45	0	1051.82	0
10	1	0.87	19.62	323.64	304.02	303.19	0	0	0	319.15	0	11.26	697.17	0
11	1	1.22	63.89	376.11	312.22	311.06	0	0	0	327.43	0	311.06	333.32	0
12	1	1.67	49.05	638.23	589.18	587.60	0	0	0	618.53	0	587.60	0	587.60
13	1	1.68	242.65	447.09	204.44	202.85	0	0	0	213.52	0	202.85	0	202.85
14	1	1.57	123.88	590.05	466.17	464.68	0	0	0	489.13	0	464.68	0	464.68
15	1	1.02	256.98	542.01	285.03	284.06	0	0	0	299.01	0	284.06	0	284.06
16	1	0.54	534.53	437.50	0	0	534.53	0.54	508.34	0	457.50	0	457.50	0
17	1	0.05	424.17	524.36	100.18	100.14	0	0	0.0	105.41	340.36	0	340.36	0
18	1	0	828.47	498.33	0	0	828.47	0	787.04	0	1048.69	0	1048.69	0
19	1	0	646.20	397.84	0	0	646.20	0	613.89	0	1601.14	0	1601.14	0
20	1	0	640.81	583.59	0	0	640.81	0	608.77	0	2148.97	0	2148.97	0
21	1	0	300.06	662.95	362.89	362.89	0	0	0	381.98	1724.46	0	1724.46	0
22	1	0	603.95	616.93	12.98	12.98	0	0	0	13.66	1709.21	0	1709.21	0
23	1	0	354.91	419.38	64.46	64.46	0	0	0	67.86	1633.74	0	1633.74	0
24	1	0	148.71	532.99	384.27	384.27	0	0	0	404.50	1184.24	0	1184.24	0

2.4. Optimal Management

Optimal management of the integrated system was performed according to the scenarios and sub-scenarios. A novel graphical approach was employed to determine the minimum enviro-economic penalty costs that are detailed in the following sections.

2.4.1. Economic Model

An economic model was developed to evaluate the economic performance of the optimal retrofitted system. The model was developed using a multi-integer linear program to obtain the total annual cost (*TAC*) of the system [40]. The overall *TAC* is a summation of the subsystems' TAC including PVs, WTs, and battery banks (BB) considering a constant TAC for the MBR as given in Equation (22).

$$TAC_{overal} = TAC_{PV} + TAC_{WT} + TAC_{BC}, \tag{22}$$

The *TAC* of each subsystem is calculated by summing the annual capital cost (*ACC*), the annual operating and maintenance cost (*AOMC*), and the annual replacement cost (*ARC*) using Equation (23) [41].

$$TAC_{Sub} = ACC_{Sub} + AOMC_{Sub} + ARC_{Sub}, \tag{23}$$

where the subscript *Sub* accounts for PV, WT or BB. *ACC* is calculated using Equation (24) [41].

$$ACC_{Sub} = CC_{Sub} \times AF, \tag{24}$$

where CC_{Sub} is the capital investment cost of each subsystem and *AF* is the amortization factor, which is calculated via Equation (25) [41].

$$AF = \frac{ir \cdot (1+ir)^{LT_{OS}}}{(1+ir)^{LT_{SS}} - 1}, \tag{25}$$

where *ir* is the interest rate, and LT_{OS} and LT_{SS} are the lifetime of the overall system and the subsystems, respectively.

Because it may be necessary to replace several subsystems during the lifetime of the system, the annual replacement cost is calculated using Equation (26) [41].

$$ARC = RC_{Sub} \times \left(\frac{LT_{OS}}{LT_{SS}} - 1\right) \times AF, \tag{26}$$

where RC_{Sub} is the replacement cost of each subsystem.

The economic constants including the capital investment cost, replacement cost, lifetime, and operating and maintenance cost of each subsystem are summarized in Table 2. The quantities were obtained using reliable literature and the uncertainties of the market were ignored [39,41].

Table 2. Economic constants.

Subsystem	Parameters	Symbol	Value	Unit	Reference
Overall	A lifetime of the overall system	LT_{OS}	20	year	[29]
PV	Capital cost	CC_{PV}	350	$/module	[29]
	Replacement cost	RC_{PV}	350	$/module	
	Lifetime	LT_{PV}	20	year	
	Operating and maintenance cost (% of capital cost)	OMC_{PV}	0	%	

Table 2. Cont.

Subsystem	Parameters	Symbol	Value	Unit	Reference
Wind turbine	Capital cost	CC_{WT}	150,000	\$/unit	[29]
	Replacement cost	RC_{WT}	130,000	\$/unit	
	Lifetime	LT_{WT}	15	year	
	Operating and maintenance cost	OMC_{WT}	2500	\$/year	
BC	Capital cost	CC_B	120	\$/kWh	[29,39]
	Replacement cost	RC_B	120	\$/kWh	
	Lifetime	LT_B	4	year	
	Operating and maintenance cost (% of capital cost)	OMC_B	1	%	

2.4.2. Environmental Model

Global warming and climate change by CO_2 emission have become an urgent environmental concern over the world. One of the main sources is the combustion of fossil fuels to produce electrical power, including coal, natural gas, and liquid gas [42]. According to the U.S. Energy Information Administration, coal combustion-related CO_2 emissions globally increased by 7% between 1990 and 2018. It is also anticipated that the combustion of coal will make a slight increase in the average rate of CO_2 emissions from 2018 to 2040 (0.1%) [43].

A linear mathematical program was developed to evaluate the environmental performance of the integrated system. The total annual amount of carbon dioxide emissions ($TACO_2$) was determined for environmental assessment of the integrated system in each sub-scenario. The present energy mix in Vietnam is shown in Figure 8. According to the figure, coal-fired plants emit the greatest proportion of CO_2, representing 34.3% of the current energy mix in Vietnam [5]. Coal was therefore assumed to be the fossil fuel burned to produce the required outsourced electricity. The following procedure was employed to obtain the $TACO_2$:

1. The required coal was obtained to generate one-kilowatt-hour electricity. The amount of coal burned in the production of electricity (kg/kWh) depends on the heat rate (HR) of the generator and the heat content (HC) of the coal as given in Equation (27) [44].

$$Amount\ of\ Coal\ used\ per\ kWh = \frac{HR}{HC}, \tag{27}$$

where HR is the heat rate of the generator, and HC is the heat content of coal. HR and HC were assumed to be 10,493 BTU/kWh and 4362.2 BTU/kg, respectively [44].

2. Assuming that 2578 kg of CO_2 are emitted while burning one short ton (2000 pound or 907.2 kg) of anthracite coal, the CO_2 emitted to generate a kilowatt-hour of electricity is calculated using Equation (28) [44].

$$Amount\ of\ CO_2\ emitted\ per\ kWh = \frac{2578}{907.2} \times \frac{HR}{HC}, \tag{28}$$

3. The total annual amount of CO_2 emitted (kg) to generate the outsourced electricity is calculated via Equation (29) [44].

$$m_{CO_2} = \frac{2578}{907.2} \times \frac{HR}{HC} \times OE, \tag{29}$$

where OE is the amount of outsourced electricity, which is the sum of the needed electricity (NE) and make-up electricity (ME).

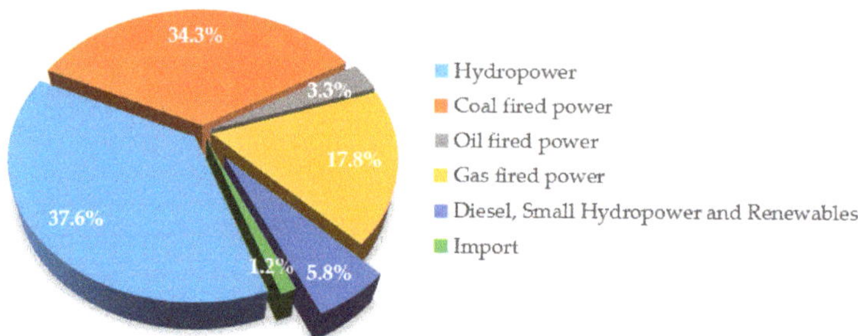

Figure 8. Vietnam energy mix (2017) [5].

2.4.3. Graphical Optimization

The economic and environmental performances of the integrated system were not included in the optimal system using PoPa. Thus, optimal management was investigated at the local scale considering all the sub-scenarios. A combined enviro-economic penalty cost was defined as the objective function given in Equation (30) [45].

$$CEEC = TAC + \alpha \times EnPC \times TACO_2, \quad (30)$$

where $CEEC$ is the combined enviro-economic penalty cost (\$/year), $EnPC$ is the environmental penalty cost of the CO_2 emission (2.65 cent/kg), and α is a constant determined by the management cases [45]. The significance of environmental penalty costs versus economic costs varies by location. Industrial air pollution is restricted in more unhealthy regions, whereas economic concerns are given priority in cleaner developing countries. Three management cases were defined according to the local environmental situation under the following assumptions [46]:

- Case 1: CO_2 emissions are not fined so that $\alpha = 0$, and only economic costs are considered for optimization. A slightly polluted environment would be considered as the best condition to apply for case 1.
- Case 2: Economic penalty costs of CO_2 emissions conform to the international agreements, and $\alpha = 1$. This case should be applied to a moderately polluted environment.
- Case 3: Air quality is sufficiently unhealthy to require that no additional pollution be emitted, and $\alpha = 10$. This case is suitable for countries and regions, especially developing countries, having hazardous high levels in environmental pollution.

The curves of $CEEC$ were drawn including all sub-scenarios in the three management cases, and minimum points were selected graphically to determine the optimal management in local settings.

3. Results and Discussion

An optimal configuration of the integrated system was obtained for all sub-scenarios in each operational month. Accordingly, the optimal sizes of PV panels, WTs, batteries, and NE were calculated for each sub-scenario, applying the SCT of PoPA. An optimal management was then performed using the combined enviro-economic costs of the optimal system considering the management cases. Results are presented in the Supplementary Materials.

3.1. Component Sizing in an Optimal Configuration

The optimal size of PVs and WTs were determined in each sub-scenario in every month to ensure that the PVs and WTs were sized for 12 months to meet the monthly average power demand in the sub-scenarios. For example, the data for the 31st sub-scenario described in Section 2.3 is illustrated graphically in Figure 9a–c. The unit power generated by each PV or WT was calculated using Equations (5) and (6), respectively. PV and WT sizes were calculated considering 12 months as detailed in Table 3. The peak demand load, which occurred in April, was approximately 4000 kWh. The optimal sizes of PVs and WTs were calculated to be 4 panels and 18 modules, respectively, to meet peak loads. Considering an annual case, the sizes of PVs and WTs should be 23 panels and 6 modules, respectively, to meet peak demand in all 12 months. The RES met 30% of total demand load in this case.

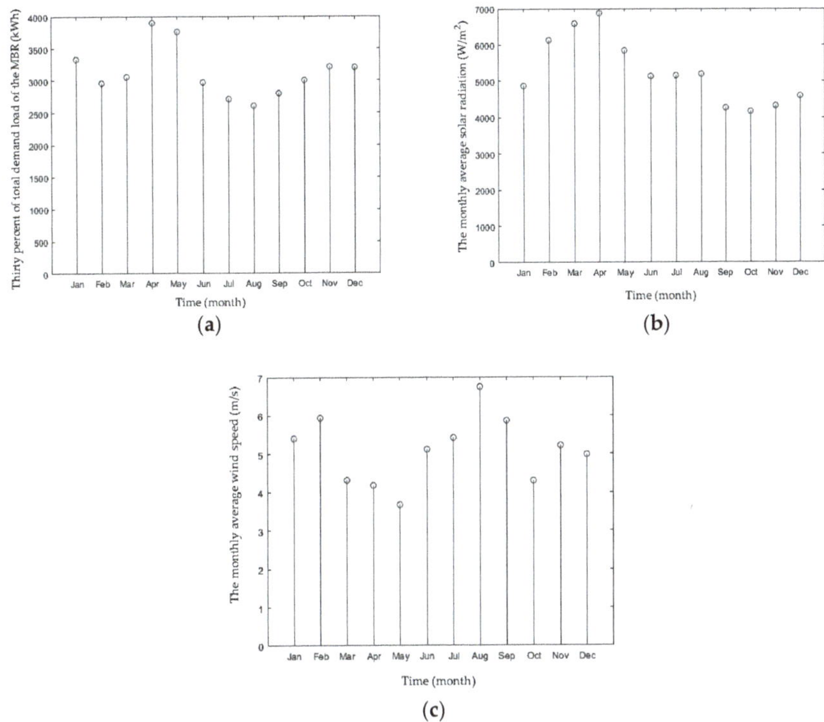

Figure 9. Column charts of the described 31st sub-scenario: (**a**) 30% of total demand load of the MBRs; (**b**) monthly average solar radiation; and (**c**) wind speed.

Table 3. Monthly optimal size of PV panels and WTs for the described 31st sub-scenario.

Month	January	February	March	April	May	June	July	August	September	October	November	December
No. WT	2	2	3	4	6	2	2	1	2	3	2	2
No. PV	21	15	14	18	20	18	16	16	20	22	23	21

All optimal sizes in each sub-scenario were calculated following the same procedure and are presented in the Supplementary Materials. Results considering 11 sub-scenarios under each scenario are also summarized in Table 4. The smallest NE was obtained in Scenario 2, in which wind sources contributed 90% of the RES. The smallest NE obtained in Scenario 2 was approximately 6 and 31 times less than those in Scenarios 3 and 1, respectively. This means that the wind energy offered superior consistency in terms of meeting MBR demand loads in Vietnam. As wind potential increased, pollution caused by fossil-fuel combustion decreases, and NE decreased as renewable penetration increases in the sub-scenarios. However, BC did not exhibit meaningful regression with the sub-scenario numbers and the scenarios. This is because of the dynamicity of the demand load and the intermittent nature of the generated power by PVs and WTs. However, a regular increase in BC was seen in Scenario 1, in which solar contribution is 90% as a result of higher DC power storage. Therefore, higher solar share led to greater storage capacity in the HRES. Furthermore, the smallest size of the solar and wind components in sub-Scenario 100 (100% RES) occurred in Scenario 2. This may be due to two reasons: either wind power has stronger potential compared with solar power in Vietnam or the losses for several AC/DC conversions mean weaker solar PV power compared with wind.

Table 4. The optimal size of PV panels, WTs, BC and NE in the scenarios.

Sub-Scenario	RES Contribution, %	No. PVs, Panel	No. WTs, Module	BC, kWh	NE, kWh
			Scenario 1		
1	0	0	0	-	3,832,303
2	1	7	1	3887	2,893,434
11	10	68	1	3928	2,878,527
21	20	135	1	3922	2,862,233
31	30	202	1	3867	2,845,988
41	40	269	1	3811	2,829,747
51	50	336	2	5551	2,171,979
61	60	403	2	5575	2,157,772
71	70	471	2	5599	2,143,399
81	80	538	2	5623	2,129,251
91	90	605	2	5646	2,115,134
101	100	672	3	5400	1,662,696
			Scenario 2		
1	0	0	0	-	3,832,303
2	1	1	1	3892	2,894,900
11	10	8	2	5435	2,242,853
21	20	15	4	6515	1,437,793
31	30	23	6	11,379	953,380
41	40	30	8	7163	640,381
51	50	38	10	5622	418,646
61	60	45	12	5568	259,490
71	70	53	14	6205	172,006
81	80	60	16	5953	106,127
91	90	68	17	6053	85,188
101	100	75	19	5227	53,571
			Scenario 3		
1	0	0	0	-	3,832,303
2	1	4	1	3890	2,894,168
11	10	38	2	5446	2,236,286
21	20	75	3	5807	1,775,114
31	30	112	4	6443	1,421,974
41	40	150	5	10,704	1,148,995
51	50	187	6	10,435	931,827
61	60	224	7	8517	761,944
71	70	262	8	7343	614,820
81	80	299	9	6496	495,026
91	90	336	10	5519	392,191
101	100	374	11	5644	304,963

The SCT table was constructed using 8760 annual hourly data objects to obtain the BC and the quantity of outsourced NE. Two days of operation are described to show how the SCT was used to determine the BC and NE. The BC and NE were calculated in the above-mentioned 31st sub-scenario as described in Section 2.3 using Table 1. BoC and NoE were calculated in every single day to obtain BC and NE for one operational year. On the first day of operation represented in Columns 12 and 13, ISC and IOE were calculated using Equations (12) and (13) using a cascade technique of PoPA. In Column 12, 0 s indicates that RES was not enough to meet the demand load, so BB was discharged completely. Thus, additional electricity was needed from the grid listed in Column 13 at those time intervals. $AEEND$ and $MOES$ were calculated using Equations (16) and (17), respectively, for every single day. There was no $MOES$ in the first day of operation. However, $AEEND$ on the first day was cascaded at 24 h and used to reduce MOES on the second day. Similarly, $AEEND$ was cascaded at 24 h on the second day and used to reduce MOES on the third day. The same procedure was followed for all operational days. If $AEEND$ was greater than $MOES$, no additional electricity was needed from the grid, and if $AEEND$ was less than $MOES$, the difference between them represented NoE from the grid. According to the SCT, $AEEND$ on the first day and $MOES$ on the second day were 1184.24 and 2267.37 kWh, given in the last row of Column 12 and the first row of Column 15, respectively. Thus, 1083.24 kWh of NoE was required on the second day considering the difference between $AEEND$ and $MOES$. $AEEND$ at 24 h was 1148 kWh on the second day of operation as given in the last row of Column 13. Thus, $AEEND$ could vary on every single day because of various solar radiation, wind speed and demand load quantities. FSC and FOE were calculated using the same cascade procedure for the remaining 364 days, but taking the $AEEND$ of a previous day into account using Equations (14) and (15), respectively. $BoCs$ were determined as the maximum value of $FSCs$ from 0 to 24 h (2148.97 kWh) on each single day. The same cascade procedure was iterated for all 365 days. BC was the maximum value of $BoCs$ and NE was the sum of all $NoEs$. BC and NE were calculated considering all sub-scenarios in scenarios and the results are given in the Supplementary Materials.

The annual OE was calculated by summing the determined needed electricity (NE) and make-up electricity (ME). The ME was the demand load not met by the RES according to the sub-scenarios. As an instance, the ME was equal to 70% of the required MBR demand load in the 31st sub-scenario described in Section 2.3. However, the NE was the demand load not met by the optimal HRES, in which the battery was totally discharged. The NE was determined using the SCT. The OE was used to calculate and graphically compare the scenarios in Figure 10. According to the figure, both NE and OE decreased as the HRES share increased. The NE has the smallest quantity in Scenario 2, which consisted of the numerical values in Table 4.

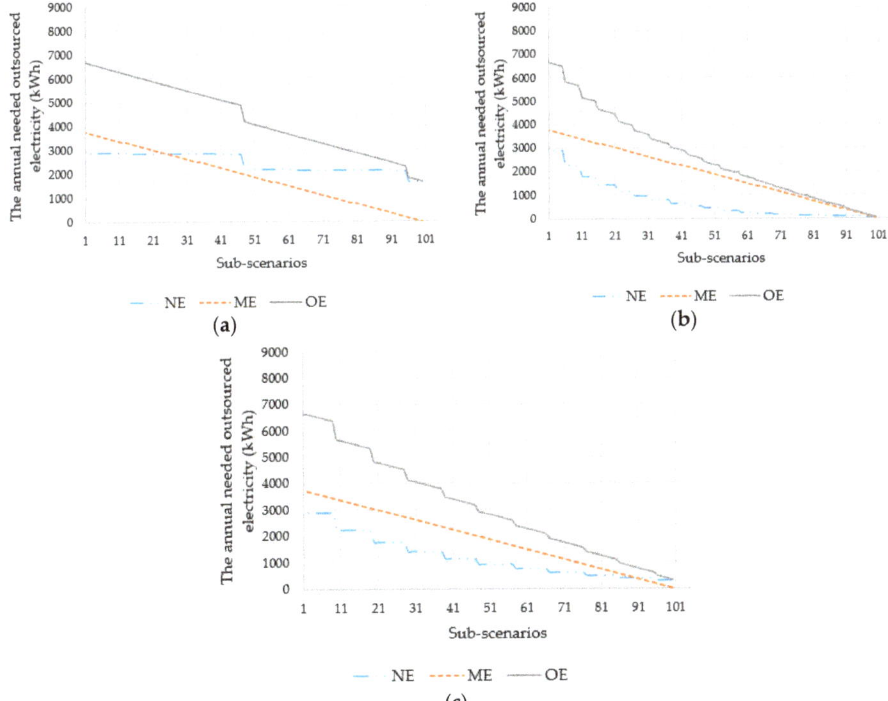

Figure 10. The annual amount of outsourced electricity for: (**a**) Scenario 1; (**b**) Scenario 2; and (**c**) Scenario 3.

3.2. Graphical Optimal Management

The TAC and $TACO_2$ of the optimal system are compared in Figure 11 for three scenarios and considering all 101 sub-scenarios. The $TACO_2$ decreased for all three scenarios as the contribution of the RES increased. On the other hand, the TAC depended on the sub-scenarios; the RES contribution increased as demand load increased. Accordingly, the size of PVs and WTs increased, but the size of battery fluctuated depending on the demand load patterns. A robust configuration should, therefore, be determined by analyzing combined CEECs. For this purpose, a graphical approach was followed.

A CEEC was obtained for all management scenarios and sub-scenarios, and the results are presented in the Supplementary Materials. The results are also shown graphically in Figure 12, excluding those for sub-Scenario 1, in which no contribution was made to the HRES. Scenarios 1–3 are depicted in continuous blue, dashed orange, and gray lines in Figure 12, respectively. As $\alpha = 0$ in Management Case 1, the CEEC was equal to TAC in Figure 12a. The smallest TAC was obtained in Scenario 1. The TAC slightly decreased in Scenario 1 until an abrupt jump occurred after sub-scenario 47, in which the CEEC was equal to \$850,710/year. The optimal configuration was determined when the HRES met 47% of the MBR demand load, considering only the economic model. Solar and wind power shares were 90%, and 10%, respectively, in the optimal HRES. This configuration is recommended for regions where the air quality is good and economic growth has a priority. The BC, NE, and number of PVs and WTs were 3773 kWh, 2818 MWh, 316, and 1, respectively, in the optimal configuration of Management Case 1.

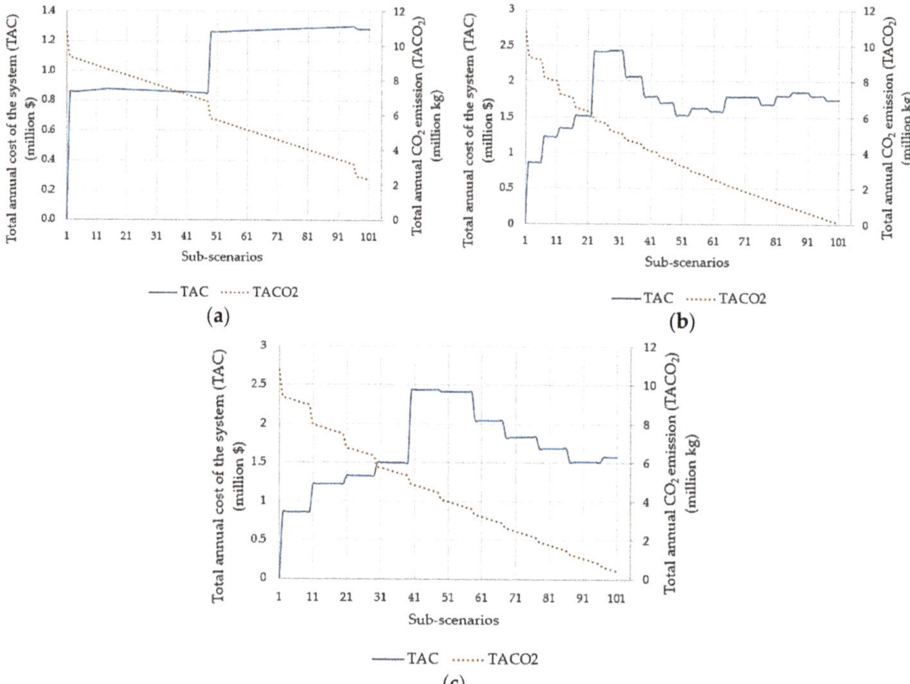

Figure 11. Total annual cost of the system and total annual carbon dioxide emissions for: (**a**) Scenario 1; (**b**) Scenario 2; and (**c**) Scenario 3.

The CEEC for Management Case 2 is shown graphically in Figure 12b. The similarity between Figure 12a,b proves the governance of economic costs in the CEEC. In other words, the economic costs were much greater than the environmental penalty costs, considering an EnPC of CO_2 of

$24/ton. The optimal configuration and the overall patterns of the graphs showed little variation when considering a global EnPC for CO_2 emissions. This shows that RES cannot compete with fossil fuels unless less-expensive green technologies are developed, higher penalty costs are assigned to greenhouse-gas emissions, or government subsidies are offered to green technologies.

Air quality varies regionally, limiting the ability to establish a common global environmental policy. Several regions, including the industrial regions in Vietnam, suffer from unhealthy air quality, even though economic conditions are mild. This condition was mentioned in the Management Case 3, and the results are shown in Figure 12c. Increased emphasis on environmental costs ($\alpha = 10$) influenced the CEEC patterns considerably in all scenarios. The CEEC diagrams had a negative slope, which means that as the contribution of the RES (sub-scenario No.) increases, the CEEC decreases. Although the obtained CEEC was much higher in Management Case 3 compared with that of Case 1, the system can be operated if the MBR demand load is fully met by the HRES. Scenario 3 produced the optimal configuration, in which solar and wind technologies had equal shares in the HRES. The TAC and $TACO_2$ were $1,693,476/year and 429,016 kg/year in the optimal configuration, respectively. The BC, NE, and number of solar PVs and WTs were 5644 kWh, 304.9 MWh, 374, and 11 modules, respectively. Although the sizes of all components were much larger in Management Case 3 compared with the other cases, the NE and subsequently the air pollution were much less.

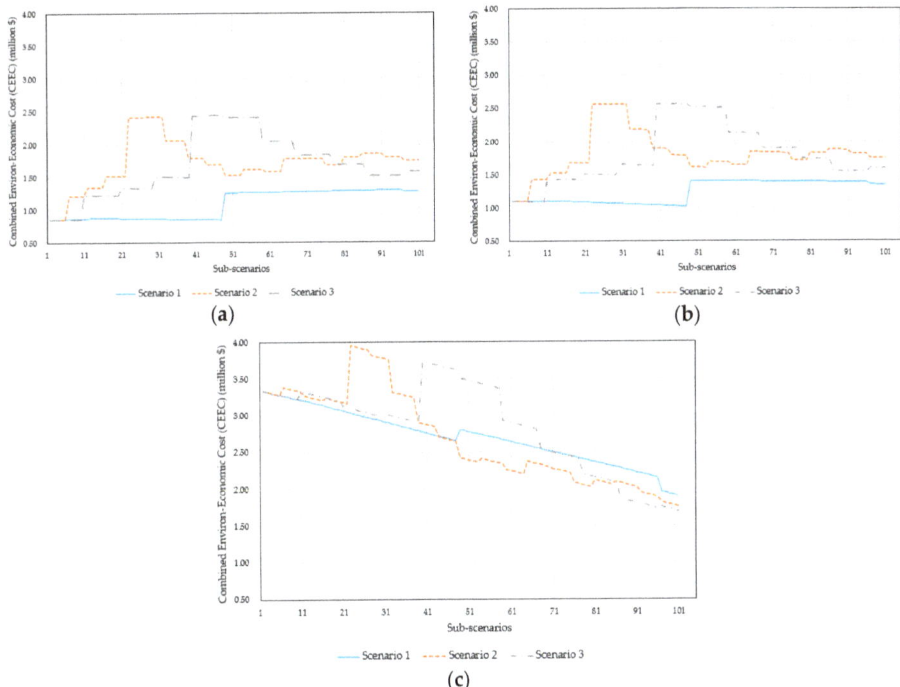

Figure 12. Combined enviro-economic costs for management of: (**a**) Case 1; (**b**) Case 2; and (**c**) Case 3.

4. Conclusions

The dynamic demand load of a MBR was met by a model-based HRES comprising solar PVs, WTs, and BBs. The model was validated using real data in Vietnam. Accordingly, a dual-scale mathematical model was assigned in which the HRES was technically optimized in the first scale and an optimal configuration was obtained in the second scale considering its enviro-economic performance. A PoPa was used for the system-scale optimization and a graphical approach was followed for the local-scale optimization. Probable design configurations were assigned to three scenarios, 101 sub-scenarios, and three management cases. According to PoPa results, wind turbines met MBR demand load more consistently and had greater potential and fewer losses compared with solar farms in Vietnam. WTs should, therefore, represent a greater share of the energy mix of a dynamic load-driven HRES in Vietnam. The performance of an optimal system varied under local conditions, and the system was optimized at the local scale. The weights assigned to the EnPCs played a significant role in the optimization, with the CEEC reaching $850,710/year, $1,030,628/year, and $1,693,476/year in three management cases with different weights. The proposed multi-scale optimization approach was validated in a coupled MBR-HRES system, but it could be efficiently used for any dynamic systems. Noting that this study was conducted using computing models, the accuracy of the results obtained by this model-based study was slightly different from those obtained by an actual situation. The followings could be the main reasons to explain: (1) the demand load of the MBR did not include electrical equipment within the MBR plant other than those given in the schematic diagram; (2) conversation efficiency, and charge and discharge efficiencies of the battery could vary from actually-installed devices; and (3) the costs of investment, replacement and operating and maintenance in systems could change according to the uncertainty of the power market.

Supplementary Materials: The following are available online at http://www.mdpi.com/2071-1050/11/1/66/s1.

Author Contributions: Conceptualization, T.-V.H., S.H. and C.Y.; Data curation, K.N.; Formal analysis, T.-V.H.; Funding acquisition, J.-M.O.; Methodology, P.I., K.N. and J.R.; Project administration, S.H.; Supervision, C.Y.; Visualization, P.I.; Writing—original draft, T.-V.H.; and Writing—review and editing, P.I. and C.Y.

Acknowledgments: This subject was supported by "Development of algae management using stream structures in the stream" (code 18AWMP-B098640-04) funded by MOLIT (Ministry of Land, Infrastructure and Transport). This work was supported by the National Research Foundation of Korea (NRF) grant funded by the Korea government (MSIT) (No. NRF-2017R1E1A1A03070713), and Korea Ministry of Environment (MOE) as Graduate School specialized in Climate Change.

Conflicts of Interest: The authors declare no conflict of interest.

References

1. Soares, N.; Martins, A.; Carvalho, A.; Caldeira, C.; Du, C.; Castanheira, É.; Rodrigues, E.; Oliveira, G.; Pereira, G.; Bastos, J. The challenging paradigm of interrelated energy systems towards a more sustainable future. *Renew. Sustain. Energy Rev.* **2018**, *95*, 171–193. [CrossRef]
2. Luong, N.D. A critical review on potential and current status of wind energy in Vietnam. *Renew. Sustain. Energy Rev.* **2015**, *43*, 440–448. [CrossRef]
3. Ifaei, P.; Karbassi, A.; Lee, S.; Yoo, C. A renewable energies-assisted sustainable development plan for Iran using techno-econo-socio-environmental multivariate analysis and big data. *Energy Convers. Manag.* **2017**, *153*, 257–277. [CrossRef]
4. Le, V.V.; Nguyen, D.C.; Dong, V.H. The Renewable Energy in Vietnam: Potential, Development Orientation. *Int. J. Sci. Technol. Res.* **2017**, *6*, 204–207.
5. EVN. *Vietnam Electricity Annual Report*; EVN: Hanoi, Vietnam, 2017; p. 12. Available online: https://en.evn.com.vn/userfile/User/huongbtt/files/2018/2/AnnualReport2017.pdf (accessed on 18 November 2018).
6. Luong, N.D. A critical review on energy efficiency and conservation policies and programs in Vietnam. *Renew. Sustain. Energy Rev.* **2015**, *52*, 623–634. [CrossRef]
7. Nguyen, K.Q. Wind energy in Vietnam: Resource assessment, development status and future implications. *Energy Policy* **2007**, *35*, 1405–1413. [CrossRef]
8. Awan, A.B.; Zubair, M.; Abokhalil, A.G. Solar energy resource analysis and evaluation of photovoltaic system performance in various regions of Saudi Arabia. *Sustainability* **2018**, *10*, 1129. [CrossRef]
9. Jianzhong, X.; Assenova, A.; Erokhin, V. Renewable Energy and Sustainable Development in a Resource-Abundant Country: Challenges of Wind Power Generation in Kazakhstan. *Sustainability* **2018**, *10*, 3315. [CrossRef]

10. Ifaei, P.; Karbassi, A.; Jacome, G.; Yoo, C. A systematic approach of bottom-up assessment methodology for an optimal design of hybrid solar/wind energy resources–Case study at middle east region. *Energy Convers. Manag.* **2017**, *145*, 138–157. [CrossRef]
11. Nguyen, D. A brief overview on assessments of wind energy resource potential in Vietnam. *J. Fundam. Renew. Energy Appl.* **2014**, *4*, 2.
12. TrueWind Solutions, LLC. *Wind Energy Resource Atlas of Southeast Asia*; World Bank: Washington, DC, USA, 2001; Available online: http://documents.worldbank.org/curated/en/252541468770659342/Wind-energy-resource-atlas-of-Southeast-Asia (accessed on 18 November 2018).
13. AWS Truepower, LLC. *Wind Resource Atlas of Vietnam*; AWS Truepower: Hanoi, Vietnam, 2011; Available online: https://www.esmap.org/sites/default/files/esmap-files/MOIT_Vietnam_Wind_Atlas_Report_18Mar2011.pdf (accessed on 18 November 2018).
14. GIZ/MoIT. *Wind Measurement for Developing Wind Power Plan and Wind Power Project*; GIZ: Hanoi, Vietnam, 2013; Available online: www.giz.de/vietnam (accessed on 18 November 2018).
15. Sakamoto, Y.; Shoji, K.; Bui, M.T.; Phạm, T.H.; Vu, T.A.; Ly, B.T.; Kajii, Y. Air quality study in Hanoi, Vietnam in 2015–2016 based on a one-year observation of NOx, O3, CO and a one-week observation of VOCs. *Atmos. Pollut. Res.* **2018**, *9*, 544–551. [CrossRef]
16. ICEM. *Analysis of Pollution from Manufacturing Sectors in Vietnam*; International Center for Environmental Management: Indooroopilly, Australia, 2007; Available online: http://icem.com.au/portfolio_category/report/ (accessed on 18 November 2018).
17. Environmental Performance Index. Available online: https://epi.envirocenter.yale.edu/epi-country-report/VNM (accessed on 18 November 2018).
18. Nakagami, K. Environmental Management and Sustainable Development in Vietnam. *Policy Sci.* **1999**, *7*, 1.
19. WB. *Vietnam–Environmental Program and Policy Priorities for a Socialist Economy in Transition*; 13200-VN; WB: Washington, DC, USA, 1995; Available online: http://documents.worldbank.org/curated/en/953301468761685706/The-supporting-annexes (accessed on 18 November 2018).
20. Sartor, M.; Kaschek, M.; Mavrov, V. Feasibility study for evaluating the client application of membrane bioreactor (MBR) technology for decentralised municipal wastewater treatment in Vietnam. *Desalination* **2008**, *224*, 172–177. [CrossRef]
21. Pabby, A.K.; Rizvi, S.S.; Sastre, A.M. *Handbook of Membrane Separations: Chemical, Pharmaceutical, Food, and Biotechnological Applications*; CRC: Boca Raton, FL, USA, 2008.
22. Joss, A.; Böhler, M.; Wedi, D.; Siegrist, H. Proposing a method for online permeability monitoring in membrane bioreactors. *Water Sci. Technol.* **2009**, *60*, 497–506. [CrossRef] [PubMed]
23. Kim, J.-S.; Lee, C.-H.; Chang, I.-S. Effect of pump shear on the performance of a crossflow membrane bioreactor. *Water Res.* **2001**, *35*, 2137–2144. [CrossRef]
24. Thunuguntla, R.; Mahboubi, A.; Ferreira, J.; Taherzadeh, M. Integration of membrane bioreactors with edible filamentous fungi for valorization of expired milk. *Sustainability* **2018**, *10*, 1940. [CrossRef]
25. Esfahani, I.J.; Rashidi, J.; Ifaei, P.; Yoo, C. Efficient thermal desalination technologies with renewable energy systems: A state-of-the-art review. *Korean J. Chem. Eng.* **2016**, *33*, 351–387. [CrossRef]
26. Esfahani, I.J.; Lee, S.; Yoo, C. Extended-power pinch analysis (EPoPA) for integration of renewable energy systems with battery/hydrogen storages. *Renew. Energy* **2015**, *80*, 1–14. [CrossRef]
27. Kantola, M.; Saari, A. Renewable vs. traditional energy management solutions—A Finnish hospital facility case. *Renew. Energy* **2013**, *57*, 539–545. [CrossRef]
28. Rozali, N.E.M.; Alwi, S.R.W.; Manan, Z.A.; Klemeš, J.J.; Hassan, M.Y. Process integration of hybrid power systems with energy losses considerations. *Energy* **2013**, *55*, 38–45. [CrossRef]
29. Li, Q.; Moya, W.; Esfahani, I.J.; Rashidi, J.; Yoo, C. Integration of reverse osmosis desalination with hybrid renewable energy sources and battery storage using electricity supply and demand-driven power pinch analysis. *Process Saf. Environ. Prot.* **2017**, *111*, 795–809. [CrossRef]
30. Jiang, T. Characterization and Modelling of Soluble Microbial Products in Membrane Bioreactors. Ph.D. Thesis, Ghent University, Ghent, Belgium, 2007.
31. Kim, M. An Integrated Prediction and Diagnosis Approach for an Understanding of Membrane Fouling Phenomena in Membrane Bioreactor (MBR): Multi-Variable Statistical and Mechanistic Models. Master's Thesis, Kyung Hee University, Global Campus, Yongin, Korea, 2012.
32. Energy Data. Available online: https://energydata.info/ (accessed on 6 July 2018).

33. International Electrotechnical Commission (IEC). *WIND TURBINES—Part 121: Power Performance Measurements of Grid Connected Wind Turbines*; IEC: Geneva, Switzerland, 2005.
34. Alex, J.; Benedetti, L.; Copp, J.; Gernaey, K.; Jeppsson, U.; Nopens, I.; Pons, M.; Rieger, L.; Rosen, C.; Steyer, J. *Benchmark Simulation Model No. 1 (BSM1)*; Lund University: Lund, Sweden, 2008; Available online: https://www.iea.lth.se/publications/Reports/LTH-IEA-7229.pdf (accessed on 18 November 2018).
35. Safder, U.; Ifaei, P.; Yoo, C. Multi-objective optimization and flexibility analysis of a cogeneration system using thermorisk and thermoeconomic analyses. *Energy Convers. Manag.* **2018**, *166*, 602–636. [CrossRef]
36. Esfahani, I.J.; Ifaei, P.; Kim, J.; Yoo, C. Design of hybrid renewable energy systems with battery/hydrogen storage considering practical power losses: A MEPoPA (modified extended-power pinch analysis). *Energy* **2016**, *100*, 40–50. [CrossRef]
37. Hocaoğlu, F.O.; Gerek, Ö.N.; Kurban, M. A novel hybrid (wind–photovoltaic) system sizing procedure. *Sol. Energy* **2009**, *83*, 2019–2028. [CrossRef]
38. Tyagi, R.K. Wind energy and role of effecting parameters. *Eur. J. Appl. Eng. Sci. Res.* **2012**, *1*, 73–83.
39. Mahmoudi, H.; Abdul-Wahab, S.; Goosen, M.; Sablani, S.; Perret, J.; Ouagued, A.; Spahis, N. Weather data and analysis of hybrid photovoltaic–wind power generation systems adapted to a seawater greenhouse desalination unit designed for arid coastal countries. *Desalination* **2008**, *222*, 119–127. [CrossRef]
40. Esfahani, I.J.; Ataei, A.; Shetty, V.; Oh, T.; Park, J.H.; Yoo, C. Modeling and genetic algorithm-based multi-objective optimization of the MED-TVC desalination system. *Desalination* **2012**, *292*, 87–104. [CrossRef]
41. Esfahani, I.J. Optimal Design of High Efficient Combined Desalination and Refrigeration System Coupled with CHP and Hybrid Renewable Energy Sources. Ph.D. Thesis, Kyung Hee University, Seoul, Korea, 2015.
42. Zhang, Z.; Li, Y.; Zhang, W.; Wang, J.; Soltanian, M.R.; Olabi, A.G. Effectiveness of amino acid salt solutions in capturing CO2: A review. *Renew. Sustain. Energy Rev.* **2018**, *98*, 179–188. [CrossRef]
43. EIA. International Energy Outlook. 2017. Available online: https://www.eia.gov/outlooks/ieo/pdf/0484(2017).pdf (accessed on 18 November 2018).
44. EIA. How Much Coal, Natural Gas, or Petroleum Is Used to Generate a Kilowatthour of Electricity? Available online: https://www.eia.gov/tools/faqs/faq.php?id=667&t=2 (accessed on 6 July 2018).
45. Ifaei, P.; Yoo, C. The compatibility of controlled power plants with self-sustainable models using a hybrid input/output and water-energy-carbon nexus analysis. *J. Clean. Prod.* **2018**. [CrossRef]
46. Loy-Benitez, J.; Li, Q.; Ifaei, P.; Nam, K.; Heo, S.; Yoo, C. A dynamic gain-scheduled ventilation control system for a subway station based on outdoor air quality conditions. *Build. Environ.* **2018**, *144*, 159–170. [CrossRef]

Article

One-Dimensional Analytical Modeling of Pressure-Retarded Osmosis in a Parallel Flow Configuration for the Desalination Industry in the State of Kuwait

Bader S. Al-Anzi * and Ashly Thomas

Department of Environmental Technology Management, College of Life Sciences, Kuwait University, P.O. Box 5969, Safat 13060, Kuwait; ashly.ann@hotmail.com
* Correspondence: bader.alanzi@ku.edu.kw

Received: 19 March 2018; Accepted: 19 April 2018; Published: 22 April 2018

Abstract: The present study deals with the application of one-dimensional (1D) analytical expressions for a parallel flow configuration in pressure-retarded osmosis (PRO) exchangers by using actual brine and feed salinity values from the Kuwait desalination industry. The 1D expressions are inspired by the effectiveness-number of transfer unit (ε-NTU) method used in heat exchanger analysis and has been developed to "size" an osmotically-driven membrane process (ODMP) mass exchanger given the operating conditions and desired performance. The driving potentials in these mass exchangers are the salinity differences between feed and draw solution. These 1D model equations are employed to determine mass transfer units (MTU) as a function of different dimensionless groups such as mass flowrate ratio (MR), recovery ratio (RR), concentration factors (CF) and effectiveness (ε). The introduction of new dimensionless groups such as the dilution rate ratio (DRR) and dilution rate (DR) would be used to relate the actual water permeation to the brine draw stream. The results show that a maximum power of 0.28 and 2.6 kJ can be produced by the PRO system using seawater or treated wastewater effluent (TWE) as the feed solution, respectively, which might be able to reduce the power consumption of the desalination industry in Kuwait.

Keywords: pressure-retarded osmosis; mass transfer units; maximum power; effectiveness; dilution rate ratio

1. Introduction

Desalination and wastewater treatment based on membrane technology comprise one of the approaches that has been extensively explored over the past two decades to tackle the challenges of increasing access to clean drinking water resulting from the rapidly-growing global population, as well as economic development [1]. It was reported that approximately 50% of the world's desalination capacity has been installed in the Middle East region [1]. Kuwait, for instance, has six multi-stage flash (MSF) and two seawater reverse osmosis (SWRO) plants with a total plant capacity of 2 million m^3/day [2,3]. This is sufficient for the 3.4 million population's daily consumption.

Although the reverse osmosis (RO) membrane process has been employed for seawater desalination without significant technical drawbacks since the 1970s, statistics have revealed that its operating cost per cubic meter of water produced is significantly decreased due to the improvements of membrane fabrication techniques and the entire operating system [4,5]. Energy consumption being a vital component that characterizes the performance of RO processes, it also influences the produced freshwater costs. Gude et al. [6] summarized that the use of energy recovery devices (ERDs), high permeable membranes, two pass-two stage RO operations, water recovery options by reuse and recycling the permeate water for multiple uses will reduce the energy consumption and production costs in small and a wide range of applications. A recent work by Karabelas et al. [7] on the analysis of specific energy consumption (SEC) in RO processes of seawater and brackish water concluded that SEC can be reduced by improving the membrane permeability and efficiency of pumps along with ERD. Management of brine effluents from desalination plants however is still a challenge that needs to be addressed. The MSF process dominates the desalination industry in Kuwait, where 1.47 million m^3/d of the total desalination capacity of 1.65 million m^3/d are provided by MSF and 0.17 million m^3/d is only supplied by reverse osmosis (RO) [3]. Desalination of seawater in Kuwait, in general, produces a brine discharge with a total dissolved solid (TDS) concentration of 70,000 ppm (i.e., almost twice the TDS of seawater) [8]. This brine of extremely high salinity that gets disposed into Kuwait's coastal waters would create a negative impact on the marine ecosystem and must be managed sustainably and efficiently [9]. Numerous studies have shown that appropriate utilization of highly concentrated brine solution with diluted treated wastewater or seawater via pressure-retarded osmosis (PRO) not only can mitigate the severe environmental impacts from brine discharge, but also generate substantial energy output from the feed and draw salinity gradient [10–12].

The first mathematical model was proposed by Loeb in 1976 [13], who studied the simulation of PRO performance with the combination of different solutions such as seawater and fresh water or highly saline bodies such as the Dead Sea and seawater to generate osmotic power. Since then, a few models have been proposed and developed by researchers to improve the performance of using PRO membranes by minimizing the effect of concentration polarization [14], as well as the salt reversal from the draw to feed side [15]. In 1981, Lee et al. [16] developed a zero-dimensional performance model that dealt with the effects of internal concentration polarization (ICP) of the PRO membrane. After that, a similar research study was conducted by McCutcheon [17] to further determine the combined effects of both internal and external concentration polarization (ECP) on the performance of forward osmosis (FO) and PRO membranes. It was reported that a zero-dimensional (0D) model can be used to investigate the effects of various operating parameters such as the salinity level of feed and draw solutions, flow velocities, hydraulic pressure and membrane characteristics on PRO power generation [18,19]. Recently, He et al. [20] developed a simple PRO model by considering the detrimental effects of ICP, ECP and reverse salt permeation (RSP) to evaluate the actual flux and power density of a PRO system and also to address the behavior of a PRO process at different applied pressures. Naguib et al. [21], on the other hand, developed a mathematical model for PRO processes in commercial length hollow fiber membranes with respect to the effects of ICP, ECP and RSP. They found that the reduction in the concentration gradient due to polarization and axial variation was proportionately increased at high membrane flux. Addressing the detrimental axial variations helps to minimize polarization, but comes with a cost of increased pumping loads. Previously, our research group [22] examined the feasibility of using PRO to generate energy from wastewater and desalination plants in Kuwait by performing the sensitivity analysis and calculating the power density using a PRO 0D model. The effects of concentration polarization (CP) and salt leakage (B') on the power density at varying applied pressures and concentration differences between the feed and draw solutions were studied. The results showed that the effects of CP and B' on the power fraction were high at lower concentration differences, which might lead to lower flux and salt rejection caused by the decreasing driving force. Therefore, CP and B' effects can be nullified at a high concentration gradient.

As much of the literature focuses on lab-scale transport through a membrane by assuming that the concentrations along the membrane for both flow streams are constant, which might result in under-sizing exchangers for use in large systems because the average osmotic driving force across a long membrane is lower than the maximum osmotic driving force in a 0D transport model [23], therefore, a large system must be sized with a model that considers the change in driving force along the membrane length. In 2013, Shaqawy et al. [24] developed a one-dimensional (1D) analytical expressions for parallel and counterflow PRO mass exchangers with respect to the recovery ratio of the membrane as a function of dimensionless parameters such as mass transfer units (MTU), mass flowrate ratio (MR) and osmotic pressure ratio (SR). They developed the first ε-MTU model for the osmotic mass exchanger, which can be used as a design tool for PRO systems. It was further extended later to estimate the required area of a PRO exchanger to determine the power production at a given feed and draw salinity by considering the effect of CP on the membrane performance [25]. Similar methods were also used by their research group to acquire the designated area of a forward osmosis (FO) and an assisted forward osmosis (AFO) system on the application of fertigation [23].

In this study, 1D model equations for parallel flow configuration are employed to determine the required mass transfer units (MTU) to achieve high permeation and power generation using actual operational data from the desalination industry in Kuwait. This study will focus on the relationship between MTU with different dimensionless groups such as mass flowrate ratio (MR), recovery ratio (RR), concentration factors (CF), effectiveness (ε), dilution rate ratio (DRR) and dilution rate (DR), which relate the actual permeation to the draw stream. Furthermore, the maximum work simulated from these analytical expressions will be discussed, as well.

2. Pressure-Retarded Osmosis Mass Exchanger Model in a Parallel-Flow Configuration

Figure 1 shows the schematic diagram of a PRO mass exchanger device [23]. Herein, the feed solution of a lower concentration flows into the channel on one side, through a semipermeable membrane. On the other side of the membrane, the draw solution of a higher concentration flows into the channel. The flow direction of the draw solution is in the same direction as that of the feed, and it is called the parallel flow configuration, as shown in Figure 1.

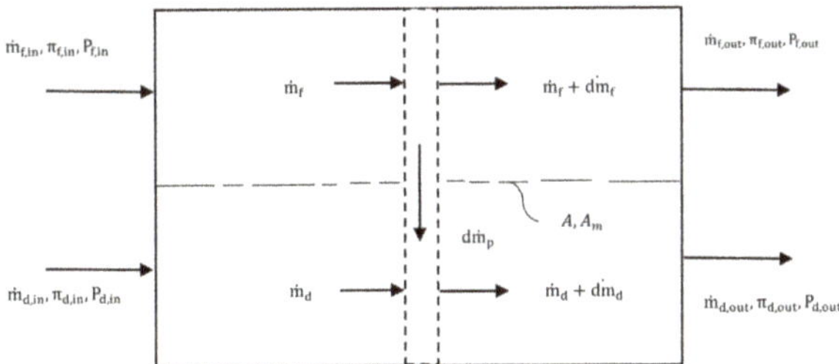

Figure 1. Schematic diagram of a parallel flow pressureretarded osmosis (PRO) mass exchanger. [20].

The differential permeate flowrate in a PRO parallel-flow configuration is given in Equation (1):

$$d\dot{m}_p = A(\Delta\pi - \Delta P)dA_m \tag{1}$$

where $\dot{m}_p$ = permeate mass flowrate through membrane (kg/s), A = water permeability coefficient of the membrane (kg/m^2.s.kPa), ΔP = difference in hydraulic pressure between the draw (P_d) and feed (P_f) solution (kPa), $\Delta\pi$ = difference in osmotic pressures between the draw (π_d) and feed (π_d) solution (kPa) and A_m = surface area of the membrane (m^2).

Applying the van't Hoff law for osmotic pressure as shown in Equation (2),

$$\Delta\pi = \pi_d - \pi_f = C(w_d - w_f) \tag{2}$$

where w_f and w_d are the salt concentration or salinity in g/kg for the feed and draw solution, respectively, while C is the modified van't Hoff coefficient. Substituting Equation (2) into Equation (1) yields:

$$d\dot{m}_p = A[C(w_d - w_f) - \Delta P]dA_m \tag{3}$$

We assume that there is 100% salt rejection and only pure water permeates through the membrane. Therefore, the salinity of the permeate is zero. Applying the conservation of solutes to the feed stream between the inlet and any arbitrary location along the flow channel yields,

$$\dot{m}_{s,f} = \dot{m}_{f,in}\dot{w}_{f,in} = \dot{m}_f w_f \tag{4}$$

For the same arbitrary control volume for the feed stream, the conservation of solution requires,

$$\dot{m}_{f,in} = \dot{m}_f + \dot{m}_p \tag{5}$$

Substituting Equation (5) into Equation (4) yields:

$$w_f = \frac{\dot{m}_{f,in} w_{f,in}}{\dot{m}_{f,in} - \dot{m}_p} \tag{6}$$

Applying the conservation of solutes and solution on the draw side of the parallel model gives,

$$w_d = \frac{\dot{m}_{d,in} w_{d,in}}{\dot{m}_{d,in} + \dot{m}_p} \tag{7}$$

Substituting Equations (6) and (7) into Equation (3) gives,

$$d\dot{m}_p = A\left[C\left(\frac{\dot{m}_{d,in} w_{d,in}}{\dot{m}_{d,in} + \dot{m}_p} - \frac{\dot{m}_{f,in} w_{f,in}}{\dot{m}_{f,in} - \dot{m}_p}\right) - \Delta P\right] dA_m \tag{8}$$

It is required that Equation (8) be cast into dimensionless form, and hence, four parameters are used for this purpose as follows:

(a) The recovery ratio (RR) is defined as the ratio of the total mass flowrate of the permeate recovered from the feed stream to the mass flowrate of the incoming feed stream. It represents the amount of pure water that can be recovered from the feed stream [25]. The RR is given by:

$$RR = \frac{\dot{m}_p}{\dot{m}_{f,in}} \tag{9}$$

(b) The mass flowrate ratio (MR) is the ratio of the mass flowrate of the draw solution to that of the feed solution at the inlet of the PRO mass exchanger. The MR is given by:

$$MR = \frac{\dot{m}_{d,in}}{\dot{m}_{f,in}} \tag{10}$$

(c) The osmotic pressure ratio (SR) is defined as the ratio of osmotic pressure at the draw or feed inlet to the difference in hydraulic pressure.

$$\text{For draw side}: SR = \frac{\dot{\pi}_{d,in}}{\Delta P} \tag{11}$$

$$\text{For feed side}: SR = \frac{\dot{\pi}_{f,in}}{\Delta P} \tag{12}$$

In a PRO operation, SR at the draw side will be greater than the SR at the feed side.

(d) Mass transfer units (MTU):

$$MTU = \frac{A_m A \Delta P}{\dot{m}_{f,in}} \tag{13}$$

$$dRR = \left(\frac{MRSR_d}{MR+RR} - \frac{SR_f}{1-RR} - 1\right)dMTU \tag{14}$$

$$\int_0^{RR} \frac{1}{((MRSR_d)/(MR+RR)) - (\frac{SR_f}{1-RR}) - 1} dRR = MTU \tag{15}$$

$$\int_0^{RR} \frac{(MR+RR)(1-RR)}{(RR-\alpha)(RR-\beta)} dRR = MTU \tag{16}$$

where:

$$\alpha = \frac{1}{2}\left[(1 + MR(SR_d - 1) + SR_f) - \sqrt{(-1 - MR(SR_d - 1) - SR_f)^2 - 4MR(SR_d - SR_f - 1)}\right] \tag{17}$$

$$\beta = \frac{1}{2}\left[(1 + MR(SR_d - 1) + SR_f) + \sqrt{(-1 - MR(SR_d - 1) - SR_f)^2 - 4MR(SR_d - SR_f - 1)}\right] \tag{18}$$

$$MTU = \frac{(\beta-1)(MR+\beta)}{(\alpha-\beta)} \ln\left(\frac{\beta-RR}{\beta}\right) - \frac{(\alpha-1)(MR+\alpha)}{(\alpha-\beta)} \ln\left(\frac{\alpha-RR}{\alpha}\right) - RR \tag{19}$$

Here, Equation (19) represents the expression to find the mass transfer units and is used to design a mass exchanger.

In order to determine the maximum transfer units allowed through a membrane when the output salinity of the brine is limited to a salinity lower than a certain limit, the concentration factor (CF) is used. The CF at the draw and the feed side are given by Equations (20) and (21).

$$\text{CF for the draw side}: \quad CF_d = \frac{w_{d,out}}{w_{d,in}} = \frac{MR}{MR + RR} \tag{20}$$

$$\text{CF for the feed side}: \quad CF_f = \frac{w_{f,out}}{w_{f,in}} = \frac{1}{1 - RR} \tag{21}$$

3. Parallel Flow PRO Effectiveness

The effectiveness of the PRO system can be defined as the ratio of the permeate flowrate to the maximum permeate flowrate [25], which occurs when MTU is increased to infinity. The maximum permeate flowrate is achieved when $\Delta\pi$ between the feed and the permeate increases to a point such that $\Delta P - \Delta\pi = 0$ at the exit of the channel. The effectiveness can also be defined as the recovery ratio divided by the maximum recovery ratio. The following is a derivation of the effectiveness of the PRO exchanger.

Using Equation (1), the maximum permeate in the case of the parallel flow configuration will occur when the hydraulic pressure difference is equal to the osmotic pressure difference at the outlet, as shown in Equation (22),

$$\Delta\pi_{out} = \pi_{d,out} - \pi_{f,out} = \Delta P \tag{22}$$

Using the van't Hoff model,

$$C(w_{d,out} - w_{f,out}) = \Delta P \tag{23}$$

Applying the conservation of solutes and solution on the draw and feed side, one can find that:

$$w_{d,out} = \frac{MR}{MR + RR} w_{d,in} \tag{24}$$

$$w_{f,out} = \frac{1}{1 - RR} w_{f,in} \tag{25}$$

Substituting Equations (24) and (25) into Equation (23) and replacing those variables with the dimensional group defined previously can yield Equation (26):

$$\frac{MR.SR_d}{MR + RR} - \frac{SR_f}{1 - RR} = 1 \tag{26}$$

Solving Equation (26) to find RR_{max}, there are two solutions that can be yielded,

$$RR_{max,1} = \alpha \tag{27}$$

$$RR_{max,2} = \beta \tag{28}$$

where α and β can be found in Equations (17) and (18), respectively. We notice from Equations (17) and (18) that α is always less than one. Since the recovery ratio must be less than one, the maximum recovery ratio is equal to α. Thus, the effectiveness can be defined as:

$$RR = \varepsilon RR_{max} = \varepsilon\alpha \tag{29}$$

By substituting Equation (29) into Equation (19), an expression for MTU as a function of the effectiveness can be obtained as illustrated in Equation (30).

$$MTU = \frac{(\beta - 1)(MR + \beta)}{(\alpha - \beta)} \ln\left(\frac{\beta - RR}{\beta}\right) - \frac{(\alpha - 1)(MR + \alpha)}{(\alpha - \beta)} \ln(1 - \varepsilon) - \varepsilon\alpha \tag{30}$$

To relate the actual water permeation to the draw stream, a new dimensionless group defined as the dilution rate ratio (DRR) is introduced, which is the ratio of the amount of water permeates, the permeate mass flowrate, through the membrane divided by the mass flowrate of the draw solution, as shown in Equation (31):

$$DRR = \frac{m_p}{m_d} \tag{31}$$

Moreover, the actual dilution rate (DR) can be defined as the ratio of the mass feed flowrate (m_f) divided by the mass draw flowrate (m_d), which is equal to the reciprocal of the mass flowrate ratio, 1/MR.

$$DR = \frac{m_f}{m_d} = \frac{1}{MR} \tag{32}$$

The determination of the power production is similar to that reported in the studies of Banchik and his co-workers [25], as shown in Equation (33),

$$W = \eta \frac{m_p}{\rho_{d,o}} \Delta P \tag{33}$$

where m_p is the mass flowrate of the permeate, $\rho_{d,o}$ is equal to the density of the diluted outlet draw stream, ΔP is the pressure drop across the hydro-turbine and η is the combined turbine and generator efficiency.

To plot the graphs using the above-mentioned numerical model, we have to input the feed and draw concentrations based on the real data from the Kuwait desalination industry, as shown in Table 1. According to Sharqawy et al. [24], for feed and draw salinities between 1 and 70 g/kg (treated wastewater effluent (TWE)) and 35 and 70 g/kg (seawater), the modified van't Hoff coefficient is determined to be 78.92 and 76.76 kPa-kg/g, respectively.

Table 1. Input data for numerical model. TWE, treated wastewater effluent.

Input	Value/Range
Ambient temperature	25 °C
Modified van't Hoff coefficient	78.92 kPa-kg/g (TWE) and 76.76 kPa-kg/g (SW)
Mass flowrate of the feed	$\dot{m}_f$ = 1 kg/s
Inlet draw salinity, $w_{d,in}$	70 g/kg (brine water)
Inlet feed salinity, $w_{f,in}$	1 g/kg (TWE), 40 g/kg (SW)
Osmotic pressure ratio	2 (TWE), 2.5 (SW), 4 (brine water)

4. Results and Discussion

Figure 2 represents the plot of RR with respect to MTU for a parallel flow configuration with different MR contours using TWE and seawater as the feed and brine as the draw solution. As observed from the figure, RR increases with increasing MR values (0.5, 0.75, 1, 2, 3, 4, 5) for both TWE and seawater feed solutions. With respect to the TWE feed stream, RR values of 0.95 and 0.4 are achieved for MR equal to five and 0.5, respectively. Furthermore, for seawater feed stream that has a higher salinity compared to that of TWE, the recovery ratio is lower with a value of 0.22. The fact that the RR value of using the TWE feed stream is higher than the seawater feed stream can be attributed to the higher salinity difference between TWE and the brine draw solution, which leads to larger osmotic driving force and a higher mass flowrate of the permeate. An increasing trend in the value of RR with higher MR values for both TWE and seawater feed streams agrees with previous findings by Banchik et al. [25]. These could be explained by using the outlet feed and draw salinity with respect to MTU, as shown in Figures 3 and 4. As observed, the draw solution concentration has smaller changes compared to the feed solution concentration at high MR due to the mass flowrate of the permeate being low compared to the mass flowrate of the draw solution. However, the feed solution will become more concentrated due to the continuous permeation of water molecules across the membrane, which reduces the overall permeation mass flowrate, and hence, RR will require greater MTU for high MR.

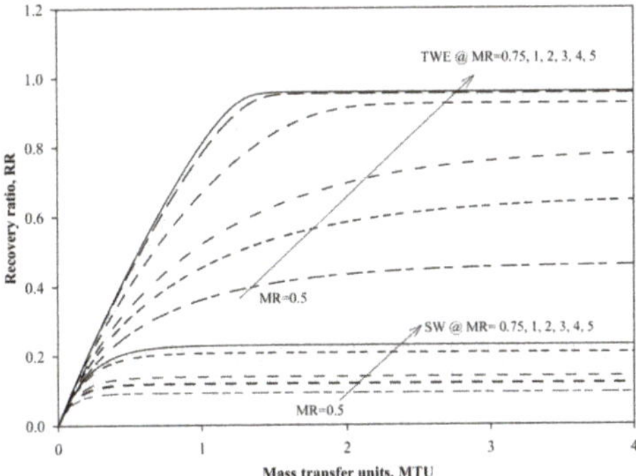

Figure 2. Recovery ratio (RR) versus mass transfer units at different mass flowrate ratio (MR) values using seawater or TWE as the feed solution and brine water as the draw solution.

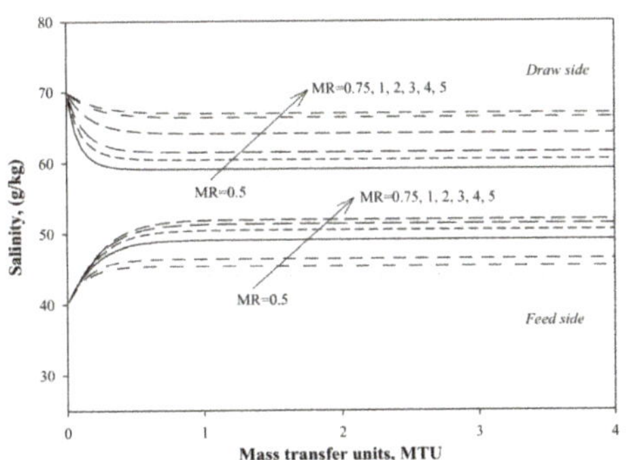

Figure 3. Salinity versus mass transfer units using seawater as the feed and brine as the draw solution for the parallel flow configuration.

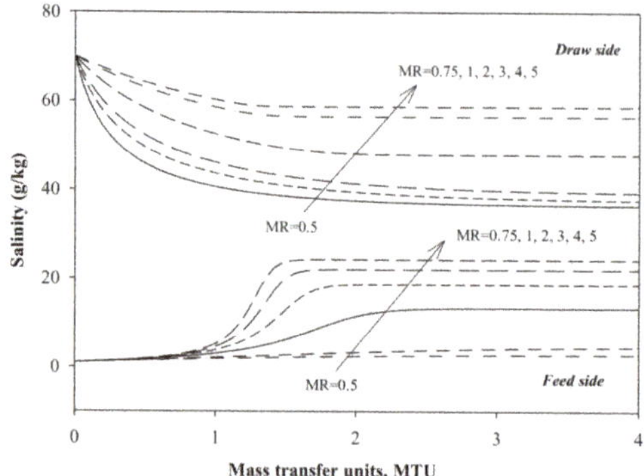

Figure 4. Salinity versus mass transfer units using TWE as the feed and brine as the draw solution for the parallel flow configuration.

Figures 5 and 6 depict the plot of the mass flowrate of the permeate versus mass transfer units at varying MR values. From Figure 5, it can be seen that m_p decreases as MR increases for MR < 1 when using seawater as the feed solution. However, for MR > 1, m_p increases with an increase in MR. More water permeate can be recovered at high MR due to the better restoration of the concentration differences between the feed and draw solution. A similar pattern is found in Figure 6 when using TWE as the feed solution. The mass flowrate of permeate (m_p) tends to be higher at lower MR for MR < 1. For MR > 1, m_p is increased proportionately with MR. As expected, the m_p of the membrane in the treatment of TWE is much higher than that using seawater due to the larger osmotic driving force produced by the higher salinity difference. Furthermore, as m_p is equal to dimensionless group RR when m_f = 1 kg/s (see Equation (9)), by taking the example of TWE as the feed and brine as the draw solution for MR < 1 and MR > 1 at the highest MTU = 4, the m_p values obtained based on RR from Figure 2 are compared with those shown in Figure 6. It is found that different m_p values are obtained for MR = 0.5 (i.e., m_p = 0.4 (Figure 2) and m_p = 0.9 (Figure 6)), while the same m_p value (i.e., 0.9 (for both Figures 2 and 6)) can be achieved for MR > 1. Hence, it can be concluded that the m_p value obtained using RR does not represent the actual permeation rate for MR < 1. It is necessary to develop another new dimensionless group to relate the actual water permeation across the membrane between the feed and draw solution, which will be discussed in the following sections.

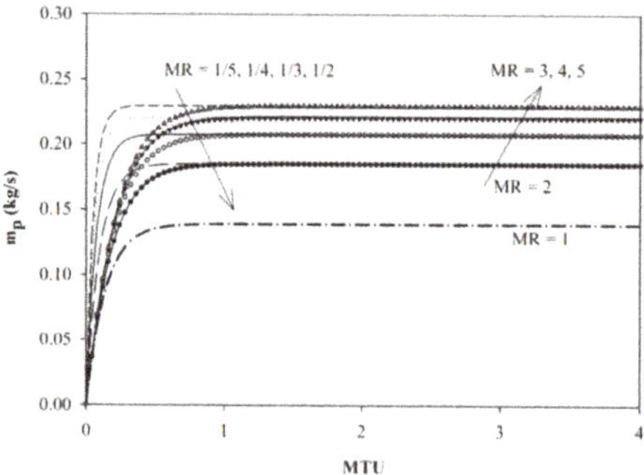

Figure 5. Mass flowrate of the permeate versus mass transfer units at different MR values using seawater as the feed and brine as the draw solution.

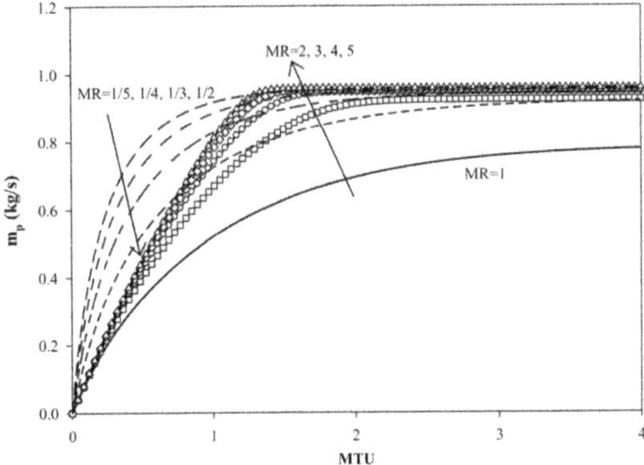

Figure 6. Mass flowrate of the permeate versus mass transfer units at different MR values using TWE as the feed and brine as the draw solution.

The concentration factor is the ratio of the outlet salinity of a stream to the inlet salinity. Owing to the salinity difference, water molecules would be transported from the feed to draw solution, and this leads to highly diluted draw effluent and more concentrated feed effluent, resulting in lower CF_d and higher CF_f at high MR, as illustrated in Figures 7 and 8. It can be observed that the CF when using TWE as the feed is much higher than those using the seawater feed stream. This can be ascribed to the higher salinity difference between TWE and brine, resulting in more water permeation across the membrane and an increase in the final concentration of the feed solution. Furthermore, CF_d becomes very small when using TWE as the feed solution regardless of different MR being able to be attributed to the large dilution of the draw solution at a higher salinity difference between TWE and the brine solution.

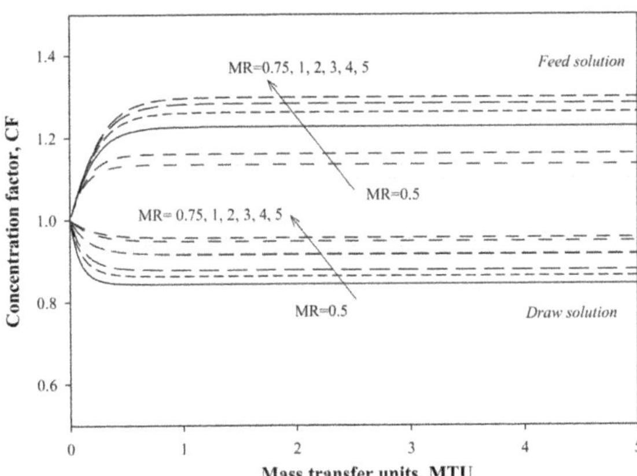

Figure 7. Concentration factor versus mass transfer units at different MR values using seawater as the feed and brine as the draw solution.

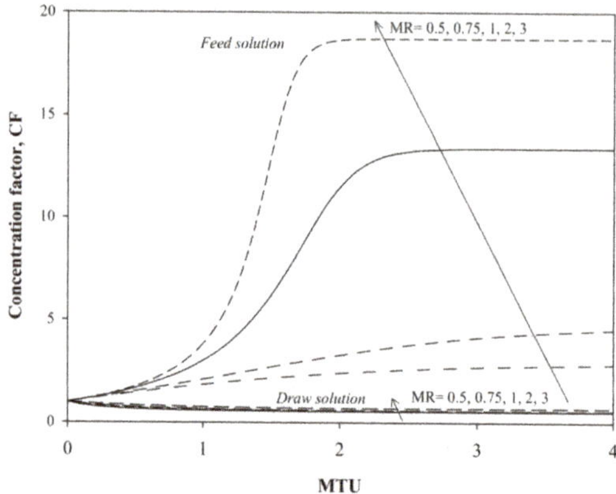

Figure 8. Concentration factor versus mass transfer units at different MR values using TWE as the feed and brine as the draw solution.

In order to show how high of a dilution rate at the draw side could be achieved, two new dimensionless groups defined as the dilution rate ratio (DRR) and dilution rate (DR) are introduced to relate the actual permeation into the draw stream. From Figures 9 and 10, it can be observed that DRR is increased with an increase in DR, owing to the large amount of feed solution diffusing through the membrane, resulting in highly diluted brine solution. As expected, DRR using TWE as the feed is higher than that using seawater, due to more water molecules being drawn from the feed to the draw side at a higher salinity difference and thus significantly decreasing the draw concentration. In addition, the maximum permeation rate would occur when the hydraulic pressure is equal to the osmotic pressure at the exit channel [24]. Thus, it is important to know the effectiveness of a PRO system in order to determine the overall PRO membrane performance.

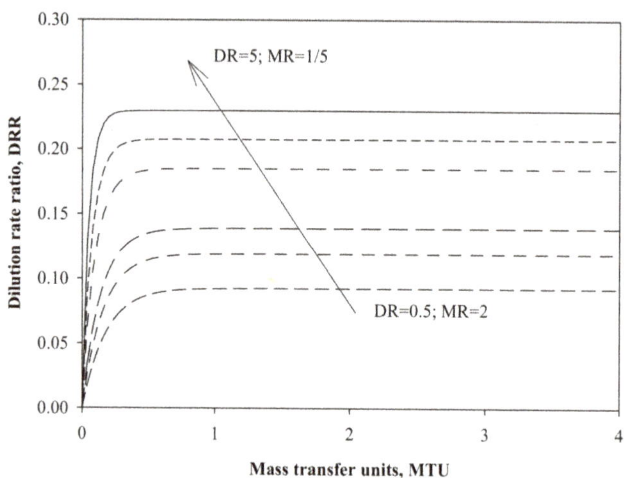

Figure 9. DRR as a function of MTU for seawater and brine at varying dilution rate values.

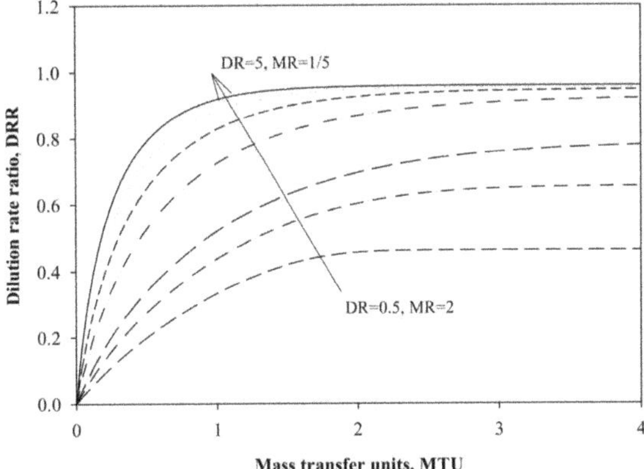

Figure 10. DRR as a function of MTU for TWE and brine at varying dilution rate values.

As illustrated in Figures 11 and 12, the effectiveness of the PRO system for the treatment of TWE requires larger MTU compared to that of treating seawater as the feed. At higher values of MR (>1), the effectiveness is gradually increased to unity at larger MTU values. High MR means a higher draw flowrate than the feed flowrate, and hence, the draw solution concentration will not experience a drastic change because the permeate flowrate is low compared to the draw flowrate at high MR. However, the feed solution will become more concentrated due to continuous water permeation, which reduces the overall permeation mass flowrate, hence requiring greater MTU. Moreover, for the treatment of TWE and brine solution, more water permeates across the membrane due to larger osmotic pressure caused by the concentration difference between TWE (1 g/kg) and the brine solution (70 g/kg), which can lead to a higher concentration of the feed solution. As a result, both the ICP effect and transport resistance increase, and a larger MTU is required to achieve the same desired effectiveness as obtained in the treatment of seawater and brine solution.

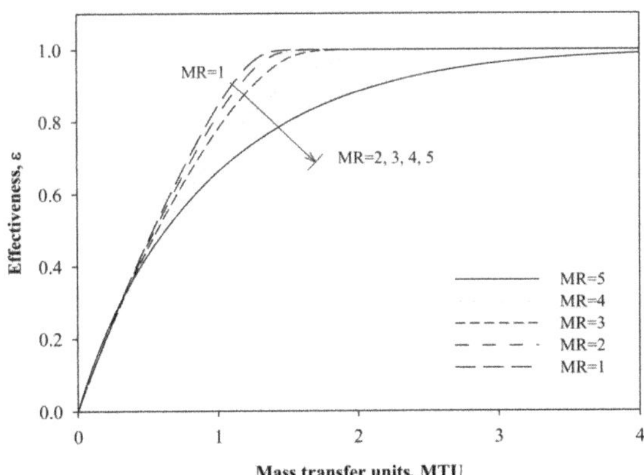

Figure 11. Effectiveness versus mass transfer units for TWE and brine solutions at different MR values.

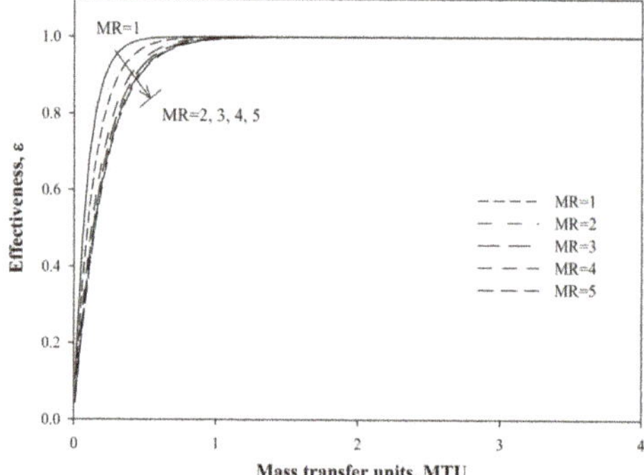

Figure 12. Effectiveness versus mass transfer units for seawater and brine solutions at different MR values.

One of the most important features in the PRO system is generating power using osmotic pressure differences. Figures 13 and 14 show the maximum work versus MTU with MR contours using TWE and seawater as the feed, respectively, and brine as the draw solution. As observed, the treatment of seawater and brine solution produced significantly lower work compared to the combination of TWE and brine solution. The maximum work is found to be increased with MTU and MR values. As seen from Equation (33), the power generated per unit membrane area is proportional to the water flux produced and the pressure drop across the membrane. With respect to the TWE feed stream, a maximum power of 2.6 kJ is attained at MR = 10. However, for the seawater feed stream, the maximum work of 0.28 kJ at the same MR is attained, which is significantly lower than that of the TWE feed stream. It is clear that higher water flux can be produced when TWE is used as the feed and brine as the draw solution due to the larger salinity difference; hence, more power can be obtained after depressurizing these permeates through a hydro-turbine [18]. A similar trend was obtained by Banchik et al. [25] that the power generation is proportionately increased with the feed salinity difference and MR values.

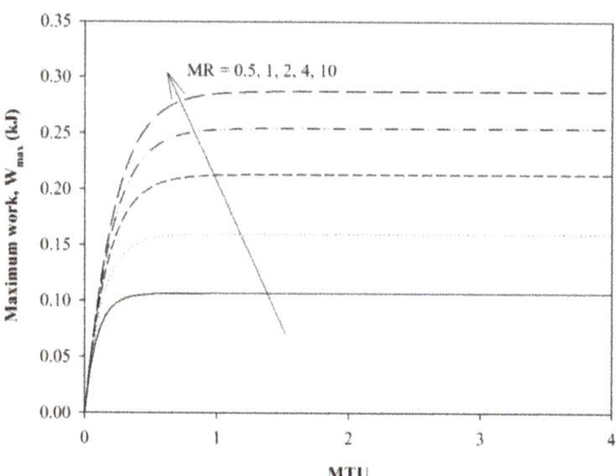

Figure 13. Maximum work versus mass transfer units for seawater and brine solution at different MR values.

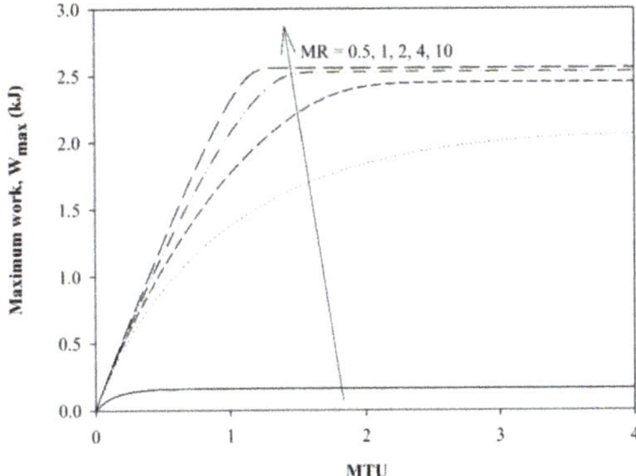

Figure 14. Maximum work versus mass transfer units for TWE and brine solution at different MR values.

5. Conclusions

One-dimensional analytical modeling was employed to design a parallel flow configuration in a PRO mass exchanger based on the effects of different dimensionless groups such as mass flowrate ratio (MR), recovery ratio (RR), concentration factor (CF), effectiveness (ε) dilution rate ratio (DRR) and dilution rate (DR) by using seawater and treated waste water effluent (TWE) as the feed and brine as the draw from the Kuwait desalination industry. The RR value of using the TWE feed stream is higher than the seawater feed stream, and the values increase with increasing MR values for MR > 1. Results indicate that the mass flowrate (m_p) of the membrane in the treatment of TWE is much higher than that using seawater as the feed stream. It has been found that the relation between RR and MR is not proportional for all MR values due to the m_p obtained using RR not giving the actual permeation. Thus, DRR and DR were developed to relate the actual permeation of feed to the draw stream. It was found that DRR values using TWE as the feed are higher than those of seawater, and hence, the effectiveness of a PRO system for the treatment of TWE requires larger MTU. Besides that, the modeling study shows that the treatment of seawater and brine solution produced significantly lower work compared to the combination of TWE and brine solution. A maximum power of 0.28 kJ and 2.6 kJ can be produced by the PRO system using seawater and TWE as the feed solutions, respectively, at MR = 10.

Acknowledgments: The authors would like to thank the Kuwait Foundation for the Advancement of Sciences (KFAS) for their financial support through Project No. P31475EC01.

Author Contributions: The work is conceived and supervised by Bader Al-Anzi, who also contributed to the study design, introduction of two new dimensionless parameters such as DRR and DR into the model. Ashly Thomas worked on plotting the figures for the model. Both the authors contributed towards the preparation and review of the manuscript.

Conflicts of Interest: The authors declare no conflict of interest.

References

1. Shahzad, M.W.; Burhan, M.; Ang, L.; Ng, K.C. Energy waster environment nexus underpinning future desalination sustainability. *Desalination* **2017**, *413*, 52–64. [CrossRef]
2. Freyberg, T. Az Zour, Kuwait Desalination Plant Recycles Power Station Cooling Water. Available online: http://www.waterworld.com/articles/2015/01/az-zour-kuwait-desalination-plant-recycles-power-plant-cooling-water.html (accessed on 5 June 2017).
3. Hamoda, M. Desalination and water resource management in Kuwait. *Desalination* **2001**, *138*, 385–393. [CrossRef]
4. Feo, J.; Jaime Sadhwani, J.; Alvarez, L. Cost analysis in RO desalination plants production lines: Mathematical model and simulation. *Desalinat. Water Treat.* **2013**, *51*, 4800–4805. [CrossRef]

5. Zhou, Y.; Tol, R.S.J. Evaluating the costs of desalination and water transport. *Water Resour. Res.* **2005**, *41*. [CrossRef]
6. Veera Gnaneswar Gude. Energy consumption and recovery in reverse osmosis. *Desalt. Water Treat* **2011**, *36*, 239–260.
7. Karabelas, A.J.; Koutsou, C.P.; Kostoglou, M.; Sioutopoulos, D.C. Analysis of specific energy consumption in reverse osmosis desalination processes. *Desalination* **2018**, *431*, 15–21. [CrossRef]
8. Wan, C.F.; Chung, T.S. Maximize the operating profit of a SWRO-PRO integrated process for optimal water production and energy. *Renew. Energy* **2016**, *94*, 304–313. [CrossRef]
9. Danoun, R. *Desalination Plants: Potential Impacts of Brine Discharge on Marine Life*; The University of Sydney: Sydney, Australia, 2007.
10. Song, X.; Liu, Z.; Sun, D.D. Energy recovery from concentrated seawater brine by thin-film nanofiber composite pressure retarded osmosis membranes with high power density. *Energy Environ. Sci.* **2013**, *6*, 1199–1210. [CrossRef]
11. Zhang, S.; Chung, T.S. Osmotic power production from seawater brine by hollow fiber membrane modules: Net power output and optimum operating conditions. *AIChE J.* **2016**, *62*, 1216–1225. [CrossRef]
12. Akram, W.; Sharqawy, M.H.; John, H.; Leinhard, V. Energy utilization of brine from an MSF desalination plant by pressure retarded osmosis. In Proceedings of the International Desalination Association World Congress on Desalination and Water Reuse 2013, Tianjin, China, 20–25 October 2013.
13. Loeb, S. Production of energy from concentrated brines by pressure retarded osmosis: I. Preliminary technical and economic correlations. *J. Membr. Sci.* **1976**, *1*, 49–63. [CrossRef]
14. Mehta, G.D.; Loeb, S. Internal concentration polarization in the porous substructure of a semi permeable membrane under pressure retarded osmosis. *J. Membr. Sci.* **1978**, *4*, 261–265. [CrossRef]
15. Sivertsen, E.; Holt, T.; Thelin, W.; Brekke, G. Modelling mass transport in hollow fibrer membranes used for pressure retarded osmosis. *J. Membr. Sci.* **2012**, *417–418*, 69–79. [CrossRef]
16. Lee, K.L.; Baker, R.W.; Lonsdale, H.K. Membranes for power generation by pressure- retarded osmosis. *J. Membr. Sci.* **1981**, *8*, 141–171. [CrossRef]
17. McCutcheon, J.R.; Elimelech, M. Influence of concentrative and dilutive internal concentration polarization on flux behavior in forward osmosis. *J. Membr. Sci.* **2006**, *284*, 237–247. [CrossRef]
18. Achilli, A.; Cath, T.Y.; Childress, A.E. Power generation with pressure retarded osmosis: An experimental and theoritical investigation. *J. Membr. Sci.* **2009**, *343*, 42–52. [CrossRef]
19. She, Q.; Jin, X.; Tang, C.Y. Osmotic power production from salinity gradient resource by pressure osmosis: Effects of operating conditions and reverse solute diffusion. *J. Membr. Sci.* **2012**, *401–402*, 262–273. [CrossRef]
20. He, W.; Wang, Y.; Shaheed, M.H. Modelling of osmotic energy from natural salt gradients due to pressure retarded osmosis: Effects of detrimental factors and flow schemes. *J. Membr. Sci.* **2014**, *471*, 247–257. [CrossRef]
21. Naguib, M.F.; Maisonneuve, J.; Laflamme, C.B.; Pillay, P. Modeling pressure retarded osmotic power in commercial length membranes. *Renew. Energy* **2015**, *76*, 619–627. [CrossRef]
22. Al-Anzi, B.; Thomas, A.; Fernandes, J. Lab scale assessment of power generation using pressure retarded osmosis from wastewater treatment plants in the State of Kuwait. *Desalination* **2016**, *396*, 57–69. [CrossRef]
23. Banchik, L.D.; Weiner, A.M.; Al-Anzi, B.; Lienhard, J.H.V. System scale analytical modeling of forward and assisted forward osmosis mass exchangers with a case study on fertigation. *J. Membr. Sci.* **2016**, *510*, 533–545. [CrossRef]
24. Sharqawy, M.H.; Banchik, L.D.; Lienhard, J.H.V. Effectiveness-mass transfer units (ε-MTU) model of an ideal pressure retarded osmosis membrane mass exchanger. *J. Membr. Sci.* **2013**, *445*, 211–219. [CrossRef]
25. Banchik, L.D.; Sharqawy, M.H.; Lienhard, J.H.V. Limits of power production due to finite membrane area in pressure retarded osmosis. *J. Membr. Sci.* **2014**, *468*, 81–89. [CrossRef]

MDPI
St. Alban-Anlage 66
4052 Basel
Switzerland
Tel. +41 61 683 77 34
Fax +41 61 302 89 18
www.mdpi.com

Sustainability Editorial Office
E-mail: sustainability@mdpi.com
www.mdpi.com/journal/sustainability

www.ingramcontent.com/pod-product-compliance
Lightning Source LLC
LaVergne TN
LVHW071302170726
843515LV00006BA/1466

* 9 7 8 3 0 3 9 2 8 9 7 1 4 *